DATE DUE

DEMCO 38-296

Using the
Biological
Literature

ADDITIONAL VOLUMES IN PREPARATION

Using the
Biological
Literature

A PRACTICAL GUIDE

Second Edition, Revised and Expanded

Elisabeth B. Davis
Diane Schmidt

University of Illinois at Urbana-Champaign
Urbana, Illinois

Marcel Dekker, Inc.   **New York • Basel • Hong Kong**

lication Data

Davis, Elisabeth B.
    Using the biological literature : a practical guide / Elisabeth B.
Davis. — 2nd ed., rev. and expanded.
        p.   cm. — (Books in library and information science ; 57)
    Includes bibliographical references and index.
    ISBN 0-8247-9477-X (hardcover : acid-free paper)
    1. Biological literature.   I. Schmidt, Diane.
II. Title.   III. Series: Books in library and information science ; v. 57
QH303.6.D38   1995
574'.072—dc20   95—11639
                                                                         CIP

The publisher offers discounts on this book when ordered in bulk quantities. For more information, write to Special Sales/Professional Marketing at the address below.

This book is printed on acid-free paper.

MARCEL DEKKER, INC.
270 Madison Avenue, New York, New York 10016

Current printing (last digit):
10 9 8 7 6 5 4 3 2 1

PRINTED IN THE UNITED STATES OF AMERICA

Preface

This book grew out of a series of handouts prepared for students using the Biology Library at the University of Illinois at Urbana-Champaign. Its purpose is to acquaint students new to the literature of biology with important primary and secondary resources of the field. Aimed toward undergraduate and graduate biology students, it is also appropriate for anyone interested in searching the biological literature and keeping up with its bibliography.

This guide to the literature of the biological sciences presents a comprehensive list of important sources that may be found in large research libraries, with emphasis on current materials in the English language. Retrospective reference works have been selected for historical perspective and to provide access to the taxonomic literature. All main fields of the biological sciences are covered; applied areas such as medicine, clinical psychology, veterinary medicine, agriculture, horticulture, nutrition, and teaching of biology are not included.

The handbook is arranged by broad subject chapters, which are subdivided by form of material. Entries are annotated only once, according to their primary focus, though they may be cross-referenced elsewhere. Every textbook listed is recommended, although not all of them are annotated. A certain knowledge of scientific literature is assumed, and there is minimal explanation of the definitions, uses, or importance of primary and secondary literature. A very brief discussion of searching strategies and indexing policies is included in Chapter 2, "Subject Access to Biological Information," but it is assumed that more detailed information on how to search for information in the biological sciences is available elsewhere. Periodical editorial policy and subject scope have been included with journal annotations as an aid to the student looking for publication possibilities. Unless otherwise cited, quotations in the body of an annotation are taken from prefatory material of the item under consideration.

New to the second edition is the addition of information on the greatly expanded universe of electronic resources for the biological sciences, including on-line databases, CD-ROMs, and most notably, Internet-accessible resources. The history of electronic information resources and the Internet is covered in Chapter 1, "Introduction to the Biological Literature," while there is a brief discussion of the types of Internet resources in Chapter 2. Materials dealing with particular subjects such as plant biology or entomology are dealt with in the subject chapters under the heading "Guides to Internet Resources." It should be remembered that the Internet universe is undergoing daily change, so that some of the information about Internet resources may be outdated by the time the guide is published; this is why more detailed information about Internet-accessible resources is not included.

The authors would like to acknowledge the assistance of John M. Stassi for his assistance in gathering information. We would also like to express our appreciation for the cooperation we received from the other life sciences librarians and staff at the University of Illinois. In particular, the assistance of the Biology Library personnel has been invaluable. And finally, this book could not have been written without the excellent collections of the University of Illinois Library.

Elisabeth B. Davis
Diane Schmidt

Contents

Using the
Biological
Literature

1

Introduction to the Biological Literature

Early History

The history of biological literature reflects the evolution of biology the discipline, and in this context it is interesting to review the progress of biology as a science to gain perspective into the development of its literature. This description of biology's milestones will be brief, with the understanding that the names, experiments, and publications singled out are representative, not enumerative, of the biological advances made over the centuries.

1

The word *biology* was first used by Jean Baptiste Lamarck and Gottfried Treviranus in the early 19th century, but the study of living things began long before that. The ancient Greeks, and Aristotle in particular, are credited with "inventing" natural science. Although there was neither a social organization for science nor a self-perpetuating scientific community, knowledge was valued for its own sake as well as for the understanding of nature. There were few attempts to manipulate nature or apply scientific knowledge systematically. The important goal was the use of rational, general, and logical explanations—in other words, Greek culture valued a new way of looking at the natural world.

Aristotle has had an immense influence on the history of biology, evoking overall a high evaluation of his biological observations. His biology was intimately linked to experience and his theories relied on his astute observations of animal structure and human anatomy. Aristotle developed a comprehensive philosophy with the major premise being that everything in nature has a purpose. Although this concept of teleology was discarded at the time of Charles Darwin, this idea of final causes was valuable to the progress of biology. For example, William Harvey's discovery of how blood circulates was based on productive, biologically useful Aristotelian questions about the purposes of the veins.

Science in the Middle Ages was basically located in three centers: the Latin West, the Latin East, and Islamic countries. In the Latin West fragments of Greek science were passed on by the scholastics in monasteries. Although basic knowledge was preserved, nothing new was added. Pliny's encyclopedic *Natural History* represented all that was known of natural history in the West. Byzantium was able to preserve more than its Western counterpart but the outcome was the same: preservation was the mode during the Dark Ages. Islamic countries, as the third center, made more rapid progress in spreading Greek science. The Arabs' political expansion and assimilation were expeditious in using Greek knowledge and promoting support for medical, astrological, and mathematical study. Islam not only preserved Greek science, it added to it. Galen, the most important ancient physician and teacher other than Hippocrates, was almost unknown outside of Islam. His work, translated into Arabic and then again into Latin, remained the authority for medical students for hundreds of years despite its many inaccuracies. In fact, Galen's work was virtually unquestioned until the 16th century when Vesalius' willingness to dissect and to experiment raised scientific investigation to a new level, signalling the beginnings of modern science.

Until the 16th century, scientific writings were preserved in manuscript and copies of books were rare and expensive to own and to duplicate. Descriptive writing, commentary, and herbals represent the incunabula left to posterity. Not until the invention of the printing press in the 16th century was scientific writing available to the common person.

17th Century

Some of the scientists important to 17th century biology were William Harvey, René Descartes, Francis Bacon, Marcello Malpighi, and Antonie Van Leeuwenhoek. Harvey epitomized experimentation in biology in contrast with Descartes, whose theories reduced everything to mechanical explanations, a point of view which ultimately failed. The name Francis Bacon is linked to the scientific revolution and to the inspiration of the Royal Society of London. Although he made no discoveries himself, he was a reformer who advocated observation, collection, and organization of data. Bacon, who had a great impact on science, promoted experimentation, emphasized systematic investigation, and encouraged publication of results. He believed science to be utilitarian and that knowledge leads to the relief of humanity's estate.

The 17th century saw the spread of the printed word and the organization of learned societies to disseminate information about scientific investigations. The first journal serving scientists, the *Philosophical Transactions of the Royal Society of London*, was begun in 1665, and by 1880 there were 100 journals serving scientific purposes. The mid-17th century saw a new value system for science based on the beliefs that science was communal and that knowledge was secular and apolitical and should be shared. There was high interest in experimentation and the mode of science was mathematical and utilitarian. In England, France, and Italy learned societies were the major transmitters of scientific values and were instrumental in the systematic communication of scientific results.

The microscope was discovered in the 17th century by Dutch lensmakers but its use, for the most part, passed into the hands of amateurs. Antonie Van Leeuwenhoek was not a trained or educated person and his observations on microscopic life allowed no biological generalizations. Robert Hooke described "cells" in cork; even so, the microscope as a scientific instrument did not demonstrate its power until the problems of light source and spherical and chromatic aberrations were resolved in the 19th century. The Italians put the microscope to the most use; for example, Marcello Malpighi used the instrument in his embryological investigations to contribute to the studies of preformation, epigenesis, ovism, and spontaneous generation, which were of interest to 17th century biologists. Unfortunately, difficulties of generation and development were too complicated and complex to reduce to simple chemical or mechanical analogies.

John Ray, the 17th century botanist, gave the first modern description of species based on morphological, structural, and biological information. Classification attempts such as his were somewhat more successful in terms of organization but there were serious problems with extant and fossil species that were for the time being insoluble. In summary, problems in biology yielded only sparingly to the various attacks of the 17th century, and there were very few lasting achievements.

18th Century

The 18th century may be characterized as an energetic age of exploration, collection, and organization. As an example of this activity, the number of described plant species is relevant: in 1700, 6,000 species were ascertained; by 1800, there were 50,000. J. J. Dillenius was employed by William Sherard of England to be, probably, the first person in history paid as a full-time taxonomist.

Carolinus Linnaeus was the dominant figure of the 18th century. To this day, he reigns supreme as a classifier and inventor of the binomial system of classification. He was a prodigious writer, an indefatigable worker, and an amazing teacher whose inspired disciples explored the world proselytizing botanists to the Linnaean system, providing descriptive standards and stimulating research. Linnaeus' most important rival was the Comte de Buffon, an extraordinary writer and the author of a great natural history for the layperson.

In the New World most of the biological work was carried on by naturalists, college professors, and physicians trained as botanists. Science was a transplant with a great dependence on Mother England, with the result that much of the biological literature was transmitted to England for publication in the *Philosophical Transactions of the Royal Society of London*. American biologists and botanists operated on personal funds, with extremely limited facilities, and with little formal training or education. Americans were busy being pioneers and science was limited to natural history, practically the only way that untrained amateurs could compete. Men such as John and William Bartram, John Bannister, and Thomas Jefferson all made contributions to biological knowledge, overcoming arduous conditions and difficulties that made science a dangerous and uncertain undertaking. The most famous early American scientist was, of course, Benjamin Franklin, whose discoveries, inventions, experiments, and commentaries are well chronicled.

Biology was in a transitional phase during the 18th century, compelled to wait on the chemical revolution began by Lavoisier in the late 1780s before it could fully develop as a scientific discipline. By the end of the 18th century the characteristics of the industrial revolution became the characteristics of science: it was middle class, nonestablishment, provincial, nonconformist, and applied to industrial use.

New forms of communication and new specialized societies were established in the 18th century. Itinerant lecturers published their lectures as scientific texts, and the appearance of greatly ambitious encyclopedias, such as Diderot's *Encyclopédie* in 1751, represented intellectually significant achievements. Numerous societies continued to publish their members' scientific papers, allowing biological events to be recorded and disseminated. There was a maturity in the development of science around 1800 as a consensus of scientific identity and method emerged. The appearance of social institutions and the gathering of a broader base of social support and interest permitted science to rise as an authority, to challenge traditional influences, and to stand on its own as an independent source of knowledge. The number of scientific journals grew from 100 in 1800 to 10,000 in 1900, an incredible increase that forced the development of the great abstracting and indexing tools of the 19th century because biologists could no longer depend on their own reading to cover the scientific literature. Botanists relied on *Botanisches Centralblatt, Just's Botanischer Jahresbericht,* and *Zeitschrift für Pflanzenkrankheiten*; zoologists trusted *Zoologischer Bericht, Berichte über Wissenschaftliche Biologie*, and *Zoological Record*; bacteriologists used *Zentralblatt für Bakteriologie.*

19th Century

Biology in the 19th century made quite striking advances based on the consolidating concepts of evolution, cellular organization, and the germ theory of disease. Rigorous methodology had developed, and biology was recognized as a unified discipline focusing on the processes of life and the functions of the organism. Natural history was discarded as physical and chemical terms were used to explain physiological processes.

The many unanswered questions emanating from the multitude of biological explanations of the 18th century culminated in making evolution the major concept of the 19th century. The works of Buffon, Erasmus Darwin, and Lamarck were precursors of evolutionary thought that set the stage for Charles Darwin's revolutionary work *On the Evolution of Species by Means of Natural Selection*, published in 1859. Darwin's theory achieved eventual acceptance based on massive evidence and a selective mechanism for evolution. Not until the turn of the century, when Gregor Mendel's work in genetics was rediscovered, was Darwin's mechanism of evolution comprehensively explained.

After the American Revolution, America's scientific connections with England and the Royal Society were largely cut off, and scientists traveled to France and then to Germany for their scientific training and education. During this period the United States had few really prominent natural historians. Science was still empirical, based on fieldwork, and not theoretical. Conflict between science and religion was fairly widespread in America. Most of the few American colleges were tiny, sectarian institutions organized to teach classical subjects and produce people of high moral character rather than to conduct research for the advancement of knowledge. It was only after the Civil War that American education was renovated, that scientific societies flourished, that Harvard was brought to scientific prominence, and that the Johns Hopkins University was founded on the German laboratory and university model. By the end of the 19th century America had scientific eminence, scientific journals, academic scientists, educational institutions, and scientific societies with social approval.

20th Century

Mendelian genetics had an important effect on many naturalists, statisticians, and experimentalists in the United States who became involved in the modern theory of evolution. By 1911 the United States was leading the world in genetic research as money poured into land grant institutions, research careers had become a reality for biologists, and interest in agriculture had become popular. Thomas Hunt Morgan, from his research on the genetics of *Drosophila melanogaster*, became the first American to win a Nobel Prize. By the 1930s the scientific forces of paleontology, geology, genetics, and natural history were all contributing to the broad assumptions of evolution.

Probably the second most important biological event after Darwin occurred in 1953 when J. D. Watson and Francis Crick published the results of their research on DNA. This revolution in molecular biology was vastly different from that of classical genetics almost a century earlier. The roots of molecular biology were founded in physical, biochemical, and structural research totally unlike the evidence collected by Charles Darwin and Alfred Wallace. Unlike in the 19th century, scholarly competition was fierce. There was a huge amount of scientific information swirling around the participants, team research was in vogue, and massive funding was available. Watson and Crick rushed to publication, unlike Darwin, who waited 20 years to publish his theory after developing it.

The spectacular events popularized by *The Double Helix: A Personal Account of the Discovery of the Structure of DNA* epitomize the significant occurrences in biological literature during this period. Specialist literature abounded; big science begat big scientific publishing programs; academic science libraries became decentralized; and library collections experienced a remarkable growth rate. The need to collect both descriptive and functional literature put an unbelievable strain on biology library budgets. Not only was it necessary to provide the voluminous cumulative literature of a discipline based on rare and costly titles, complete runs of periodicals, and a large core of reference books, but it was also necessary to purchase new journal subscriptions and the latest editions of monographs in order to have available all the latest information.

Almost every library currently faces a crisis in serials prices. More and more journals are published each year, and established journals often increase their frequency or page count and with that, their price. In the early 1990s, it was not unusual for journal prices to go up 20 percent in a single year. Clearly, library budgets cannot hope to keep up with these increases, and thus since the mid-1980s, almost every library has been forced to cancel journals, many of them cancelling a number of titles each year. There are many causes for this budget crisis, including the publish-or-perish dilemma of untenured faculty, the ever-increasing amount of research being done, the attendant proliferation of specialist journals, the effect of the changing strength of the dollar on titles published elsewhere. The scientists themselves are frequently well aware of the problem, having seen journals of importance for their own research get canceled. Various remedies have been proposed, including efforts to get promotion and tenure decisions based on quality of research rather than quantity (thus avoiding the "salami syndrome" in which a project is sliced into many small articles rather than one major one). Most solutions, however, are too little, too late. The best hope for the scientific literature probably lies in electronic publishing, but this transition will be a lengthy one and its results will not be known until well in the future.

Characteristics of Biological Literature

Characteristics of the biological literature are typical of other scientific disciplines. Bonn and Smith report an estimated 55,000 sci/tech titles worldwide, with about 9,500 titles from the United States. About 95% of the cited literature in the sciences is published in serials. Thus, biological literature is international in scope, dependent on serial publications for dissemination, interdisciplinary and overlapping, and complex in origin, with the primary periodical as the most important source of new information. Biological literature is distinguished by its broad spectrum, its volume, its generator who is also its user, and its appeal to the public interest.

In general, the number of scientific, scholarly periodicals has doubled every 10- to 15-year interval during the 20th century. Since World War II, however, there has been logarithmic growth of serial publications, and the growth of biological journals has paralleled this burst. The increase during the last decade has been especially explosive for the life sciences, making it one of the fastest growing disciplines. The 1994 "New Journals" supplement to *Nature* listed 50 new, English-language scholarly journals in the sciences (excluding clinical medicine, engineering, and pure mathematics). Of these 50 new titles, one-half were in the biological sciences.

The source of all this activity can be attributed to learned societies, commercial publishers, and university presses. Types of literature issued from points of origin are primary, secondary, and tertiary publications. Primary publications report original research, the first published record of scientific investigations, and are represented by periodicals, preprints, research reports, patents, dissertations, and trade technical bulletins. Secondary literature publications are reference works that derive from primary sources. Examples are encyclopedias, handbooks, treatises, bibliographies, reviews, abstracting and indexing serials, and translations. Tertiary sources, which discuss science rather than contribute to it, are textbooks, directories, and guides to the literature.

The geographic origin of biological literature may be fairly described using the information provided by BIOSIS, publisher of *Biological Abstracts*, the largest, most comprehensive biological abstracting service in the English language. A survey of the literature sources monitored by BIOSIS in 1993 reports that 50% were from Europe and the Middle East, 30% from North America, 14% from Australia and Asia, 3% from Central and South America, and 2% from Africa. Europe and North America had slight growth in their percentages, while the other regions reported a decrease over the past 5 years. Growth in biological literature since World War II has been particularly great from Japanese, Russian, and Spanish-Portuguese publications.

In 1962 Bourne reported that English was the dominant language of scientific literature, and this is still true today, as more and more journals are being published in, or being translated into, English. Garfield and Welljams-Dorof reported in 1992 that 95% of the microbiology articles indexed in *Science Citation Index* in 1991 were in English, with Russian the next most frequent at a mere 5%. The results are similar, but not identical, in other indexes. A recent search of Biological Abstracts on CD for July–September 1994 showed that 87% of all articles were English, followed by Russian at 3% and Japanese at 2%. Even those journals which publish in other languages usually have English abstracts. An article in *Nature* reported that in 1992 a French researcher charged that he was turned down for a promotion because he did not publish in English, showing the perceived importance of English as the language of science.

The most popular frequency of biological serials is irregular, followed, in decreasing order, by quarterly, monthly, and annual publication. Subject coverage and emphasis of biological literature are on basic and applied research that is interdisciplinary. The biological literature is scattered, fragmented, and dispersed, with 65–68% of cited articles appearing in less than 20% of the biological journals. Over the last decade the interfacing, overlap, and integration of disciplines have been typical of biology, with tremendous growth especially in areas relevant to endocrinology, immunology, and the neurosciences. Biological investigations often have more immediate, apparent impact on society, and because of this characteristic, research in the life sciences is heavily reported and critiqued in the popular press. The power of the biological revolution is visible through articles on biotechnology, aging, molecular genetics and the genetics of diseases such as breast cancer, and computational aspects of various disciplines such as computational neuroethology, just to name a few topics of wide interest.

Electronic Biological Literature

The vast proliferation of biological literature has made the computer an indispensable part of any biologist's toolkit. Although abstracts and indexes have been computerized since the early 1970s, through the mid-1980s they were still searched by trained intermediaries such as librarians and information specialists. Beginning in the mid-1980s, CD-ROMs (compact disk—read only memory) widened the access of electronic databases beyond the expert searcher by allowing the end-user (the person actually using the information) to perform his or her own searches. In addition, many libraries offer locally mounted databases for their patrons. These databases are loaded onto local mainframes and are freely accessible to registered users.

The early wave of computerization made secondary tools such as abstracts and indexes more widely available and more easily used. The next wave, which is still ongoing, is to improve access to the primary literature, particularly journal articles. There are a number of services currently in place which provide computer-readable copies of articles to users, usually for a fee which is much higher than photocopying. This fee does include fees for copying material beyond the copyright law's fair-use provisions, however. These services, and more elaborate versions which are under development, are useful since most libraries have been forced to cancel journal subscriptions, often resulting in much reduced access to less-used titles.

Both the first and second wave of computerization have basically built upon the earlier paper formats—indexes and journals—rather than replacing them. Libraries may have canceled less-used paper indexes and replaced them with online or CD-ROM databases, but most libraries still have both paper copies and electronic versions of the most heavily used items. The few electronic journals at this time are largely CD-ROM versions of major journals, such as the *Journal of Biological Chemistry* or the journals published by the American Society for Microbiology, and are not published as frequently as their print equivalents. Electronic journals are still rare enough and new enough that librarians are uncertain about their utility and archiving; thus few have canceled print copies of electronic journals. Some publishers offer price breaks for keeping print subscriptions as well as electronic versions, providing another incentive for keeping the old familiar print copy. There are few journals which are available in only electronic form; the first commercially backed strictly electronic scientific journal, *Online Journal of Current Clinical Trials*, had a slow and difficult start. Subscriptions were fairly good, but few authors were willing to write for it. More electronic journals are in the works, however.

Computer networks such as the Internet burst upon the scene in the early 1990s and profoundly changed the way in which researchers communicated. Electronic mail (email) is an inexpensive and simple way to collaborate with colleagues from across the country, even around the world. Large data files can be transferred quickly and easily, and files mounted on distant computers can be searched. The most common uses of the Internet are still for communication and file transfer, although with its growing popularity has come an astonishing variety of other applications.

In many ways, the Internet is often used as an expansion of the eternal invisible college. It makes networking and brainstorming with colleagues from distant areas a daily occurrence, rather than something which only occurred at conferences or symposia. In addition to using email, many researchers subscribe to one or more discussion groups such as USENET groups, listservs, or bulletin boards. These discussion groups make it easy to communicate with other people with similar interests, whether it is in *Arabidopsis* genetics or conspiracy theories. They also make it easy to tap into the vast amount of expertise held by other discussion group members. If one's local library is unable to find a vaguely remembered article, perhaps someone on the Net knows the reference. Requesters are dependent on the goodwill and reliability of strangers, however, so it is best to double-check responses received from these sources.

There is still a rather limited amount of officially sanctioned data available through the Internet. Much of what is available is non-copyrighted information posted by volunteers or is rather narrow in scope. With the greater access of commercial ventures to the Internet, this is rapidly changing but at present the Internet is still the domain of rugged individualists. Learning of the existence of resources is still very much a hit-or-miss or word-of-mouth proposition.

There are exceptions to this rule, of course. GenBank, the major gene sequence database, is freely accessible through the Internet, as are a number of other gene or protein sequence databases. Academic institutions are beginning to accept the idea that Internet resources are valuable and are supporting the development of authoritative files sanctioned by societies or departments.

While electronic resources seem likely to change the manner in which biological research and study is performed, the basics will remain the same. Internet-accessible sources, at the present time, cannot begin to replace the authoritative print resources annotated in this book. As the Internet becomes less the province of enthusiastic volunteers and more the province of committed official bodies, whether commercial or non-profit, this situation will likely change. As cyberspace becomes more transparent to the casual user, too, its usefulness will increase. For now the Internet acts as a complement or supplement and is no replacement for the old-fashioned paper publication.

Bibliography

Allen, D. E. 1976. *The Naturalist in Britain: A Social History*. London: Allen Lane.
Anonymous. 1994. New journals. *Nature*, 371(6496): 439–458.

Arber, A. 1938. *Herbals: Their Origin and Evolution.* New, enl. ed. Cambridge, England: Cambridge University Press.

Bakewell, D. 1992. Publish in English, or perish? *Nature*, 356(6371): 6348.

Bonn, George S. and Linda C. Smith. 1992. Literature of science and technology. *McGraw-Hill Encyclopedia of Science and Technology*, 7th ed. New York: McGraw-Hill, pp. 128–134.

Bourne, C. P. 1962. The world's technical journal literature: An estimate of volume, origin, language, field, indexing and abstracting. *American Documentation*, 13: 159–168.

Garfield, E. 1980. Has scientific communication changed in 300 years? *Current Contents/Life Sciences*, 23(8): 5–11.

Garfield, E. and A. Welljams-Dorof. 1992. The microbiology literature: Language of publication and their relative citation impact. *FEMS Microbiology Letters*, 100(1): 33–38.

Kronick, D. A. 1992. The scientific journal: Devant le deluge. *Current Contents/Agriculture, Biology, and Environmental Sciences*, 23(27): 6–10.

Steere, W. C. 1976. *Biological Abstracts/BIOSIS: The First Fifty Years. The Evolution of a Major Science Information Service.* New York: Plenum.

Thompson, K. S. 1994. Scientific publishing: An embarrassment of riches. *American Scientist*, 82(6): 508–511.

Wood, E. H. 1993. Online Journal of Current Clinical Trials. *Bulletin of the Medical Library Association*, 81(1): 89–90.

2

Subject Access to Biological Information

Introduction

Unless a researcher is already completely familiar with all books and articles in a particular subject (and who is?), it is necessary to do subject searches. Finding material on a particular subject is often more complicated than it seems. However, whether searchers are looking for books or articles, and whether they are looking in the good old paper card catalog or index, or in a complicated new computer public terminal or database, some basics stay the same. Along with looking in the right places, using the right terms to look up a subject is one of the most critical steps in finding useful material. Naturally, there are many aids to help researchers find the material they are looking for. The vocabulary used by the people who classify books and create article indexes is one important aid.

This chapter will deal with some topics which are important when doing subject searches in the biological sciences. Of course, many of these topics are important when doing searches in any subject, but some are more important to the sciences. Controlled vocabulary, for instance, is more useful in fields such as biology, which use a standardized vocabulary, and is less important in some of the humanities and social sciences. Indexes such as *Index Medicus* and *Biological Abstracts* use controlled vocabulary extensively.

Definitions

Controlled vocabulary describes any terms which are decided upon by indexers or catalogers and used to describe the subject of a book or article. These terms may or may not be the same words that an author uses to describe his or her writings. The form of controlled vocabulary which is most familiar to biologists is the use of Latin names. *Felis concolor* is the "controlled vocabulary" term for the animal that may be called puma, cougar, mountain lion, screamer, panther, painter, catamount, or lion.

Subject headings are the traditional way of describing the subject of a book. Anyone who has looked up a subject in a card catalog has used subject headings. They may consist of a main heading and one or more subheadings, such as "Flowers—North America—Identification." Usually, only a few subject headings are used to describe the entire book. The most commonly used subject headings at large libraries are the Library of Congress Subject Headings (LCSH). Smaller libraries may use the *Sears List of Subject Headings*, which is similar but less detailed.

While subject headings are usually used to describe whole books, **descriptors** are used to identify individual concepts covered in an article or, sometimes, a book. They usually define much narrower topics than subject headings do, and many more are used to describe the article, often up to 15 or 20. These descriptors are usually added to a record by indexers, who decide which words are used as descriptors.

The opposite of controlled vocabulary is **natural language** or **free text**, in which no attempt is made to standardize subject terms, and the author's own words are used to describe the subject. One example of free text searching might be flipping through the F's in a card catalog, hoping that titles beginning with "flowers" will include books on identifying North American flowers.

One term which is often used to describe words used for natural language searching is **keyword**, which is usually just a term used to search for a topic anywhere in a record, including title, subject headings, abstract, and so on. Unfortunately, sometimes "keyword" is also used as a synonym for "descriptor."

Advantages and Disadvantages of Controlled Vocabulary

Controlled vocabulary terms offer many advantages when doing searches, but there are also disadvantages. One advantage of controlled vocabulary has already been mentioned. Scientific names are an example of the advantages of using controlled subject terms to pull together all possible synonyms of a term. One does not need to think of all possible ways in which an author might name *Felis concolor*. Controlled terms also usually control for the use of plurals and alternate spellings. Searching for "puma" may miss items in which the word "pumas" is used. Controlled terms added to a record can also eliminate ambiguity in an author's selection of words. Is the cat in a title such as *The Ghost Cat of the Rockies* a cougar, or just a figment of the author's imagination?

Controlled vocabulary terms may also be more specific than the associated natural language terms. "Stress," for instance, means one thing to a psychologist, another to an ecologist, and still another to a civil engineer. Using a subject heading such as "Stress, psychological" would eliminate those articles dealing with metal fatigue, and thus reduce the stress on the searcher.

Of course, there are disadvantages as well to using controlled vocabulary terms. One problem familiar to users of the LCSH is matching your idea of the subject with the LCSH. It is not intuitively obvious, for instance, that "Moving-pictures" is the proper term for movies, film, and video. Under previous editions, the proper heading for beehives was "Honeybee—Housing," though "Beehives" is now correct. For books on wolverine behavior, is the proper heading "Wolverines—Behavior" or "Animal behavior—Wolverines"?

Another disadvantage to controlled vocabulary is that subject headings change slowly, if at all, so that hot new fields may be poorly covered. The field of computational neuroethology, for instance, is not indexed in LCSH as such. Headings now in use include "Computers—Biology—Data processing" and "Animal behavior—Computer simulation." In addition, card catalogs and online systems may not be updated every time a change occurs, so unless cross-references are included (and used), searchers may not find materials indexed under earlier or later headings.

Yet another serious problem is that different systems use different terms. LCSH uses "Tumors" while *Index Medicus* uses "Neoplasms," and either term will work in *Biological Abstracts*. Indexes such as the *Biology and Agriculture Index* and the *Reader's Guide to Periodical Literature* use subject headings which are similar, but not identical, to the headings used by the Library of Congress. For instance, these indexes use "Tumors" rather than "Neoplasms," as does LCSH, but they use "Cancer—Therapy" whereas LCSH uses "Cancer—Treatment." Each time a researcher uses a new system or index, he or she must figure out which of several possible terms is the "proper" one.

Important Indexing Systems for Biology

The three most important indexing systems for biologists are the Library of Congress Subject Headings (LCSH), the Biological Abstracts subject codes, and the Medical Subject Headings (MeSH). LCSH is used to find books, the Biological Abstracts codes are used to find articles in the print or computer versions of *Biological Abstracts*, and MeSH is used to locate either books or articles in medical areas, including the most important medical index, *Index Medicus*. For a more detailed description of *Biological Abstracts* and *Index Medicus* and their computer versions, see Chapter 4, "Abstracts and Indexes." An additional important indexing scheme which deserves mention is that used by *Chemical Abstracts*.

Library of Congress Subject Headings are updated irregularly and have been in use since 1897. The most recent edition is the sixteenth, which was published in 1993. LCSH subdivides subject headings in several ways, including topical (such as "Flowers—Identification"), form of publication (such as "Flowers—Bibliography"), chronological period (such as "Flowers—Pre-Linnean works"), and location (such as "Flowers—North America"). Several subheadings may be strung together, as in the earlier example of "Flowers—North America—Identification." In previous editions, and to some extent in the present edition, subject headings were often in inverted form. Thus, extinct birds were formerly listed under "Birds—Extinct," but are now listed as "Extinct Birds." Each edition of LCSH is published by the Library of Congress in the form of a thesaurus of headings. Currently there are four volumes of headings. There are many cross-references in these big red subject heading lists, including pointers leading to correct subject headings, as well as to broader, narrower, and related terms. The subject heading lists also include selected subheadings, although not all possible subheadings will be listed for each main heading. It is very helpful to consult the subject heading lists to find the proper heading or for suggestions for further searches for books.

LCSH headings are used in computer catalogs as well as in card catalogs. Some online systems have extensive cross-references leading to the proper LCSH term, while others do not. Where these cross-references are included, searching for material is made much easier. Whether searched in paper or on computer, however, using the correct LCSH heading will greatly improve the retrieval of useful materials.

Biological Abstracts, the most important index for biologists, has some very useful and advanced subject indexing in its computerized version, which is known as BIOSIS Previews. BIOSIS uses both keywords (natural language) and concept codes (controlled vocabulary). The keywords are taken from the language used by authors, and are listed in the *BIOSIS Previews Search Guide*. These keywords are intended to give searchers search hints, but any term may be searched as a keyword. The concept codes are a very powerful method of pulling together broad subjects, such as pollution or limnology. The Master Index in the front of the *BIOSIS Previews Search Guide* lists keywords and their associated concept codes, if any. For instance, under "estuary" possible synonyms such as aquatic, bay, brackish, limnology, river, and stream are listed, as well as the concept code CC07510 for oceanography and limnology. The concept code section of the search guide lists the concept codes and includes examples of searches using that concept code as well as strategy hints. The concept codes are useful when combining broad subjects with narrower ones, such as the effect of pollution on the Chesapeake Bay. In this case, a searcher might combine the concept codes for oceanography and limnology (CC07510) and air, water, and soil pollution (CC37015) with keywords such as Chesapeake. This is better than trying to come up with a list of terms dealing with all possible types of pollution and all possible terms dealing with estuaries.

BIOSIS also uses special codes for families of organisms, the Biosystematic Codes. These are useful for the searcher who wants to find articles dealing with all canines, for instance, but not as helpful for the one who is only interested in red wolves. A similar, but broader, taxonomic grouping is the Supertaxonomic Group. These are broader hierarchical groups listed for each species discussed in an article. An article on red wolves, for instance, would also be indexed under carnivores, nonhuman mammals, mammals, nonhuman vertebrates, vertebrates, and animals. The Supertaxonomic Groups are particularly useful when searching for broad groups of organisms, such as all plants or all birds. For more assistance in searching *Biological Abstracts*, see Allen and Strickland-Hodge's *How to Use Psychological Abstracts and Biological Abstracts*.

Medical Subject Headings (MeSH) are the subject headings created by the National Library of Medicine (NLM) for use in medical library card and on-line catalogs, the print index *Index Medicus*, and the database MEDLINE, as well as the other publications of NLM. The MeSH headings are updated every year and are absolutely essential when using these sources.

The MeSH headings for use with the printed *Index Medicus* are published each year as volume 1 of the *Cumulated Index Medicus*. A much more detailed listing for users of the online MEDLINE is also published each year. The online aids consist of three separate sections, of which the *Medical Subject Headings: Annotated Alphabetic List* is the most frequently used. This document lists all of the subject headings, as well as the allowable subheadings. This list is designed for the use of indexers, but it includes many notes which are equally useful for interested searchers. Another MeSH listing is the *Permuted Medical Subject Headings*. This volume takes each significant word from the

MeSH terms and lists all MeSH headings, including cross-references, in which that term is used. Thus, under "lobe" the list includes "Epilepsy, frontal lobe", "Epilepsy, temporal lobe," "Frontal lobe," "Middle lobe syndrome," "Occipital lobe," "Optic lobe," "Parietal lobe," and "Temporal lobe."

MeSH headings divide up medical subjects very narrowly, while listing only broad subject headings for related but non-medical subjects. As well as the main subject headings, MeSH also includes subheadings of various types, including publication types, geographic subheadings, and common descriptors which are added to every article to which they apply. These common descriptors (or "check tags") include terms such as "animal" (for animal studies), "female," and "in-vitro," and age groups such as "adolescence" and "aged-80-and-over." There are also topical subheadings, which further describe the subject heading. These subheadings include terms such as "adverse effects" and "diagnostic use."

A related medical subject classification used in MEDLINE is the tree structure, which is listed in *Medical Subject Headings: Tree Structures*. The tree structure is a hierarchical arrangement of the MeSH headings, so that narrower subjects can be combined and searched using broader categories. Thus, A is anatomy, C is diseases, C4 is neoplasms, C4.557.337 is leukemia, C4.557.337.428 is lymphatic leukemia, and C4.557.337.428.511 is acute lymphotic leukemia. It is possible to search a broad subject by truncating the tree number (known as "exploding" the tree). Exploding C4.557.337 will retrieve all forms of leukemia. For help in using *Index Medicus*, see Strickland-Hodge's *How to Use Index Medicus and Excerpta Medica*.

Chemical Abstracts does not use the same names for biologically or medically important chemicals as the other biological or medical indexes, which can have a significant impact on subject searches. Guanine, for instance, is accepted by *Index Medicus* and *Biological Abstracts*, but not by *Chemical Abstracts*. The *Chemical Abstracts Index Guide*, which is updated each year and with each *Cumulative Index*, lists the proper headings used in *Chemical Abstracts* for chemical compounds, classes of substances, applications, properties, and other subjects. The *Index Guide* also lists the CAS (Chemical Abstracts Service) Registry Number for chemicals. These Registry Numbers are useful since they are often less complicated than their related chemical names and pull together all synonyms of a particular chemical. They are also used in many other databases, including all other CAS databases, MEDLINE and other medical databases and indexes, and BIOSIS Previews through some vendors. An example of an *Index Guide* entry is the one for guanine, in which a searcher is directed to see 6H-Purin-6-one,2-amino-1,7-dihydro- or the Registry Number 73-40-5. Schulz and Georgy's *From CA to CAS Online: Databases in Chemistry* includes more information on how to use *Chemical Abstracts* and the other chemistry databases on STN (see Chapter 3, "General Sources").

Other Important Systems

While the previously discussed indexing systems are the most important for the field of biology, there are several others which merit mention. Each scientific index has its own subject headings list, with the exception of the paper *Science Citation Index*, which uses only keywords from titles. To use the other indexes to their fullest, it is useful to find the subject listing, often called a thesaurus, and give it some study. Indexes such as the *Bibliography of Agriculture* and *CAB (Commonwealth Agriculture Bureau) Abstracts* (which both use the *CAB Thesaurus*), the *Life Sciences Collection, Psychological Abstracts*, and *Zoological Record* all have subject listings of varying importance. These thesauri, like the LCSH and MeSH lists, usually provide the approved subject terms as well as broader, narrower, and/or related terms.

Internet Resources

Unlike library materials, which generally have good subject access, information available through the Internet is in a state of chaos. At present, there is little subject access for the many information sources, although this situation is improving. In this section, we will discuss general strategies for finding useful information. The situation is constantly changing, so the information provided here should be taken as only a general guide. Three major sources of information will be discussed here: discussion groups such as USENET groups, Gopher-accessible information, and the World Wide Web.

The best way for the uninitiated would-be Internet user to find materials of interest is to acquire one of the excellent books about the Internet, such as Krol's *The Whole Internet*, and simply check things out. There are also a number of guides to the Internet which are available via the Internet at no charge. Their availability and currency will vary, so be careful. Another source is Una Smith's *A Biologist's Guide to the Internet*, which has much information about Internet-accessible resources for biologists and is freely available via the Internet in a variety of ways.

There are literally thousands of **discussion groups** which have been created to allow like-minded enthusiasts to talk to each other. Many, though not all, are Usenet groups. Most Usenet groups regularly post FAQs, which are lists of Frequently Asked Questions and often have good pointers to other sources of information. The BIOSCI/bionet Frequently Asked Questions FAQ is general in nature and provides information on a variety of discussion groups for biologists. It is also available in several formats through several mechanisms. There are also a number of LISTSERV lists. Information about these LISTSERV lists can be obtained from Diane Kovacs' directory.

Gopher is a program which simplifies access to a wide variety of information sources which are available on computers from around the world. There are a variety of Gopher "clients" (i.e., software versions), some of which are mounted on individual personal computers and others of which are on mainframes. Most Gophers have some kind of subject arrangement. One strategy to use to find information from a Gopher server is to look for institutions which have a strong program in the area you are interested in. For instance, the Smithsonian Institution Gopher is a good starting point for taxonomy and biodiversity. The Indiana University IUBio Archive for Biology is another good starting point since it offers access to an extensive list of other biological gophers.

Another method of finding Gopher-accessible information is to use either Archie or Veronica, two search systems for locating files. Archie searches for FTP files (files which can be sent from a remote computer to your local machine using the FTP protocol). These files include software and text files of all types. Veronica searches for keywords in Gopher directories, and is usually the most useful system for Gopher searches.

The World Wide Web (WWW) is a group of hypertext-linked resources which has recently become very popular. Programs such as Mosaic or Lynx offer a relatively simple way to navigate the WWW. Programs such as Mosaic also offer access to Gopher and FTP servers, so they are a powerful tool. Mosaic, currently the most popular WWW client, can be searched in two ways, by doing a keyword search or by browsing various home pages which offer lists of other Internet resources. Two other sources are the "Starting Points for Internet Exploration" document and the "Internet Resources Meta-Index," which are both available under the "Navigate" menu in Mosaic.

Bibliography

Allan, Barbara and Barry Strickland-Hodge. *How to Use Psychological Abstracts and Biological Abstracts*. Brookfield, VT: Gower, 1987. 76 p. $76.00. ISBN 0566035359.

BIOSCI/bionet Frequently Asked Questions (FAQ). USENET bionet.announce. Available via Gopher, anonymous ftp, and e-mail from many archives. For a free copy via email, send the command "info faq" to biosci-server@net.bio.net.

BIOSIS Previews Search Guide. 1977– . Philadelphia: Biological Abstracts. 2 v. Updated annually.

CAB Thesaurus. 1990 ed. Wallingford, England: CAB International, 1990. 2 v. Updated irregularly. ISBN 0851986870.

Chemical Abstracts Index Guide. v. 69– , 1968– . Columbus, OH: American Chemical Society. Updated annually. ISSN 0009-2258.

CSA Life Sciences Collection Thesaurus, 3rd ed. Bethesda, MD: Cambridge Scientific Abstracts, 1991. 163 p. Updated irregularly.

Kovacs, Diane. Directory of Scholarly Electronic Conferences. For a free copy via e-mail, send the text "get acadlist readme" to listserv@kentvm.kent.edu.

Krol, E. *The Whole Internet: User's Guide and Catalog*, 2nd ed. Sebastopol, CA: O'Reilly and Associates, 1994.

Library of Congress Subject Headings, 16th ed. Washington, DC: Cataloging Distribution Service, Library of Congress, 1993. 4 v. Updated irregularly. ISSN 1048-9711.

Medical Subject Headings: Annotated Alphabetic List. 1975– . Bethesda, MD: Medical Subject Headings Section, Library Operations, National Library of Medicine. Updated annually. ISSN 0147-5711.

Medical Subject Headings: Tree Structures. 1972– . Bethesda, MD: Medical Subject Headings Section, Library Operations, National Library of Medicine. Updated annually. ISSN 0147-099X.

Permuted Medical Subject Headings. 1976– . Bethesda, MD: Medical Subject Headings Section, Library Operations, National Library of Medicine. Updated annually. ISSN 1045-2338.

Schulz, Hedda and Ursula Georgy. *From CA to CAS Online: Databases in Chemistry*, 2nd, completely rev. and enl. ed. New York: Springer-Verlag, 1994. 311 p. ISBN 0387574832.

Sears List of Subject Headings, 14th ed. Edited by Martha T. Mooney. New York: Wilson, 1991. 731 p. Updated irregularly. ISBN 082420803X.

Smith, Una R. 1993. A Biologist's Guide to Internet Resources. Usenet sci.answers. Available via Gopher, anonymous ftp, and e-mail from main archives. For a free copy via email, send the text "send pub/usenet/sci.answers/biology/guide/*" to the e-mail address mail-server@rtfm.mit.edu. Approximately 45 pages.

Strickland-Hodge, Barry. *How to Use* Index Medicus *and* Exerpta Medica. Brookfield, VT: Gower, 1986. 60 p. $32.00. ISBN 0566035324.

Thesaurus of Psychological Index Terms, 6th ed. Arlington, VA: American Psychological Association, 1991. Updated irregularly. ISBN 1557981116.

Zoological Record Search Guide. 1985– . Philadelphia: BioSciences Information Service. Updated irregularly.

3

General Sources

Introduction

This chapter describes selected sources that are relevant to biology in general, with no attempt to be comprehensive. These titles were chosen as especially appropriate for undergraduates needing an introduction to the field, or for anyone requiring sources covering the broad spectrum of the biological sciences. Knowledge of most of the publications annotated in this chapter is helpful in effectively utilizing the more specialized chapters that follow. Arrangement is by topic, presenting publications that acquaint readers to the field of biology from the viewpoint of the history of the life sciences, mathematical and statistical sources, and pertinent techniques, just to name a few of the sections that follow. These general sources may be used as a base upon which to expand or define more specific subjects, to open up the literature as a beginning, not an end.

Audiovisual Materials and Sources

A-V ONLINE. Online database accessible from DIALOG as file 46. Bibliographic information, updated quarterly. Albuquerque, NM: Access Innovations, Inc. 1964 to the present, selective earlier coverage.

This database provides comprehensive coverage of educational AV materials for preschool to professional/graduate school levels. Films, filmstrips, overhead transparencies, audio tapes, video tapes, phonograph records, motion picture cartridges, and slides are included. Also, see *Science Books and Films* in the Bibliographies section (p. 25).

AV Source Directory: A Subject Index to Health Science AV Producer/ Distributor Catalogs. 1977– . Chicago: Midwest Health Science Library Network, Management Offices, The John Crerar Library.

This index to AV producers' catalogs assigns *MeSH* subject headings and discipline headings for each source. Listing of producers is alphabetical; there is a subject index. This index is useful for biologists interested in the basic medical sciences.

The Video Librarian. v. 1– , 1986– . Bremerton, WA: Randy Pitman. Monthly. ISSN 0887-6851.

This review periodical provides evaluative, dependable reviews for video recordings. It is indexed in *Library Literature.*

The Video Source Book. 1st– ed.; 1979– . Syosset, New York: National Video Clearinghouse.

This comprehensive catalog to television programs and video tapes is arranged by title with subject access. Major producers and 1,500 distributors are included with indexes for subtitle, credits, videodisc, 8-mm, and captioned videos. Complete bibliographic information is provided for each title.

Other sources include catalogs from major publishers and distributors such as:

Ambrose Video Publishing, Inc., New York, NY;
The Annenberg/Corporation for Public Broadcasting, Washington, DC;
Britannica Films & Video, Chicago, IL;
Coronet/MTI Film & Video, Simon & Schuster, Deerfield, IL;
Films for the Humanities & Sciences, Princeton, NJ;
Films Incorporated Video, Chicago, IL;
Insight Media, New York, NY;
National Film Board of Canada, New York, NY;
National Geographic Educational Services Catalog, Washington, DC;
Voyager Video, Darien, CT.

Bibliographies

The following bibliographies, catalogs, and book review indexes may be used to verify book titles, to update and evaluate editions, and to provide purchasing and availability information for books, microforms, serials, and other materials. General bibliographies for current and retrospective journal articles are discussed in Chapter 4, "Abstracts and Indexes."

Book Review Index (BRI). 1969 – . Detroit, MI: Gale Research Inc. Bimonthly. $195.00/yr. ISSN 0524-0581.

Available in print and online from DIALOG. *BRI* contains references to more than two million reviews of approximately one million books and periodical titles, scanning more than 500 magazines and newspapers.

Books in Print. File 470, online from DIALOG.

Updated monthly, this file is provided by Bowker, New York and is the major source of bibliographic information on books currently published and in print in the United States. Scientific, technical, medical, scholarly, and popular works, as well as children's books, are included in the file. Also, see *Scientific & Technical Books & Serials in Print*, annotated on p. 25, which is a subset of on-line File 470.

British Books in Print. File 430, online from DIALOG.

Updated monthly, this directory is provided by J. Whitaker & Son, London. The database provides comprehensive indexing of books published in the United

Kingdom as well as those published throughout the world that are printed in the English language and are available in the United Kingdom.

Clewis, Beth. *Index to Illustrations of Animals and Plants.* New York: Neal-Schuman, 1991. 217 p. $49.95. ISBN 15557-00721.

This is useful as a supplement to Munz and Slauson (below), and Thompson (p. 26), and it follows a similar format. Clewis covers books published in the 1980s and lists 62,000 entries for access to 142 books with illustrations for plants and animals from around the world. The book is arranged by common name, with indexes for scientific name and book title.

Guide to Microforms in Print. 1975– . Germany: K. Saur, distributed by R. R. Bowker. *Author/Title-U.S., Canada, Mexico and South America.* Annual. ISSN 0164-0747. *Subject*, Annual. $325.00. ISSN 0163-8386. *Supplement*, 1979– . $235.00. ISSN 0164-0739.

Includes books, journals, and other materials in microform, with the exception of dissertations and theses.

LC MARC-Books. File 426 available from DIALOG.

Provided by the U.S. Library of Congress, this bibliographic file is updated weekly and covers 1968 to the present. The database contains complete bibliographic records for all books cataloged by the Library of Congress beginning with books in English and adding coverage of books in other languages from 1970–1979. The database is comprehensive and can be searched by author, title, subject, series, publication date, and other access points.

Meisel, Max. *A Bibliography of American Natural History. The Pioneer Century.* 1994 reprint. 3 v. $195.00. ISBN 0945345704.

"The Role played by the Scientific Societies; Scientific Journals; Natural History Museums and Botanical Gardens; State Geological and Natural History Surveys; Federal Exploring Expeditions in the Rise and progress of American Botany, Geology, Mineralogy, Paleontology, and Zoology."

Munz, Lucile Thompson and Nedra G. Slauson. *Index to Illustrations of Living Things Outside North America; Where to Find Pictures of Flora and Fauna.* Hamden, CT: Archon Books, 1981. 441 p. $57.50. ISBN 0208018573.

A companion volume to John W. Thompson's *Index to Illustrations . . .* (p. 26). Arrangement is by common name to illustrations of plants, birds, and animals. There is a scientific name index and a bibliography of sources. Updated by Clewis (above).

National Union Catalog. Washington, DC: U.S. Library of Congress.

Complete bibliographic records for books, manuscripts, audiovisual and carto-graphic materials, etc., cataloged at the Library of Congress, and other U.S. libraries. Records are available since the late 1800s, issued in various sections, in hundreds of volumes. In essence, this is the United States national biblio-graphy emanating from the national library. Available in print, as online data-bases from DIALOG (files 421–426), and on the Internet through Gopher as LC MARVEL. Although not specifically cited here, other countries maintain their own national bibliographies, available in print in larger public and university libraries.

Plant, Animal & Anatomical Illustration in Art & Science: A Biblio-graphical Guide from the 16th Century to the Present Day. Compiled by Gavin D. R. Bridson and James J. White. Win-chester: St. Paul's Bibliographies in association with Hunt Institute for Botanical Documenta-tion; Detroit: Omnigraphics, 1990. 450. p. $150.00. ISBN 0906795818.

This very complete guide covers illustrations for natural history, medicine, botany, and zoology. Historical bibliography is well served in this volume to include sources important since antiquity. Also, see Clewis (p. 24), Munz (p. 24), and Thompson (p. 26).

Science Books and Films. v. 1– , 1965– . Washington, DC: American Association for the Advancement of Science. Monthly. $35/yr. ISSN 0098-342X.

Includes reviews of new science books and science education films, providing more comprehensive coverage of science and technology than any other review publication.

Scientific & Technical Books & Serials in Print. 1972– . New Provi-dence, NJ: R. R. Bowker. Annual. $295/yr. ISSN 0000-054X.

Vol. 1: Subject index to books. Vol. 2: Title and Author indexes. Vol. 3: Subject and title indexes to serials; vendor listings; serials online; key to publishers' and distributors' abbreviations. These volumes provide comprehen-sive bibliographic information for scientific and technical books and serials pub-lished or distributed in the United States. Also available online as *Books in Print* from CDP and DIALOG.

Smit, Pieter. *History of the Life Sciences*.

See this entry in the Biography and History section (p. 30), for complete information.

Thompson, John W. *Index to Illustrations of the Natural World.*
    Where to Find Pictures of the Living Things of North America.
    Hamden, CT: Shoe String Press, 1983 reprint of the 1977 ed.
    265 p. $48.50. ISBN 0208020381.

This book indexes illustrations to more than 9,000 species of animals and plants from 206 books. Arrangement is by common name, with a scientific name index. Companion to Munz, and updated by Clewis.

Ulrich's International Periodicals Directory. 1932– . Annual since 1980.
    New Providence, NJ: R. R. Bowker. $395/yr. 5 vols. ISSN 0000-
    0175.

A classified list of serials covering periodicals and annuals published worldwide, U.S. newspapers, refereed serials, serials available on CD-ROM, vendor lists, cessations, and indexes to publications of international organizations, ISSN and titles. *Ulrich's* provides complete bibliographic information including beginning date, frequency, price, publisher and address, ISSN, circulation, and brief description when available. Available online from CDP and DIALOG.

Biography and History

Also, see the Directories section (p. 38) for current biographical information about scientists.

The American Development of Biology. Edited by Ronald Rainger, et al.
    Philadelphia: University of Pennsylvania Press, 1988. 380 p.
    $18.00. ISBN 081228092X.

This book, like *The Expansion of American Biology*, was commissioned by the American Society of Zoologists to celebrate its centennial (1889–1989).

Annual Bibliography of the History of Natural History. v. I– , 1985– .
    Annual. London: Natural History Museum. ISSN 0268-9986.

This publication, covering life and health sciences, aims to provide a comprehensive record of literature relating to the history of natural history. Includes biographies and obituaries.

Archives of Natural History. v. I– , 1981– . London: Natural History
    Museum. Irregular. ISSN 0260-9541. $170/yr.

Published by the Society for the History of Natural History and devoted to publishing papers within the broad field of the interest of the Society. Supersedes the *Journal of the Society for the Bibliography of Natural History.* v. 1–9, 1968–1980.

Asimov, Isaac. *Asimov's Biographical Encyclopedia of Science and Technology: The Lives and Achievements of 1510 Great Scientists from Ancient Times to the Present Chronologically Arranged*, 2nd rev. ed. Garden City, New York: Doubleday, 1982. 941 p. $29.95 (paper). ISBN 0385177712.

Biographical guide to the history of science concentrating chiefly on the last two centuries.

Asimov, Isaac. *A Short History of Biology*. Westport, CT: Greenwood Press, 1980 reprint of the 1964 ed. 189 p. $35.00. ISBN 0313225834. Index.

This brief history is by one of the most prolific and well-known popular science writers.

The Biographical Dictionary of Scientists, 2nd ed. Edited by Roy Porter. New York: Oxford University Press, 1994. 1,024 p. $85.00.

Contains biographical information for over 1,200 scientists from astronomy, chemistry, physics, biology, mathematics, engineering, technology, and geology. Seven chronological reviews of significant developments in various scientific areas, 150 illustrations, a list of Nobel laureates, an extensive glossary, and a comprehensive index are included.

Biographical Encyclopedia of Scientists, 2nd ed. Edited by John Daintith et al. Philadelphia: Institute of Physics, 1994. 2 v. $190.00. ISBN 0750302879 (set).

Contains a biography section of over short 2,000 entries. Name and subject indexes.

Borell, Merriley. *The Biological Sciences in the Twentieth Century*. (Album of Science). New York: Scribner's, 1988. 306 p. $75.00. ISBN 0684164833.

This is a pictorial record of growth of the scientific enterprise with attention to the social matrix and implications of biological research and discovery. A guide for further reading, an index, and a list of picture sources and credits are included.

Bowler, Peter J. *Charles Darwin: The Man and His Influence*. Oxford: Blackwell, 1990. 250 p. $34.00. ISBN 0631168184. Index.

Biographical history of Charles Darwin and his milieu.

Bowler, Peter J. *The Mendelian Revolution: The Emergence of Hereditarian Concepts in Modern Science and Society*. Baltimore: Johns Hopkins University Press, 1989. 207 p. $29.95. ISBN 0801838886.

The Correspondence of Charles Darwin. Edited by Frederick Burk
    hardt and Sydney Smith. New York: Cambridge University
    Press, 1985– . Multivolume. Vol. 8 (1993): 1860. 800 p.
    $59.95. ISBN 0521442419.

A monumental work of scholarship that includes a host of famous biolo-
gist/scientist correspondents.

The Darwin CD-ROM [computer file]. San Francisco, CA: Lightbinders,
    1992. Multimedia created by Pete Goldie. $100.00.

Database containing a wide variety of material by, and about the life and work
of, Charles Darwin. Complete texts for the final editions of *The Voyage of the
Beagle, The Origin of Species,* and *The Descent of Man.* A Darwin bibliography, critiques, study guides, and more than 650 color and black & white images
are included.

Dictionary of the History of Science. Edited by W. F. Bynum et al.
    Princeton, NJ: Princeton University Press, 1985. 494 p. $75.00.
    ISBN 0691082871. Biographical Index.

Seven hundred signed articles discuss the key ideas and core features of recent
Western science in all fields. A useful briefly annotated bibliography and
biographical index are included.

Dictionary of Scientific Biography. 1970– . New York: Macmillan.
    Beginning with vol. 15, issued as supplements. 3 v. $80.00
    each. ISBN 0684101149 (v. 3).

"Published under the auspices of the American Council of Learned Societies."
Covers all areas of science with lengthy, detailed entries and citations to portraits
for the most famous scientists. Includes bibliographies and index.

The Expansion of American Biology. Edited by Keith R. Benson et al.
    New Brunswick: Rutgers University Press, 1991. 357 p. $42.00.
    ISBN 0813516501.

This is the second in the two-part series commissioned by the American Society
of Zoologists of original, unpublished articles discussing 20th century American
biology. (Also, see *The American Development of Biology,* p. 26.)

Ford, Brian J. *Images of Science. A History of Scientific Illustration.*
    New York: Oxford University Press, 1993. 208 p. $45.00. ISBN
    0195209834.

Historical Studies in the Physical and Biological Sciences. v. 16– ,

1986– . Semiannual. Berkeley: University of California Press. $36/yr. ISSN 0890-9997.

Continues: *Historical Studies in the Physical Sciences* and covers history of physics, biology, and other sciences.

Hughes, Arthur Frederick William. *The American Biologist Through Four Centuries*. Springfield, IL: Thomas, 1982. 386 p. $64.50. ISBN 03980445984.

Biographies of American biologists.

Isis. An International Review Devoted to the History of Science and its Cultural Influences. v. 1– , 1913– . Chicago: University of Chicago Press for the History of Science Society. Quarterly. $110/yr. ISSN 0021-1753.

Provides review articles, research notes, documents, discussions, news, and critical bibliographies of the history of science and its cultural influences. An excellent source for keeping up with the historical literature of the biological sciences.

Journal of the History of Biology. v. 1– , 1968– . Norwell, MA: Kluwer Academic. Triennial. $126/yr. ISSN 0022-5010.

Scholarly articles and book reviews.

Kronick, David A. *Scientific and Technical Periodicals of the Seventeenth and Eighteenth Centuries: A Guide*. Metuchen, NJ: Scarecrow, 1991. 332 p. $39.50. ISBN 0810824922.

Early works to 1800.

Magner, Lois N. *A History of the Life Sciences*, 2nd ed. New York: Dekker, 1994. 496 p. $59.75. ISBN 0824789423. Index.

Introduction to the main themes of biology through the post–Watson-Crick period.

Maienschein, Jane. *Transforming Traditions in American Biology 1880–1915*. Baltimore: Johns Hopkins University Press, 1991. 366 p. $48.00. ISBN 0801841267.

A snapshot of the history of the development of the biological sciences in the United States, focusing on Edmund Beecher Wilson, Thomas Hunt Morgan, Edwin Grant Conklin, and Ross Granville Harrison.

Moore, John Alexander. *Science as a Way of Knowing: The Foundations of Modern Biology*. Cambridge, MA: Harvard University Press, 1993. 530 p. $29.95. ISBN 067479480X.

"The story of the development of concepts in the biological sciences" (from the Preface) told in four parts: Understanding Nature, Growth of Evolutionary Thought, Classical Genetics, and the Enigma of Development.

Olby, Robert C. *The Path to the Double Helix*. Seattle: University of
   Washington Press, 1974. 510 p. ISBN 0295953594.

". . . an overview of the intellectual and institutional movement in experimental biology which has yielded a physical and chemical account of the gene." Foreword was written by Francis Crick, winner, with James Watson and Maurice Wilkins, of the 1962 Nobel Prize in Physiology or Medicine.

*The Origins of Natural Science in America: The Essays of George Brown
   Goode*. Edited by Sally Gregory Kohlstedt. Washington, DC:
   Smithsonian Institution Press, 1991. 432 p. $45.00. ISBN
   1560980982.

A reprint of five essays by the American naturalist, curator, and museum administrator George Brown Good (1851–1896).

Overmier, Judith A. *The History of Biology: A Selected, Annotated
   Bibliography*. New York: Garland Publishing, 1989. (*Bibliographies of the History of Science and Technology*, vol. 15). 157 p.
   $21.00. ISBN 0824091183.

This award winning bibliography includes 619 annotated entries providing access to the history of biology. There are 17 pages of author and subject indexes which are especially useful for the beginning historian of science.

Singer, Charles. *A History of Biology to about the Year 1900: A General Introduction to the Study of Living Things*, rev. ed. Ames, IA:
   Iowa State University Press, 1989 reprint. 616 p. $22.95 (paper). ISBN 081380937.

A classic history.

Smit, Pieter. *History of the Life Sciences; An Annotated Bibliography*.
   New York: Hafner Press, 1974. 1071 p. ISBN 0028525108.
   Index.

This historical bibliography contains more than 4,000 entries with full bibliographical information plus a summary review of the work cited. The work is divided into: general references and tools; historiography of the life and medical sciences; a selected list of biographies, bibliographies, etc., of famous biologists, and medical men; and an index of personal names.

Watson, James D. *The Double Helix: A Personal Account of the
   Discovery of the Structure of DNA*. Edited by Gunther S.

Stent. New York: Macmillan, 1980. 298 p. $11.95 (paper).
ISBN 068970602.

A new critical edition of the classic account of the discovery of DNA, including text, commentary, reviews, and original papers.

Classification

The sources in this section provide general information for classification schemes for living organisms. For more specific details consult individual chapters on botany, entomology, microbiology, or zoology, or look at:

Bergey's Manual of Systematic Bacteriology. Williams & Wilkins, 1989.

Radford, Albert E. et al. *Fundamentals of Plant Systematics*. Harper & Row, 1986.

Jones, Samuel B. *Plant Systematics*, 2nd ed. McGraw-Hill, 1986.

Zoological Record, annual publication from BIOSIS.

BIOSIS Previews Search Guide. Philadelphia, PA: BIOSIS, issued annually. Variously paged. $115.00. ISBN 0916246248 (1995).

This essential guide for online searchers provides a taxonomic overview of the hierarchical structure used for microorganisms, plants, and animals.

Cladistics: A Practical Course in Systematics, by Peter L. Forey et al. Oxford, UK: Clarendon Press, 1992 reissue. (Systematics Association Publications, 10). $23.40 (paper). ISBN 0198577664.

An up-to-date account of the techniques of modern cladistics, a method of systematic classification that has become a method of choice for comparative studies in all fields of biology.

Margulis, Lynn, Karlene V. Schwartz, and Michael Dolan. *The Illustrated Five Kingdoms: A Guide to the Diversity of Life on Earth*. New York: HarperCollins, 1994. 229 p. ISBN 006500843X. Index.

Arranged by the five kingdoms of Monera, Protoctista, Fungi, Animalia, and Plantae, this reference discusses classification schemes and the general features of each kingdom, and includes an illustrative phylogeny, and a bibliography of suggested reading for selected members of each phylum. An appendix lists genera, including genus, phylum, and common name for each. There is a glossary. See p. 394 for an electronic version.

Minelli, Alessandro. *Biological Systematics: The State of the Art*. New York: Chapman & Hall, 1993. 387 p. $69.00. ISBN 0412364409.

A comprehensive essay on modern biological systematics. The book includes history, molecular methods, theory and practice, methodology of phylogenetic reconstruction, and discussion of cladistics and computers. There are 23 appendices providing various important classification schedules, an extensive bibliography, and references.

Molecular Systematics, 2nd rev. ed. Edited by David M. Hillis and Craig
    Moritz. Sunderland, MA: Sinauer, 1990. 588 p. $70.00; $43.95
    (paper). ISBN 0878932836; 0878932828 (paper).

Overview of molecular systematics. There is a new edition forthcoming in 1995.

Panchen, Alex L. *Classification, Evolution, and the Nature of Biology*.
    New York: Cambridge University Press, 1992. 403 p. $80.00.
    ISBN 0521305829.

A historical treatise on the relationships between classifications and patterns of phylogeny.

Quicke, D. L. J. *Principles and Techniques of Contemporary Taxonomy*.
    London: Chapman & Hall, 1993. 200 p. $50.00. ISBN 07514-
    0019X.

A survey of the arguments and techniques of systematics as they are applied to all groups of organisms, including principles of nomenclature and classification, and the practice of cladistics.

Synopsis and Classification of Living Organisms. Sybil. P. Parker, Editor
    in Chief. New York: McGraw-Hill, 1982. 2 v. ISBN 0070790310
    (set). Index.

Treats higher-level taxonomy. The systematic positions and affinities of all living organisms are presented in synoptic articles for all taxa down to the family level. Linnaean classifications and citations are included as a guide to the specialized literature. An appendix discusses the history and role of nomenclature in the taxonomy and classification of organisms, and provides classification tables.

A Synoptic Classification of Living Organisms. Edited by R. S. K.
    Barnes. Oxford: Blackwell Scientific Publications, 1984. 273 p.
    ISBN 0632011459.

A dictionary/mini-encyclopedia of classification and diversity that presents an outline, synoptic account of the classification of living organisms from prokaryotic bacteria, through protists, to the multicellular fungi, plants and animals. There are references and suggestions for further reading, and an index to taxa is provided.

Dictionaries and Encyclopedias

Acronyms, Initialisms, and Abbreviations Dictionary. 1976– . Annual
    edition or supplement. Detroit: Gale Research Co. 3 v. $895/set.
    ISSN 0270-4404.

Over 520,000 definitions for all subjects including biology. Vol. 1: Acronyms,
Initialisms & Abbreviations Dictionary. Vol. 2: New Acronyms, Initialisms &
Abbreviations. Vol. 3: Reverse Acronyms, Initialisms & Abbreviations Dictionary.

Blinderman, Charles. *Biolexicon: A Guide to the Language of Biology*.
    Springfield, IL: Thomas, 1990. 363 p. ISBN 0398056714.
    $39.75. Bibliography. Index.

This guide to the vocabulary of biology and medicine helps students decipher
the language by providing insights into philosophy, religion, history, mythology,
theories of evolution, Renaissance anatomy, and spooky obsessions. The book
is arranged by broad topic rather than by presenting lists of terms.

Borror, Donald Joyce. *Dictionary of Word Roots and Combining Forms
    Compiled from the Greek, Latin, and Other Languages, with
    Special Reference to Biological Terms and Scientific Names*.
    Mountain View, CA: Mayfield, 1960. 134 p. $8.95 (paper).
    ISBN 087484053.

This is of particular value to the beginning student or taxonomist.

Cambridge Dictionary of Biology. Editor, Peter M. B. Walker. New
    York: Cambridge University Press, 1990. 324 p. $15.95 (paper).
    ISBN 0521397642.

Previously published under the title *Chambers Biology Dictionary*, this volume
defines 10,000 terms in zoology, botany, biochemistry, molecular biology, and
genetics.

Cambridge Encyclopedia of Life Sciences. New York: Cambridge Uni-
    versity Press, 1985. 432 p. ISBN 0521256968.

This encyclopedia presents the science of biology as linked biological phenomena. The volume is divided into broad topics discussing processes and organization, environments, evolution, and the fossil record. There is a classification of
living organisms, and a species and subject index are included.

Cambridge Illustrated Dictionary of Natural History. Edited by R. J.
    Lincoln and G. A. Boxshall. Cambridge, MA: Cambridge University
    Press, 1987. 413 p. $18.95 (paper). ISBN 0521305519.

Content reflects the popular image of natural history: life on Earth. Over 700
line drawings supplement the definitions.

Concise Dictionary of Biology. Oxford: Oxford University Press, 1985.
    256 p. ISBN 0198661444. $7.95.

This dictionary is derived from the *Concise Science Dictionary*, published by Oxford University Press in 1984.

Dictionary of Chemical Names and Synonyms, compiled by Philip H.
    Howard and Michael Neal. Boca Raton, FL: Lewis, 1991. Variously paged. $149.95. ISBN 0873713966.

Identification for over 20,000 chemicals, including the appropriate chemical name, registry number, molecular weight, and structure for 95% of the chemicals people use the most. There are indexes for chemical synonyms and formulae, using the Hill system.

A Dictionary of Life Sciences, 2nd ed. rev. Edited by E. A. Martin.
    New York: Pica Press, distributed by Universe Books, 1984. 396
    p. ISBN 0876637403. $22.50.

There is an increase of some 300 new entries in this revised edition.

Dictionary of Light Microscopy, compiled by the Nomenclature Committee of the Royal Microscopical Society. (Microscopy handbooks,
    15.) Oxford: Oxford University Press for the RMS, 1989. 139 p.
    ISBN 0198564139. $10.50 (paper). Cover title: *RMS Dictionary of Light Microscopy*.

Terms were chosen to conform to current usage, international standards, and the word's origin and meaning. Incorrect, obsolete, and inconsistent terms are indicated; RMS hopes that the recommended terms will be used by manufacturers, for teaching purposes, and for scientific papers. Appendixes include figures and tables, and English-French-German equivalent terms.

Dictionary of Natural Products. J. Buckingham, Executive Editor. New
    York: Chapman & Hall, 1994. 7 v. $4,995.00 (set). ISBN 04124-
    66201 (set). Annual supplements.

Contains the structures, bibliographies and physical properties of 100,000 natural products grouped within approximately 35,000 entries.

Dictionary of Steroids: Chemical Data, Structures, and Bibliographies,
    Edited by R. A. Hill et al. New York: Routledge, Chapman & Hall,
    1991. 2 v. $1,350 (set). ISBN 0412270609. Index.

This unique synthesis provides bibliographic, structural, and chemical data for over 10,000 steroids, aiming to document all known naturally occurring steroids of plant and animal origin, plus the most important synthetic steroids.

Dictionary of the History of Science.

Annotated in the Biography and History section (p. 28).

Dictionary of Theoretical Concepts in Biology, compiled by Keith E. Roe
    & Richard G. Frederick. Metuchen, NJ: Scarecrow Press, 1981.
    267 p. ISBN 081081353X. $32.50.

Provides access to the literature through 1979 on 1,166 named theoretical concepts by citing original sources and reviews illuminating those concepts. Both plant and animal biology are represented.

Elsevier's Dictionary of Microscopes and Microtechniques: in English,
    French, and German. Compiled by Robert Serre. New York:
    Elsevier, 1993. 286 p. $142.75. ISBN 0444889736.

A trilingual dictionary covering 1,827 terms found in contemporary microscope and microtechniques publications. Definitions, synonyms, cross-references, and a complete bibliography to sources are provided.

Encyclopedia of Bioethics. Warren T. Reich, Editor-in-Chief. New York:
    Macmillan Reference, 1994. 5 v. $425.00. ISBN 0028973550
    (set).

A complete revision of the award-winning encyclopedia, this set contains 460 authoritative, original articles, extensive bibliographies and a comprehensive index. Coverage: ethical and moral dimensions of scientific and technological innovations; basic concepts of life, death, health, etc.; principles and virtues used as guides for human behavior; ethical theories; the changing nature of ethics; worldwide religious traditions; history of medical ethics; and disciplines contributing to bioethics from anthropology to the sociology of science.

Encyclopedia of Human Biology. Editor-in-Chief, Renato Dulbecco. San
    Diego, CA: Academic Press, 1991. 8 v. $2,100 (set). ISBN
    0122267478 (set). Index (v. 8).

Over 600 articles, each averaging about ten pages, provides authoritative, up-to-date information on human biology, covering anthropology, behavior, biochemistry, biophysics, cytology, ecology, evolution, genetics, immunology, neurosciences, pharmacology, physiology, toxicology, etc. There are 2,000 illustrations; 50 color plates; 200 tables; 5,500 bibliographic entries, and 4,800 glossary entries. The advisory board reads like the who's who of the scientific world to include 11 Nobel laureates. There is a 40,000-term subject index.

The Facts on File Dictionary of Biology, Rev. and expanded ed. Edited
    by Elizabeth Tootill. New York: Facts on File, 1988. 326 p.
    $24.95. ISBN 0816018650.

This volume defines the most commonly used biological terms including those in genetics and molecular biology.

Glossary of Microscopical Terms and Definitions, 2nd ed. 1989.

Chicago: McCrone Research Institute for the New York Microscopical Society, 1989. (The Microscope series, v. 51). 78 p. ISBN 0904962121. Bibliography.

The purpose of this dictionary is to clarify meanings of microscopically oriented terms. Includes people, instrumentation, laws, and terms, all arranged alphabetically.

HarperCollins Dictionary of Biology. Edited by W. G. Hale and J. P. Margham. New York: Harper Perennial, 1991. 569 p. $13.00 (paper). ISBN 0064610152.

Originally published 1988 in Great Britain by William Collins, Son & Co. under the title *Collins Dictionary of Biology*. More than 5,600 entries and 300 diagrams provide in-depth explanations and illustration. All major biological subjects are covered including biographies of important biologists.

Henderson, Isabella Ferguson. *Henderson's Dictionary of Biological Terms*. Edited by Eleanor Lawrence, 10th ed. New York: Wiley, 1989. 637 p. $57.95. ISBN 0470214465.

Pronunciation, derivation, and definition of terms in biology, botany, zoology, anatomy, cytology, genetics, embryology, physiology. An excellent dictionary and a fine place to start.

Illustrated Glossary of Protoctista. Vocabulary of the Algae, Apicomplexa, Ciliates, Foraminifera, Microspora, Water Molds, Slime Molds, and Other Protoctists. Edited by Lynn Margulis, Heather I. McKhann, and Lorraine Olendzenski. Boston: Jones & Bartlett, 1993. 288 p. $50.00. ISBN 0867200812.

An abbreviated version of the *Handbook of Protoctista. . . .* This glossary focuses on terminology and systematics for this group of organisms.

International Dictionary of Medicine and Biology. Editorial Board, E. Lovell Becker et al; editor-in-chief, Sidney I. Landau. New York: Wiley, 1986. 3 v. $400.00. ISBN 047101849X.

This unabridged dictionary of 159,000 definitions covers the fields of the traditional basic and clinical medical sciences, technological specialties, and the delivery of health care.

Jaeger, Edmund C. *A Source-Book of Biological Names and Terms*, 3rd ed. Springfield, IL: Thomas, 1978. 360 p. $48.50. ISBN 0398009163.

The classic guide to the vocabulary of biology.

McGraw-Hill Dictionary of Biology. Sybil P. Parker, Editor-in-Chief.
New York: McGraw-Hill, 1984. 384 p. $15.95 (paper). ISBN
0070454191.

Material previously published in the 1984 3rd ed. of the *McGraw-Hill Dictionary of Scientific and Technical Terms*, currently in its 45th ed., 1989. There is a 5th edition of this *Dictionary of Scientific and Technical Terms*, 1993, $110.50.

McGraw-Hill Encyclopedia of Science and Technology, 7th ed. Sybil P.
Parker, Editor-in-Chief. New York: McGraw-Hill, 1992. 20 v.
$1,900.00. ISBN 0079092063 (set).

A comprehensive science and technology encyclopedia that deserves its continued excellent reputation. *The McGraw-Hill Scientific and Technical Reference Set* on CD-ROM combines the *McGraw-Hill Concise Encyclopedia and Science and Technology* with the *McGraw-Hill Dictionary of Scientific and Technical Terms*.

Medawar, Peter Brian and J. S. Medawar. *Aristotle to Zoos: A Philosophical Dictionary of Biology.* Cambridge, MA: Harvard University Press, 1983. 305 p. $18.50. ISBN 0674045351.

Confined to biological topics, this book is not a dictionary in the usual sense. It is arranged alphabetically, but in its authors' words, it is for browsing and reflection among theoretical concepts and ideas. The entries range from several lines to several pages; there is an index.

The Merck Index: An Encyclopedia of Chemicals, Drugs, and Biologicals, 11th ed. Rahway, NJ: Merck, 1989. 1 v. $35.00. ISBN 091191028X.

This is an invaluable dictionary for more than 10,000 significant drugs, chemicals, and biologicals. Available online from CDP.

Multilingual Illustrated Dictionary of Aquatic Animals and Plants. Commission of the European Communities. Oxford, England: Fishing News Books, 1993. 900 p. ISBN 0852382065.

Scientific name, family name, and name in each of nine languages for over 1,400 species of fish, crustaceans, molluscs, seaweeds, and other fishery products landed for commercial purposes.

Stedman's Medical Dictionary, 25th ed. Baltimore: Williams & Wilkins, 1990. 1,784 p. $43.00. ISBN 0683079166.

This and *Dorland's Illustrated Medical Dictionary*, 27th ed., 1988 ($43.25) are excellent sources for medical terminology useful for biological scientists.

USAN and the USP Dictionary of Drug Names. Rockville, MD: United
States Pharmacopeial Convention, Inc., 1992. 824 p. ISSN
0090-6816.

A compilation of the United States Adopted Names (USAN) selected and re-
leased from June 15, 1961, through June 15, 1991, current USP and NF names
for drugs, and other nonproprietary drug names. This essential reference lists
U.S. adopted name, molecular formula, chemical name, registry number, phar-
macologic and/or therapeutic activity, brand name, manufacturer, code desig-
nation, and graphic formula.

Directories

For more information about specific societies, consult the Societies section (p.
66) and the multivolume *Encyclopedia of Associations* (p. 39). Also, check out
specific societies or subjects using Gopher or the World Wide Web on the Internet.

Agricultural Information Resource Centers: A World Directory 1990.
Edited by Rita C. Fisher et al. Urbana, IL: International Association
of Agricultural Librarians and Documentalists, 1990. 641 p.
$100.00. ISBN 0962405205.

A directory to 2,531 world agricultural information resource centers. There are
institution and subject indexes.

*American Men and Women of Science; A Biographical Directory of
Today's Leaders in Physical, Biological and Related Sciences
1995–96,* 19th ed. New Providence, NJ: R. R. Bowker, 1994. 8
v. $850.00. ISBN 0835234630.

Information includes birthplace and date, spouse and children's names, scientific
field, education, honors and awards, experience, research focus, professional
membership, e-mail, fax, and mailing address. Available online from DIALOG
as part of File 236, BOWKER BIOGRAPHICAL DIRECTORY.

American Type Culture Collection.
For information, see the Handbooks and Manuals section (p. 47).

*Directory of Electronic Journals, Newsletters, and Academic Discussion
Lists,* 4th ed. Compiled by Diane K. Kovacs. Washington, DC:
Association of Research Libraries, 1994. 588 p. $59.00. Avail-
able for no charge on the Internet.

Directory to 1,800 mailing lists and 440 electronic journals.

Directory of Research Grants. Phoenix, AZ: Oryx Press. Annual.
     $135/yr. ISSN 0146-7336.

Includes almost 6,000 sources for research grants and scholarships in the United States offered by federal, state, and local governments; commercial organizations; associations; and private foundations. Online access is provided by the *GRANTS* Database from DIALOG, File 85; also available on CD-ROM.

Encyclopedia of Associations. 1st ed.– , 1956– . Detroit: Gale
     Research Co. 29th ed.: 3 v. $1,085 (set).

Provides detailed entries for over 22,000 associations, organization, clubs, and other non-profit membership groups in all fields in the United States. Vol. 1, in 3 parts: National Organizations of the United States. ISBN 0810392461. $415.00; Vol. 2: Geographic and Executive Indexes. ISBN 081039250X. $320.00; Vol. 3: Supplement. ISBN 0810392518. $350.00. Vol. 4: International Organizations, 29th ed.; 2 v. plus supplement. ISBN 0810389568, $510.00. Also, available online from DIALOG, as File 114, and on CD-ROM.

Foundation Directory. 1st ed.– , 1960– . New York: The Foundation
     Center. Annual. ISSN 0071-8092.

Comprehensive directory of 32,500 active grant funding foundations, community foundations, operating foundations, and corporate grant makers. Available online from DIALOG, File 26.

Guide to Biological Field Stations: Directory of Members, 1992. Edited
     by J. F. Merritt and C. J. Hannakan. Tyson Research Center,
     Washington University, P.O. Box 258, Eureka, MO: Published by
     the Organization of Biological Field Stations, 1992. $10.00.

Information about 150 biological field stations in North and Central America, Mexico, and the Caribbean, including location, environment, facilities, and ongoing research and educational programs.

"Guide to Scientific Products, Instruments and Services"

Published annually, ISSN 0036-8075, as a supplement to *Science.* This directory includes product listings for animals, cages, equipment, biologicals, chemicals, instruments and accessories, computers, software, databases, laboratory furniture, glassware, hardware and equipment, and services. There are indexes for manufacturers, products, and advertisers.

Life Sciences Organizations and Agencies Directory, compiled by Mar-
     garet Labash Young. Detroit, MI: Gale Research Co., 1988. 864
     p. ISBN 0810318261. $175.00.

A guide to approximately 8,000 organizations and agencies providing information in the agricultural and biological sciences worldwide, including: associations, botanic gardens, computer information services, consulting firms, educational institutions, libraries and information centers, research centers, standards organizations, state and federal agencies.

Linscott's Directory of Immunological and Biological Reagents.
See the annotation (p. 185) in Chapter 8, "Microbiology and Immunology," for the "world's resource for locating immunological and biological reagents".

Peterson's Guide to Graduate Programs in the Biological and Agricultural Sciences, 1966– . Princeton, NJ: Peterson's Guides. Annual. $40.95/yr. ISSN 0894-9360.
Comprehensive guide to postbaccalaureate programs in the United States and Canada. Available on CD-ROM, and online from DIALOG as File 273, **Peterson's Gradline**.

Research Centers Directory. 1st– . 1960– . Detroit: Gale Research Co. ISSN 0080-1518. 2 v. $455 (set).
The 1993 edition covers 12,800 university related, independent, and non-profit research centers in the United States and Canada. This comprehensive guide encompasses life sciences, physical sciences and engineering, private and public policy affairs, sociology and cultural studies, and multidisciplinary and research-coordinating centers. Arrangement is by subject with indexes for alphabetized associations and keywords, geographic locations, and subjects. This *Directory* is available online from DIALOG as file 225; diskette and magnetic tape may also be purchased. *New Research Centers*, the supplement to the 1993 edition, includes 500 centers not previously listed (ISSN 0028-6591). Also, refer to *Research Centers and Services Directory*, file 115 from DIALOG, that includes information from 27,800 research centers in 147 countries.

World Directory of Biological and Medical Sciences Libraries. Edited by Ursula H. Poland. New York: K. G. Saur, 1988. 203 p. (IFLA publications, 42). $30.00. ISBN 3598217722.
This questionnaire-generated directory covers libraries with collections in the biological and medical sciences, dentistry, veterinary sciences, and pharmaceutical sciences. All relevant libraries in developing countries were included, but only the top 25 major resource libraries in developed countries were listed. In addition, three appendixes consist of a bibliography of national and regional directories of biological and medical sciences libraries; a list of associations of biological and medical sciences librarians; addresses of union lists and cooperative service centers.

World Guide to Scientific Associations and Learned Societies, 6th ed.
    New Providence, NJ: K. G. Saur, 1994. 560 p. $245.00. ISBN
    3598205805.

All areas of academic study, culture, and technology are covered in this
directory of 17,000 national and international societies. The arrangement is
alphabetical by country, then by association name. Listings include organization
name, address, phone, telex, cable address and telefax, year of foundation,
number of members, President, General Secretary, area of activity, and publication information. There are indexes for name, subject, publications, and
German-English concordance to areas of specialization.

World Guide to Special Libraries, 3rd ed. New Providence, NJ: K. G.
    Saur, 1994. 2 v. $325.00. ISBN 3598222343.

Covers more than 35,000 libraries in 183 countries and provides access to vital
and important archives dedicated to special subjects or classifications. Data
includes name and address, holdings, loan policy, electronic data network
connections, memberships in professional and specialist associations, and the
like. Name and subject indexes.

Field Guides

There are a large number of excellent field guides available. Rather than providing a comprehensive list, this section annotates some of the more popular
field guide series. For more information, consult specific chapters depending on
subject matter, e.g., see Chapter 10, "Plant Biology," for botanical identification
manuals.

A Field Guide for the Identification of Field Guides at UIUC Libraries,
    compiled by Diane Schmidt. Urbana, IL: Diane Schmidt, privately
    published, 1991. 145 p.

This guide to field guides is useful for verification and for finding the appropriate guide to animals, plants, rocks, etc. It is arranged by subject and geographic
location.

Series

Audubon Society Field Guides

Published by Knopf in New York, this highly regarded series is useful for amateurs as well as professional plant biologists. The Audubon series presents color
photographs rather than drawings.

Golden Field Guides

Published in New York by Golden Press, these field guides cover all of North America rather than isolated sections of the United States. These guides are particularly convenient to use because they provide descriptions, illustrations, and maps of each animal or plant on the same two-page spread.

Peterson Field Guide Series

Published by Houghton Mifflin in Boston, this well-known series covers almost every group of plants and animals in almost every part of North America.

Pictured Key Series

Published by William C. Brown in Dubuque, IA, this very successful series includes introductory sections on how to look, where to look, how to collect, and how to use keys. Written for the amateur, each book provides a glossary, index, and numerous illustrations.

Guides to Internet Resources

The Internet is a superstore of information, from directories to genetic sequence databases. There are electronic catalogs, newsgroups, mailing lists, and electronic journals, just to name a few categories that biologists will find interesting. Following is a list of sources, or points of access, for the rapidly evolving, ever-changing electronic information highway.

A Biologist's Guide to Internet Resources

A free 45-page guide containing an overview and lists of free resources for biologists is available over the Internet via Usenet, Gopher, anonymous file transfer protocol (FTP), and e-mail. For more information, see Chapter 2, "Subject Access to Biological Information."

Directory of Electronic Journals, Newsletters, and Academic Discussion Lists

Look at the Directories section (p. 38) for information.

Li, Xia and Nancy B. Crane. *Electronic Style: A Guide to Citing Electronic Information*. Westport, CT: Meckler, 1993. 65 p. $15.00. ISBN 088736909X. Index.

Guidance for proper citation of electronic information, including citation style recommended by the American Psychological Association for full-text information files, bibliographic databases, Internet-accessible electronic journals and discussion lists, electronic conferences, electronic bulletin boards, and e-mail. There is a glossary.

"Top Ten List of Good Gophers." *Library Management Quarterly* v. 17(2): 13-15, Spring 1994.
There are brief descriptions for each Gopher.

Guides to the Literature

Alston, Y. R. and J. Coombs. *Biosciences: Information Sources and Services*.
Annotated in Chapter 7. "Genetics and Biotechnology."

Ambrose, Harrison W. and Katharine Peckham Ambrose. *A Handbook of Biological Investigation*, 4th ed. Winston-Salem, NC: Hunter Textbooks, 1987. 204 p. $14.95. ISBN 0887250742.
This book is useful for the beginning student in biology who needs an introduction to experimental design, biological research, writing a scientific paper, and the biological literature. The fifth edition is forthcoming.

The Author's Guide to Biomedical Journals; Complete Manuscript Submission Instructions for 185 Leading Biomedical Periodicals. New York: Mary Ann Liebert, Inc., 1993. 600 p. $175.00 (spiral). ISBN 0913113611.
Contains complete, unabridged submittal instructions for each of 185 leading biomedical journals. There is a special directory section that includes a complete list of publishers' names, addresses, and phone numbers.

Beynon, Robert J. *Postgraduate Study in the Biological Sciences: A Researchers Companion*. Brookfield, VT: Portland Press, 1993. 150 p. ISBN 1855780097. $20.00.
Intended for biological and biomedical postgraduate researchers, this guide provides basic advice and information on postgraduate study, experimental skills, research projects, scientific writing, public speaking, computers, safety and legislative control, teaching, and beginning career moves. Smith (discussed in this section, p. 46) is also a useful, if somewhat more elementary, guide.

BIOSIS Previews Search Guide. Philadelphia, PA: Biological Abstracts, issued annually. Variously paged. ISBN 0916246248. 1995. $115.00.
This guide is essential for anyone searching the BIOSIS databases. Divided into

seven sections plus a glossary and index, this volume discusses editorial policies, search system information, master index, controlled keywords, concept code directory, biosystematic code directory, and searching fundamentals.

Day, Robert A. *How to Write and Publish a Scientific Paper*, 4th ed. Phoenix, AZ: Oryx, 1994. 240 p. $25.00. ISBN 0897748646.

The author, with a nice sense of humor, focuses on the unchanging principles of good scientific writing.

Guide to Sources for Agricultural and Biological Research. Edited by R. Richard Blanchard and Lois Farrell. Berkeley, CA: University of California Press, 1981. Sponsored by the United States National Agricultural Library. 735 p. $60.00. ISBN 0520032268.

Primarily a guide to the literature of agriculture, this reference book is also relevant to the biological sciences.

Illustrating Science: Standards for Publication. Scientific Illustration Committee, Council of Biology Editors. 1988. 296 p. $49.95. ISBN 0685218937. Index.

The purpose of the book is to develop specific standards and guidelines for publication of illustrated scientific material. The Association of Medical Illustrators, the Guild of Natural Science Illustrators, the Biological Photographic Association, and the Graphic Artists Guild collaborated on this project. Also, refer to Zweifel later in this section (p. 47).

Introduction to Reference Sources in the Health Sciences, 3rd ed. Compiled by Fred W. Roper and Jo Anne Boorkman. Metuchen, NJ: Scarecrow Press, 1994. 301 p. Price not reported. ISSN 0810828898.

Although aimed at library students and librarians, this is an excellent guide to the medical literature that serves well for broader purposes.

Malinowsky, H. Robert. *Reference Sources in Science, Engineering, Medicine, and Agriculture*. Phoenix, AZ: Oryx, 1994. 328 p. $49.95. ISBN 0897747429.

More than 2,400 bibliographic entries covering science, engineering, medicine, and agriculture provide complete descriptions of important reference materials for the student and/or librarian.

Morehead, Joe and Mary Fetzer. *Introduction to United States Government Information Sources*, 4th ed. Englewood Cliffs, NJ: Libraries Unlimited, 1992. 474 p. $32.50 (paper). ISBN 1563080664.

This classic textbook serves as a guide to sources of government publications in print, nonprint, and electronic format.

Nyberg, Cheryl Rae, Maria A. Porta and Carol Boast. *Laboratory Animal Welfare: A Guide to Reference Tools, Legal Materials, Organizations, Federal Agencies*. Twin Falls, ID: BN Books, 1994. 389 p. $100.00. ISBN 0961629398.

". . . partially supported by the U.S. Dept. of Agriculture, National Agricultural Library, Agreement NO. 58-32U4-7-031, and by the University of Illinois at Urbana-Champaign" (from title page verso). This valuable guide includes chapters on reference materials, databases, legal materials, United States Government agencies, organization, information centers, consultants, acronyms, and current law dealing with animal welfare.

Opportunities in Biology. Committee on Research Opportunities in Biology, Board on Biology, Commission on Life Sciences Staff, National Research Council. Washington, DC: National Academy Press, 1989. 448 p. $42.50. ISBN 0309039274.

This evaluation of research opportunities in biology discusses new technologies, molecular structure, genes and cells, development, the nervous system and behavior, the immune system, evolution and diversity, ecology, advances in medicine, the biochemical process industry, plant biology, agriculture, and the biology research infrastructure.

The Oxford Dictionary for Scientific Writers and Editors. New York: Oxford University Press, 1991. 389 p. $36.00. ISBN 0198539207.

Provides a guide for presenting scientific information for scientists, science writers, and editors. The dictionary covers physics, chemistry, botany, zoology, biochemistry, genetics, immunology, microbiology, astronomy, medicine, mathematics, and computer science.

Pechenik, Jan A. *A Short Guide to Writing about Biology*, 2nd ed. New York: HarperCollins, 1993. 240 p. $10.00 (paper). ISBN 0673521281.

This book, aimed at college students and teachers, covers aspects of biology writing and related topics such as laboratory reports, term papers, preparing research proposals, letters of application and resumes, and giving oral presentations. Both content and style of writing are covered.

Scientific Style and Format; The CBE Manual for Authors, Editors, and Publishers, 6th ed. New York: Cambridge University Press, 1994.

704 p. $34.95. ISBN 0521471540. Formerly, the *CBE Style Manual: A Guide for Authors, Editors, and Publishers in the Biological Sciences.*

"Now covering all the physical sciences and mathematics as well as the life sciences." This standard for style in the biological sciences has been completely reorganized, with coverage expanded to all sciences and with a new focus on publication style and formats for scientific papers, journals, and books, International in scope, it recognizes both American and British preferences. Also, see Zeiger (p. 47) for writing biomedical research papers.

Smith, Robert V. *Graduate Research: A Guide for Students in the Sciences*, 2nd ed. New York: Plenum, 1990. 292 p. $24.95. ISBN 0306434652.

Problems of developing and improving research skills and preparing for professional careers are addressed in this step-by-step, detailed guide. Smith may be used alone, or as a precursor for Beynon (p. 43); topics range from "Getting Started" to "Getting a Job," with chapters on ethics; time management; library work; and choosing an advisor, a research problem, role models, just to mention a few topics discussed in this unusual and very useful guide.

Walford's Guide to Reference Material, 6th rev. ed. London: Library Association, distributed by UNIPUB, 1993. $195.00 (v. 1). ISBN 1856040151 (v. 1). Volume 1 covers science and technology.

Classic, standard resource to the literature of science and technology.

Wiggins, Gary. *Chemical Information Sources*. New York: McGraw-Hill, 1991. 352 p. $35.94. ISBN 0079099394. Includes two 3½" disks for IBM, or 100% compatible, personal computer.

This highly recommended textbook is designed to provide access to the chemical literature for chemists, librarians, and students. The printed volume describes in detail a number of basic sources, both printed and computer readable. The supplementary disks provide a computer-readable file of over 2,150 records, called the Chemistry Reference Sources Database (CRSD), which lists the complete reference collection of the Indiana University Chemistry Library plus entries from chemistry libraries at Purdue University, the University of Illinois, and the University of Michigan. The CRSD records full bibliographic information for the works discussed in the printed book and also serves as a location device for interlibrary loan. Updated information about resources for chemists is available on the Internet via the Indiana University Gopher.

Winter, Charles A. *Opportunities in Biological Science Careers*. Lincolnwood, IL: VGM Career Horizons, 1990. 149 p. $12.95. ISBN 0844286265. Glossary.

The aim of this book is to provide useful information to young people seeking career guidance. Periodic updates are planned. Chapters include basic information on biology, education requirements, and careers in biology, biomedicine, and applied biology.

Zeiger, Mimi. *Essentials of Writing Biomedical Research Papers*. New York: McGraw-Hill, Health Professions Division, 1992. 422 p. ISBN 007072833X.

The goal of this book is teach professionals how to write clearly for journal articles in the biomedical sciences. Sections discuss word choice, sentence and paragraph structure, text of the research paper, figures and tables, references, and choosing a good title. Exercises are included as an integral part of the discussion and demonstration. Also, see *Scientific Style and Format* (p. 45) for related information.

Zweifel, Frances W. *A Handbook of Biological Illustration*, 2nd ed. Chicago: University of Chicago Press, 1988. 160 p. (Chicago Guides to Writing, Editing, & Publishing). $9.95 (paper). ISBN 0226997014.

The objective is to assist the non-artist biologist produce useful, and even aesthetically pleasing, illustrations. Written by a well-known, free-lance biological illustrator. See also *Illustrating Science* (p. 44).

Handbooks and Manuals

The American Type Culture Collection (ATCC) is a nonprofit, private organization that acquires, preserves, and distributes well-characterized biological cultures for the international research community. It produces culture catalogs, lab manuals and guides, newsletters, workshops, educational products, and a culture depository service. Printed catalogs are free of charge in the United States; electronic access is provided via Gopher on the Internet, diskette, or CD-ROM. Some examples of the catalogs and manuals: *Algae and Protozoa, Bacteria and Bacteriophage, Recombinant DNA Materials, Quality Control Methods for Cell Lines, Preservation Methods: Freezing and Freeze-Drying*. For more information, contact the American Type Culture Collection, 12301 Parklawn Drive, Rockville, MD 20852. Phone: 301-881-2600; Fax: 301-231-5826.

Biomass Handbook. Edited by Osamu Kitani and Carl W. Hall. New York: Gordon and Breach, 1989. 963 p. $349.00. ISBN 2881242693.

The aim is to provide comprehensive knowledge of biomass and related systems, including recent technological developments in biotechnology, production, conversion, transportation, and utilization. Numerous table, charts, graphs, and statistics; author and subject indexes.

Biosafety in the Laboratory: Prudent Practices for the Handling and Disposal of Infectious Materials. Committee on Hazardous Biological Substances in the Laboratory, Board on Chemical Sciences and Technology, Commission on Physical Sciences, Mathematics, and Resources, National Research Council. Washington, DC: National Academy Press, 1989. 222 p. $29.95. ISBN 0309039754.

Chapters present an overview and summary of the recommendations, epidemiology of occupational infections, safe handling of infectious agents, good laboratory practices, safe disposal, safety management, references. Appendixes include information on biosafety, HIVs, zoonotic pathogens, transport regulations, teaching aids, regulation, and accreditation.

Briggs, Shirley Ann and the staff of Rachel Carson Council. *Basic Guide to Pesticides: Their Characteristics and Hazards.* Washington, DC: Hemisphere Pub. Corp., 1992. 283 p. $39.50. ISBN 1560322535. Bibliography. Index.

Appropriate for either the novice or specialist, this valuable compilation covers 700 pesticides and their contaminants. There is an index of common, trade, chemical names and CAS registry numbers; charts of pesticide characteristics covering names, class of chemical, chief use, persistence, effects in mammals, adverse effects on other species, and physical properties; information on chemical classes of pesticides; references to the literature; and six appendices on uses, alternatives, carcinogenicity testing, environmental impact, economic impact, and U.S. federal regulations 1910–1988.

CRC Handbook of Chemistry and Physics, 75th ed. Boca Raton, FL: CRC Press, 1994. 2,608 p. $99.50. ISBN 084930475X.

Indispensable, reliable source for chemical, physical, and engineering data including mathematical tables, elements and inorganic compounds, general chemical tables, physical constants, etc.

CRC Handbook of Laboratory Safety, 3rd ed. Boca Raton, FL: CRC Press, 1990. 728 p. $110.00. ISBN 849303532.

Authoritative reference providing information on protective equipment, chemical reactions, hazards of ventilation, fire, toxic substances, radiation, infection, microbiological techniques, etc.

Geigy Scientific Tables, 8th rev. ed. West Caldwell, NJ: Ciba-Geigy Corp. 4 v. 1981–1986. Includes bibliographies and index. ISBN 0914168509.

Tables for units of measurement, body fluids, nutrition, statistics, mathematical formulae, physical chemistry, blood, somatometric data, biochemistry, inborn error of metabolism, pharmacogenetics and ecogenetics.

Guidelines for Acquisition and Management of Biological Specimens. Lawrence, KS: Association of Systematics Collections, Museum of Natural history, University of Kansas, 1982. 42 p. ISBN 0942924029.

"A Report of the Participants of a Conference on Voucher Specimen Management." This report was published to remedy the general lack of understanding of the need for quality voucher specimens for reliable, accurate, good research, and to provide guidelines for the management of voucher specimens as well as the proper methods of collection, identification, preparation (including required data elements), and selection of an appropriate repository. Legislation, scholarly publication policies, costs, fees, and funding responsibilities are also considered.

Handbook of Pesticide Toxicology. Edited by Wayland J. Hayes, Jr., and Edward R. Laws, Jr.. San Diego: Academic Press, 1991. 3 v. $495.00 (set). v. 1: General Principles, ISBN 01233416612; v. 2–3: Classes of Pesticides, ISBN 0123341620 and 0123341639, respectively.

A review in *Environmental Research* (v. 56, Dec. 1991, p. 214–15) called this handbook "encyclopedic and indispensable as a reference guide to pesticide toxicology."

Handbooks and Tables in Science and Technology, 3rd ed. Edited by Russell H. Powell. Phoenix, AZ: Oryx, 1994. 359 p. $95.00. ISBN 0897745345.

Compilation of over 3,600 handbooks and tables in science, technology, and medicine, completely indexed by subject, keyword, author/editor, and title. The main section is arranged alphabetically by title and includes complete bibliographic information, and a brief annotation. Popular "how to" handbooks, dictionaries, encyclopedias, field guides, maps, directories, biographical sources, indexing, abstracting, and current awareness sources are generally excluded.

The Handbooks of Aging: Consisting of Three Volumes: Critical Comprehensive Reviews of Research Knowledge, Theories, Concepts, and Issues, 3rd ed. Editor-in-Chief, James E. Birren. San Diego: Academic Press, 1990. 3 v. V. 1: *Handbook of the Biology of Aging*; V. 2: *Handbook of the Psychology of Aging*; V. 3: *Handbook of Aging and the Social Sciences*.

Richardson, James H. *Handbook for the Light Microscope: A User's Guide*. Park Ridge, NJ: Noyes, 1991. 522 p. $79.00. ISBN 0815512694.

Complete and concise information on how to use a wide variety of optical microscopes correctly and easily. There are thorough descriptions of each type of optical microscope, including applications, setup procedures, accessories, and appendixes listing immersion fluids and long-working distance lenses.

Watterson, Andrew. *Pesticide Users' Health and Safety Handbook: An International Guide*. New York: Van Nostrand Reinhold, 1988. 504 p. ISBN 0442234872. Glossary, bibliography, index.

Written for lay people concerned about human and environmental health hazards, the book covers key issue and considerations of pesticide use. Part I discusses pesticide usage and hazards; Part II includes pesticide data sheets; Part III provides a variety of information on acute poisoning, carcinogenicity, mutagenicity, reproductive health hazards in animals, disease, banned and restricted pesticides, organizations and contact lists, etc.

Mathematics and Statistics

This array of resources was selected to give choices to the beginning student and the more experienced researcher.

Bailey, Norman T. J. *Statistical Methods in Biology*, 2nd ed. New York: Cambridge University Press, 1992, reprint of the 1981 ed. 216 p. (paper). ISBN 0521427495. Index.

There is a continuing demand for this elementary text that is specifically and successfully designed to provide workers in the biological and medical sciences with the necessary basic statistical techniques. There are suggestions for more advanced reading, a summary of statistical formulae, and appendixes of abridged statistical tables.

Batschelet, Edward. *Introduction to Mathematics for Life Scientists*, 3rd ed. New York: Springer-Verlag, 1979. 643 p. (Biomathematics, v. 2). $39.50 (paper). ISBN 0387096485.

A classic text for life scientists needing to brush up on their math.

Biomathematics. v. 1– , 1970– . New York: Springer-Verlag. Irregular. Price varies. ISSN 0067-8821.

Volume 19 of this series is *Mathematical Biology*, by J. D. Murray. It provides a toolkit of modelling techniques with examples from population ecology, reaction kinetics, biological oscillators, developmental biology, evolution, and other areas, with emphasis on practical applications.

Biometrics; Journal of the Biometric Society. v. 1– , 1945– . Alexandria, VA: Biometric Society. Quarterly. $80/yr. ISSN 0006-341X.

This international journal promotes and extends the use of mathematical and statistical methods in pure and applied biological sciences by publishing original papers, review articles, or critiques.

Brown, David and P. Rothery. *Models in Biology: Mathematics, Statistics and Computing.* New York: Wiley, 1993. 688 p. $39.95. ISBN 0471933228.

An introduction to the use of mathematical models in biology, statistical techniques for fitting and testing them, and associated computing methods.

CRC Standard Mathematical Tables and Formulae, 29th ed. Edited by William H. Beyer. Boca Raton, FL: CRC, 1991. 609 p. $39.95. ISBN 0849306299.

Covers constants and conversion factors; algebra; combinatorial analysis; geometry; trigonometry; logarithmic, exponential, and hyperbolic functions; analytic geometry; calculus; differential equations; linear algebra; special functions; numerical methods; probability and statistics; and financial tables.

Campbell, R. C. *Statistics for Biologists*, 3rd ed. New York: Cambridge University Press, 1989. 464 p. $24.95 (paper). ISBN 0521369320.

A clear introduction to the principles and elementary techniques of statistical reasoning.

Causton, David R. *A Biologist's Advanced Mathematics.* London: Allen & Unwin, 1987. 326 p. $37.95. ISBN 0045740364. Index.

This book is designed for advanced undergraduates, postgraduates, and research workers in the bio- and agrosciences. Topics discussed are linear algebra, integrals, trigonometric and related functions, and differential equations. Answers to exercises are included.

Clarke, Geoffrey M. *Statistics and Experimental Design: An Introduction for Biologists and Biochemists*, 3rd ed. New York: Wiley, 1994. 160 p. $27.95. ISBN 0470234091.

Describes how to use statistical methods, interpret the results, and illustrate reports with suitable graphics. Numerous examples and exercises.

Cleveland, William S. *Visualizing Data*. Summit, NJ: Hobart Press, 1993. 360 p. $40.00. ISBN 0963488406.

The central theme of this book is that visualization is critical to data analysis. Cleveland illustrates how graphing data and fitting mathematical functions to the data can illuminate information and reveal unexpected patterns.

Elston, Robert C. and William D. Johnson. *Essentials of Biostatistics*, 2nd ed. Philadelphia: F. A. Davis, 1994. 328 p. $31.95 (paper). ISBN 0803631235.

An overview of statistical principles, with emphasis on concepts rather than computations.

Fowler, Jim and Louis Cohen. *Practical Statistics for Field Biology*. Bristol, PA: Open University Press, 1990. 227 p. $88.00. ISBN 033509208X. Bibliography. Index.

Not just another statistics text, the authors promise that this book is written specifically to serve field biologists "whose data are often non-standard and frequently 'messy.'"

Fry, J. C. *Biological Data Analysis; A Practical Approach*. Oxford: IRL Press at Oxford University Press, 1993. 418 p. $80.00. ISBN 0199633401.

Aimed at advanced undergraduates as well as professional biologists, this book shows how to analyze biological data with commonly available software packages. There are appendices containing software packages and statistical tables.

Hassard, Thomas H. *Understanding Biostatistics*. St. Louis, MO: Mosby, 1991. 292 p. plus statistical tables appendix, index, and solutions to problems. $29.95. ISBN 0801620783.

Intended for graduate students in the health sciences, although the subject matter is directly applicable to the life sciences in general.

IMA Journal of Mathematics Applied in Medicine and Biology. v. 1– , 1984– . Oxford: Oxford University Press. Published for the Institute of Mathematics and Its Applications. $210/yr. ISSN 0265-0746.

Aims of the journal: embrace uses of mathematics in medical and biological research, and encourage mathematicians to solve mathematical problems arising in medicine and biology.

Journal of Mathematical Biology. v. 1– , 1974– . New York: Springer-International. Monthly. $515/yr. ISSN 0303-6812.

Publishes papers in which mathematics is used for a better understanding of biological phenomena, mathematical papers inspired by biological research, and papers which yield new experimental data bearing on mathematical models.

Manly, Bryan F. J. *The Design and Analysis of Research Studies*. New York: Cambridge University Press, 1992. 353 p. $89.95. ISBN 0521414539.

A practical guide on how to design research studies, analyze quantitative data, and interpret the results.

Mead, J. *Statistical Methods in Agriculture and Experimental Biology*, 2nd ed. London: Chapman and Hall, 1992. 352 p. $99.95. ISBN 0412354802.

This introductory text includes basic statistical methods appropriate to agriculture and experimental biology. The level is suitable for students and research workers in biology.

Menell, Ann C. and Michael J. Bazin. *Mathematics for the Biological Sciences*. New York: Halsted Press, 1988. 231 p. $27.50. ISBN 0745804144.

An introduction to mathematics for biology students to give them the mathematical tools necessary to understand and investigate on a quantitative and theoretical basis.

Milton, J. Susan. *Statistical Methods in the Biological and Health Sciences*, 2nd ed. New York: McGraw-Hill, 1991. 526 p. ISBN 007042506X.

Intended for a first course in statistical methods, or for those with little experience with statistical methods. Emphasis is on understanding the principles as well as practicing them. Specifically chosen for biology and health sciences; knowledge of calculus is not assumed.

Numerical Computer Methods, Part A and B. San Diego, CA: Academic Press, 1994. (*Methods in Enzymology*, vol. 240). Part B: 904 p. $115.00. ISBN 0121821412.

The contributions in Part B emphasize numerical analysis of experimental data and analytical biochemistry, using modern data analysis methods developed with computer hardware.

Roberts, E. A. *Sequential Data in Biological Experiments: An Introduction for Research Workers*. London: Chapman & Hall, 1992. 240 p. $67.50. ISBN 0412414104. Index.

The aim is "to provide research workers with methods of analyzing data from comparative experiments with sequential observations and to demonstrate special features of the design of such experiments" (from the Preface).

Rohlf, F. James and Robert R. Sokal. *Statistical Tables*, 3rd ed. San Francisco: Freeman, 1994. 220 p. $20.00 (paper). ISBN 071672412X.

Ancillary text to *Biometry* (Sokal and Rohlf, below), providing statistical tables for the biological sciences.

Rosen, Dennis. *Mathematics Recovered for the Natural and Medical Sciences*. London: Chapman & Hall, 1992. 277 p. ISBN 0412410400. Index. Answers to exercises.

The title says it all: the book was written for those students, with little time to recoup, needing to recover the mathematics learned earlier in their academic careers.

Sokal, Robert R. and F. James Rohlf. *Biometry: The Principles and Practice of Statistics in Biological Research*, 3rd ed. San Francisco: Freeman, 1994. 880 p. ISBN 0716724111.

Companion to Rohlf and Sokal *Statistical Tables*. *Biometry* emphasizes practical applications of science, from physical and life science examples.

Sokal, Robert R. and F. James Rohlf. *Introduction to Biostatistics*, 2nd ed. New York: Freeman, 1987. 363 p. $34.95. ISBN 0716718057.

Useful for the student with a minimal knowledge of mathematics.

Statistical software suppliers:
>   Abacus Concepts, Inc., Berkeley, CA
>   BMDP Statistical Software, Inc., Los Angeles, CA
>   GraphPad Software, Inc. San Diego, CA
>   SAS Institute, Inc., Cary, NC
>   SPSS, Inc., Chicago, IL
>   StatSoft, Inc., Tulsa, OK
>   Systat, Inc., Evanston, IL

Methods and Techniques

Arnold, Zach M. *A Technical Manual for the Biologist. A Guide to the Construction and Use of Numerous Tools of Value to the Economy-Minded Student of Microscopic and Small Macroscopic Free-*

Living Organisms. Pacific Grove, CA: Boxwood Press, 1993. $25.00. ISBN 0940168243.

This manual is based on marine protozoological techniques, tools, and apparatus that were developed for marine geobiology. More general discussions demonstrate its wider potential: field and laboratory procedures and equipment, pedagogical methods, working with plastics, safety, and references to the original literature.

Cruz, Yolanda P. *Laboratory Exercises in Developmental Biology*. San Diego: Academic Press, 1993. 241 p. $34.95 (spiral). ISBN 0121983900.

Laboratory exercises suitable for college classes using live organisms: sea urchins, frogs, chicks, mice, and fruit flies. Questions following each exercise and brief references to the literature are provided.

Dykstra, Michael J. *Biological Electron Microscopy: Theory, Techniques, and Troubleshooting*. New York: Plenum Press, 1992. 360 p. $49.50. ISBN 0306442779.

Written for the beginning student, this textbook covers all the basic approaches utilized in transmission and scanning electron microscopy. It presents principles of fixation, specimen preparation, cytotechniques, and immunolabeling.

Dykstra, Michael J. *A Manual of Applied Techniques for Biological Electron Microscopy*. New York: Plenum Press, 1993. 252 p. $35.00. ISBN 0306444496.

Laboratory procedures for preparation of biological samples for scanning and transmission electron microscopy. Principles for preserving specimen integrity, tips for formulating solutions, procedures for preparation, sectioning techniques, post-staining procedures, and basic approaches are all included.

Hayat, M. A. *Stains and Cytochemical Methods*. New York: Plenum Press, 1993. 474 p. $89.50. ISBN 0306442949.

Considers both theoretical aspects and practical details of cytochemical stains as applied to the detection of macromolecules and heavy metals. Includes staining methods for high-resolution and high-voltage electron microscopy; procedures for staining thin cryosections; step-by-step procedures for staining lipids, nucleic acids, proteins, glycogen, cartilage, muscle and nervous tissue; preservation of samples; and subcellular localization of metals.

Hunter, Elaine. *Practical Electron Microscopy; A Beginner's Illustrated Guide*. New York: Cambridge University Press, 1993. 192 p. $34.95. ISBN 0521385393.

Extensively illustrated laboratory manual of transmission electron microscopy techniques.

Knoche, Herman W. *Radioisotopic Methods for Biological and Medical Research.* New York: Oxford University Press, 1991. 432 p. $42.50. ISBN 0195058062.

This book presents theoretical aspects of radioisotopic methods. Particular attention is given to 1) principles and practices of radiation protection; 2) radioactivity measurement, standardization and monitoring techniques, and instruments; 3) mathematics and calculations basic to the use and measurement of radioactivity; and 4) biological effects of radiation.

Light Microscopy in Biology: A Practical Approach. Edited by A. J. Lacey. New York: IRL Press, 1989. (Practical Approach Series). 329 p. $62.00. ISBN 0199630364.

The aim of this book is to give guidance at the practical level to people who wish to make more extensive use of their microscopes and also to offer encouragement . . . ." (from the Preface). Contents include sections on the principles and aims of light microscopy, rendering transparent specimens visible, image recording, immunohistochemistry, histochemistry, fluorescence microscopy, micrometry image analysis, video microscopy, and chromosome banding. There is a list of suppliers, and an index.

MacFarlane, Ruth B. Alford. *Collecting and Preserving Plants for Science and Pleasure.* New York: Arco, 1985. 184 p. $13.95. ISBN 0668060093.

The purpose of this book is to provide information on collecting plants for scientific and ornamental purposes. Chapters include collecting, preserving, identifying, labeling, mounting and packaging, storage and display, city botany, the herbarium and ornamental uses. Two appendixes contain tables of abbreviations, equivalents, temperature conversion, and names of ornamental plants.

Martin, Bernice M. *Tissue Culture Techniques; An Introduction.* Basel, Switzerland: Birkhaeuser, 1994.

See the annotation (p. 126) in Chapter 6, "Molecular and Cellular Biology".

Mason, W. T. *Fluorescent and Luminescent Probes for Biological Activity: A Practical Guide to Technology for Quantitative and Real Time Analysis.* San Diego: Academic Press, 1993. (Biological Techniques Series). 433 p. $112.00. ISBN 0124778291.

Some of the latest advances in methodology are introduced: the use of optical probe technology, problems and limitations, applications to practical questions, and areas of future development. The focus in on hardware requirements, principles of operation, and applications to living material.

Methods. v. 1– , 1990– . San Diego: Academic Press. Bimonthly.
$132/yr. ISSN 1046-2023.

This companion journal to *Methods in Enzymology* presents new methods applicable to a number of disciplines. Each issue is devoted to a specific approach or technique and describes its theoretical basis with emphasis on clear descriptions of protocols. Other companion journals are *Genomethods* and *Immunomethods*.

Methods in Enzymology. v. 1– , 1955– . San Diego: Academic
Press. Irregular. About $69.00/vol. ISSN 0076-6879.

This is an indispensable monographic series reporting methods of importance to a wide range of biologists. Each volume is devoted to a single topic to provide comprehensive, current coverage of pertinent techniques. Vol. 224: *Molecular Evolution: Producing the Biochemical Data*, 1993. 719 p. $95.00. Also, see *Methods* (above), a companion to *Methods in Enzymology*.

*Methods of Preparation for Electron Microscopy; An Introduction for the
Biomedical Sciences*, by D. G. Robinson et al. New York: Springer-Verlag, 1987. 190 p. $55.00. ISBN 03717592X. Index.

Chapters include an introduction to electron microscopy, methods for transmission electron microscopy and scanning electron microscopy, and evaluation of micrographs.

Molecular Techniques in Taxonomy. Edited by Godfrey M. Hewitt et al.
New York: Springer-Verlag, 1991. (NATO ASI series. Series H,
Cell biology, v. 57). 410 p. $166.00. ISBN 0387517642.

Proceedings of the NATO Advanced Study Institute on Molecular Techniques in Taxonomy. Organized to provide information and practice in the molecular techniques that have had major impact in taxonomic and evolutionary studies. Three main sections discuss overviews of molecular taxonomy, case studies of the successful application of molecular methods, and protocols for a range of generally applicable methods. The editors suggest that for computational analysis of taxonomic data and the reconstruction of phylogenetic trees, *Molecular Systematics* (edited by D. M. Hillis and C. Moritz, Sinauer, 1990) be consulted.

Optical Microscopy; Emerging Methods and Applications. Edited by
Brian Herman and John J. Lemasters. New York: Academic Press,
1993. 441 p. $79.95. ISBN 0123420601.

A synopsis of the most recent and comprehensive descriptions of new techniques, with emphasis on the life sciences.

Pankhurst, Richard J. *Practical Taxonomic Computing.* New York:
    Cambridge University Press, 1991. 202 p. $45.00. ISBN
    0521417600. Index.

This book, an update of *Biological Identification* (Pankhurst, 1978), is aimed toward students as well as professional biologists who need a working knowledge of computer techniques for identification. Chapters cover databases, classification, conventional and computerized identification methods, history, biological applications, and expert systems.

Photobiological Techniques. Edited by Dennis Paul Valenzeno et al.
    (NATO ASI series. Series A, Life sciences ; vol. 216). New York:
    Plenum Press, 1991. 381 p. $110.00. ISBN 0306440571.
    Index.

Each of 21 chapters discusses one of the major subdisciplines of photobiology complete with a general overview and one or two experiments, including a list of required materials. Supplementary reading is listed, and review questions and answers are included.

Procedures in Electron Microscopy. Edited by A. W. Robards and A. J.
    Wilson. New York: Wiley, 1994. (Wiley Looseleaf Publications).
    Variously paged. $575.00, price includes 1994 core volume plus
    2 updates of 100 pages, each. ISBN 0471928534.

A lab manual that includes tested techniques and preparation methods, descriptions of expected results, troubleshooting tips, safety notes, special equipment, interpretation of methods, and references.

Sanderson, J. B. *Biological Microtechnique.* Oxford, England: BIOS,
    1994. (Royal Microscopical Society, Microscopy Handbooks, 28).
    224 p. $22.32. ISBN 1872748422.

A practical guide, this manual provides instruction for biological specimen preparation for light microscopy. Contents cover collection material for specimen preparation, fixation and embedding, microtomy and other preparative methods, staining and dyeing, mounting, finishing, and restoring preparations.

Scanning and Transmission Electron Microscopy: An Introduction, by
    Stanley L. Flegler et al. New York: Freeman, 1993. 225 p.
    $38.95. ISBN 0716770474.

This authoritative introduction to practical and theoretical fundamentals of electron microscopy discusses essentials, including operation, image production, analytical techniques, and potential applications.

Science of Biological Specimen Preparation for Microscopy and Micro-

analysis. Edited by Ralph M. Albrecht et al. Chicago: Scanning Microscopy International, 1989. 302 p. $45.00. ISBN 0931288444.

This and earlier editions record the proceedings of the Pfefferkorn Conferences on the Science of Biological Specimen Preparation for Microscopy and Micro-analysis.

Smith, Robert F. *Microscopy and Photomicrography: A Working Manual*, 2nd ed. Boca Raton, FL: CRC, 1994. 162 p. $35.00 (spiral bound). ISBN 0849386829. Index.

This manual is intended for students in the life sciences, medical technologists, histologists, and pathologists, with the aim of providing practical working information without complicated mathematics. There are numerous illustrations and photographs to assist the readers in obtaining maximum performance and image quality from their optical systems. Lists of sources and references are included.

Storage of Natural History Collections: Ideas and Practical Solutions. Edited by Carolyn L. Rose and Amparo R. de Torres. Pittsburgh, PA: Society for the Preservation of Natural History Collections, 1992. (Published with a grant from the Institute of Museum Services). 346 p. Price not reported. ISBN 09633547607. Indexes.

This book presents 113 ideas and practical solutions for the storage for a complete range of natural history collections, including those for anthropology, herbaria, taxidermy, microscopy, living cells, etc. The manual is organized into six sections, according to storage configurations to encourage people to transfer and adapt methods: supports, covers, containers, environmental control, labels, general guidelines. Appendices include a glossary, conversion tables, materials and supplies, suppliers and manufacturers. A companion volume *Collections: Basic Concepts* is forthcoming.

Watson, James D. et al. *Recombinant DNA*, 2nd ed. New York: Scientific American Books, distributed by W. H. Freeman, 1992. 626 p. $32.95. ISBN 0716719940.

Authoritative introduction to the concepts and techniques of the cutting edge of DNA technology, making it accessible to the beginner.

Three-Dimensional Confocal Microscopy: Volume Investigation of Biological Specimens. Edited by John K. Stevens, Linda R. Mills, Judy E. Trogadis. San Diego, CA: Academic Press, 1994. 507 p. $85.00. ISBN 0126683301.

"The goal of this book is to familiarize the reader with these new technologies and to demonstrate their applicability to a wide range of biological and clinical problems."

X-Ray Microanalysis in Biology: Experimental Techniques and Applications. Edited by David C. Sigee et al. New York: Cambridge University Press, 1993. $89.95. ISBN 0521415306.

An up-to-date look at the use of x-ray microanalysis in biology. There are four main themes: detection and quantification of x-rays, associated techniques, specimen preparation, and applications for many different areas of biology, including animal and plant cell physiology, medicine, pathology and studies on environmental pollution.

Nomenclature

Refer to individual subject chapters for more specific information.

Biochemical Nomenclature and Related Documents; A Compendium, 2nd ed. Edited by C. Liebecq. Published for the International Union of Biochemistry and Molecular Biology. Brookfield, VT: Portland Press, 1992. 350 p. $36.00. ISBN 1855780054.

This is an essential reference containing current recommendations on nomenclature issued by the nomenclature committees of the International Union of Biochemistry. It also includes text from, and references to, other important nomenclature documents.

Biological Nomenclature Today: A Review of the Present State and Current Issues of Biological Nomenclature of Animals, Plants, Bacteria and Viruses: Papers presented July 9, 1985, Brighton, UK. Oxford: IRL on behalf of the ICSU Press for the International Union of Biological Sciences, 1986. (IUBS Monograph Series, no. 2). 70 p. $12.00 (paper). ISBN 1852210168.

Chapters discuss botanical, zoological, bacterial, viral, and fungal nomenclature.

A Draft Glossary of Terms Used in Bionomenclature, compiled by D. L. Hawksworth. Paris: The International Union of Biological Sciences, 1994. 74 p. Price not reported. ISBN 9494060806.

Over 1,000 terms are included in the glossary that has the aim of harmonizing the various biological (botanical, zoological, etc.) codes. This is a working document but it can be useful in conjunction with recognized nomenclatural codes, such as botany.

*Enzyme Nomenclature 1992: Recommendations of the Nomenclature
    Committee of the International Union of Biochemistry and Molecu-
    lar Biology on the Nomenclature and Classification of Enzymes,*
    prepared for NC-IUBMB by Edwin C. Webb. San Diego: Academic
    Press for the IUBMB, 1992. 862 p. ISBN 0122271645.

3,196 enzymes are listed with their recommended name, their EC number, type
reaction, and the systematic name based on the classification of the enzyme.
Non-recommended names are included as are comments and references to justify
each entry. A glossary, abbreviations, and an index are also provided.

Periodicals

This section lists journals and accompanying serials that are of use to the
general biologist. For general subject specific periodicals, see the section to
which they pertain; e.g., *Biomathematics* is included in the Mathematics and
Statistics section (p. 50). For subject specific journals, refer to the particular
subject chapter.

Some of the most valuable aids in verifying or identifying serial titles are
the lists of serials indexed by database publishers. Three of the most important
serials lists, corresponding to three of the most prominent databases for biologists, (*Biological Abstracts*, *Chemical Abstracts*, and MEDLINE) precede the
listing of significant general biological serial publications.

Serial Titles

Chemical Abstracts Service Source Index (CASSI). 1969– . Easton,
    PA: American Chemical Society. $200. ISSN 0001-0634.

Includes invaluable information on all source publications covered by *Chemical
Abstracts*. Included are title, abbreviated title, coden, former titles, language,
history, frequency, price, and publisher. The extent of the information and sheer
number of publications included makes this the one of the most valuable serial
lists for the biological/chemical sciences. Available as a quarterly, an annual,
or cumulative volumes. The most recent cumulation covers 1907–1989.

List of Serials Indexed for Online Users. 1980– . Bethesda, MD:
    National Library of Medicine. Annual. $25.00. ISSN 0736-7139.

Information for sources in MEDLINE, *Health Planning & Administration*, and
POPLINE. Each entry gives abbreviated title, full title, publication city, NLM
call number, NLM title control number, ISSN number, special list indicator,
journal title code, and brief notes when indicated. An annual companion pub-

lication from the National Library of Medicine is *List of Journals Indexed in Index Medicus*, ISSN 0093-3821.

Serial Sources for the BIOSIS Previews Database. 1938– . Philadelphia: BioSciences Information Service. Annual. (Formerly, *Biosis List of Serials*.) $95.00/yr. ISSN 1044-4297.

All sources scanned for *Biological Abstracts* and *Biological Abstracts/RRM* are listed in this key publication, including information on complete title, standard abbreviation, coden, ISSN, frequency, publisher, and publishers' addresses. New serial titles, ceased and changed serial titles are noted.

Ulrich's International Periodicals Directory
See the annotation in the Bibliographies section (p. 26).

Zoological Record Serial Sources. Philadelphia: BioSciences Information Service. Annual. $40.00/1992. ISSN 1041-4657.

Includes information on more than 6,500 serial publications scanned for inclusion in *Zoological Record.* Also lists monographs, reports, and conference literature.

Periodical Titles

American Scientist. v. 1– , 1913– . New Haven, CT: Sigma Xi, The Scientific Research Society of North America. Bimonthly. $45/yr. ISSN 0003-0996.

Articles of interest to a wide range of scientists; book reviews.

Biologist. v. 16– , 1969– . London: Institute of Biology. Quarterly. $53/yr. ISSN 0006-3347.

Continues the *Institute of Biology Journal.* This journal presents overview articles, news, and reviews for professional biologists in the areas of biomedical, environmental, agricultural and educational sciences.

BioScience. v. 1– , 1951– . Baltimore, MD: Williams and Wilkins. $138/yr. ISSN 0006-3568. Continues the *AIBS Bulletin.*

This official publication of the American Institute of Biological Sciences contains research articles, news and reports from the Institute, information about people, places, new products, books, and professional opportunities. Recent special issues: July/August, 1991: Marine Biological Diversity; Jan. 1992: Phylogeny of Plant–Animal Interactions; July/August, 1992: Crop Productivity for Earth and Space; Dec. 1992: Stability and Change in the Tropics.

Computer Applications in the Biosciences (CABIOS). v. 1– , 1985– .
Cary, NC: Oxford University Press. Bimonthly. $260/yr. ISSN
0266-7061.

Forum for the exchange of information on the uses of computing in the molecular biosciences, including reviews, full papers, communications, and applications notes.

Computers and Biomedical Research; An International Journal. v. 1– ,
1967– . San Diego: Academic Press. Bimonthly. $215/yr. ISSN
0010-4809.

Endorsed by the American Medical Informatics Association, and the American College of Medical Informatics, this journal issues original reports concerning computer applications and methods to biomedical research.

Computers in Biology and Medicine; An International Journal. v. 1– ,
1971– . Oxford, England: Pergamon. Bimonthly. $604/yr. ISSN
0010-4825.

Communication of the revolutionary advances being made in the applications of the computer to the fields of bioscience and medicine, including research, instruction, ideas and information on all aspects of computer usage in these fields.

Current Biology. v. 1– , 1991– . London: Current Biology, LTD. Bimonthly. $300/yr. ISSN 0960-9822.

News and reviews from the frontiers of biology.

Experientia. v. 1– , 1945– . Basel: Birkhauser. Monthly. $600/yr.
ISSN 0014-4754.

General science journal.

FASEB Journal. v. 1– , 1987– . Bethesda, MD: The Federation.
Monthly. $360/yr. ISSN 0892-6638.

Official publication of the Federation of American Societies for Experimental Biology. Formerly *Federation Proceedings*. Multispeciality journal.

Micron; the International Research and Review Journal for Microscopy.
v. 25– , 1994– . Tarrytown, NY: Pergamon Press. Incorporates
Micro and Microscopica Acta and *Electron Microscopy Reviews*.
$565/yr. ISSN 0968-4328.

Covers original and review articles for all work involving design, application, practice or theory of microscopy and microanalysis in biology, medicine, agriculture, metallurgy, materials sciences and physical sciences.

Nature. v. 1– , 1869– . London: Macmillan. Weekly. $400/yr.
    ISSN 0028-0836.

This influential general scientific journal reports a large component of biological research. It accepts research articles, review articles, and brief reports of research. It also summarizes news and comment of interest for a wide scholarly audience; reviews of significant books are included.

New Scientist. v. 1– , 1956– . London: New Science Publications.
    Weekly. $130.00. ISSN 0262-4079. Merged with *Science Journal.*

General British periodical publishing science and technology news, commentary, feature articles, and book reviews.

Philosophical Transactions: Series B (Biological Sciences). v. 178– ,
    1887– . London: Royal Society. Monthly. $900/yr. ISSN 0962-
    8428.

A multidisciplinary journal in the biological sciences publishing fundamental, classic papers for leading scientists in the world's longest-running scientific journal.

Proceedings of the National Academy of Sciences. v. 1– , 1915– .
    Semimonthly. $420/yr. ISSN 0027-8424.

A leading scientific journal of great import to the biological sciences with papers that are contributed by, or communicated to, a member of the Academy for transmittal in the *Proceedings.* Papers are accepted in all areas of science and must report theoretical or experimental research of "exceptional importance or novelty."

Science. v. 1– , 1880– . Washington, DC: American Association for
    the Advancement of Science. Weekly. $205/yr. ISSN 0036-
    8075.

Prestigious general scientific journal with a majority of biological articles reporting original research, news, comments, book reviews. Full text available online from DIALOG. Special issues devoted to "Computers '94" (Aug. 12, 1994), "Careers '94" (Sept. 23, 1994), "Genome Issue" (Sept. 30, 1994), "Frontiers in Biology: Development" (Oct. 28, 1994).

The Scientist. "The Newspaper for the Science Professional." v. 1– ,
    1986– . Philadelphia: The Scientist, Inc. Biweekly (except in
    August and December). $58.00/yr. ISSN 0890-3670. Available
    electronically on NSFnet and the Internet.

This publication reports general sciences news, opinion on controversial topics, information on new tools and technology, details of interest to the profession, and brief notes for career opportunities, equipment, letters, obituaries, etc.

Systematic Biology. v. 1– , 1952– . Washington, DC: Society of
    Systematic Biologists, National Museum of Natural History.
    Quarterly. $60/yr. ISSN 1063-5157.

Papers on systematic biology theory and practice, covering all taxa and all taxonomic levels. The journal includes phylogenetics, biogeography, classification, morphological and molecular data, and evolutionary theory.

Reviews of the Literature

*Advances in Computer Methods for Systematic Biology; Artificial Intelli-
    gence, Databases, Computer Vision.* Edited by Renaud Fortuner.
    Baltimore: Johns Hopkins University Press, 1993. 584 p.
    $70.00. ISBN 0801844924.

Explores the application of artificial intelligence and other computer methods to biological systematics.

Advances in Experimental Medicine and Biology. 1967– . New York:
    Plenum Press. Irregular. About $115.00/vol. ISSN 0065-2598.

This is an important monographic series, with each volume organized around a distinctive topic relevant to experimental medicine and biology.

Advances in Marine Biology. v. 1– , 1963– . San Diego: Academic
    Press. Irregular. Vol. 28, 1992. $118.00. ISBN 0120261286.

A high-quality review series dealing with various aspects of marine biology.

Cold Spring Harbor Symposia on Quantitative Biology. Proceedings. v.
    1– , 1933– . Cold Spring Harbor, New York: Cold Spring Harbor
    Laboratory. Annual. Approximately $200/yr. ISSN 0091-7451.

Excellent review papers from symposia, each on a particular biological topic. Vol. 59, 1995: *Molecular Genetics of Cancer.*

Current Topics in Developmental Biology. v. 1– , 1966– . San Diego:
    Academic Press. Annual. Volumes priced separately. ISSN
    0070-2153.

This important review series presents reviews and discussions on topics of interest and import at the cellular, biochemical, and morphogenetic levels.

Harrison, Lionel G. *Kinetic Theory of Living Pattern.* New York: Cambridge University Press, 1992. 354 p. $69.50. ISBN 0521306-914.

Review of development and philosophy of the shapes of living organisms.

Index to Scientific Reviews. v. I– , 1974– . Philadelphia: Institute for Scientific Information. $830/yr. ISSN 0360-0661.

"An international interdisciplinary index to the review literature of science, medicine, agriculture, technology, and the behavioral sciences." For more complete information, see Chapter 4, "Abstracts and Indexes."

Perspectives in Biology and Medicine. v. 1– , 1957– . Chicago: University of Chicago Press. Quarterly. $35/yr. ISSN 0031-5982.

This excellent review journal serves as a vehicle for interpretive and innovative essays, new hypotheses, biomedical history, and humorous pieces.

Quarterly Review of Biology. v. 1– , 1926– . Chicago: University of Chicago Press. Quarterly. $70/yr. ISSN 0033-5770.

"Critical reviews of recent research in the biological sciences." At least half of each issue is devoted to book reviews of interest to life scientists; new biological software is also reviewed.

Scientific American. v. 1– , 1845– . New York: Scientific American, Inc. Monthly. $39/yr. ISSN 0036-8733.

Scholarly review articles written for the educated layperson; also includes news, comments, games, book reviews.

Seminars in Developmental Biology. v. 1– , 1990– . London: Academic Press. Bimonthly. $185/yr. ISSN 1044-5781.

Each issue is devoted to a state-of-the-art review of a selected topic, edited by a guest editor.

Societies

Refer to *The Encyclopedia of Associations* (p. 39) for more information. Also, society directories, or information about them, is often available on the Internet via Gopher or World Wide Web.

American Association for the Advancement of Science (AAAS)
1333 H St., N. W., Washington, DC 20005. (202) 326-6400.

This the largest general scientific organization representing all fields of science. Membership includes 135,000 individuals and 296 scientific societies, professional organizations, and state and city academies. Objectives are to further the work of the scientist and to improve the effectiveness of science in the promotion of human welfare. AAAS conducts seminars and colloquia on scientific issues, and presents scientific awards. Publications include *Science*, *Science Books and Films*, *Science Education News*, symposium volumes, and general reference works. There is an annual meeting; annual symposia and lectures are held, also.

American Institute of Biological Sciences (AIBS)
    730 11th St., N.W., Washington, DC 20001. (202) 628-1500.

AIBS is an organization of professional biological associations and laboratories whose approximately 8,000 members have an interest in the life sciences. The society conducts symposium series, supports speaker and consultant lists, and operates a placement service. The association maintains a computerized service, BioTron (Biologists Electronic Network), publishes a monthly newsletter, an annual August meeting program, and the journal *BioScience*.

Association of Systematics Collections
    730 11th St., N.W. 2nd fl., Washington, DC 20001.
    (202) 347-2850.

The purpose of this association is to foster the care, management, preservation, and improvement of systematics collections and to facilitate their use in science and society. Membership of about 100 members includes grant researchers, educational institutions, museum personnel, and biologists. *ASC Newsletter* is published bimonthly with articles on systematics collections, natural history museums, and herbaria. Other publications: *Amphibian Species of the World*, *Biogeography of the Tropical Pacific*, *Guidelines to Acquisition and Management of Biological Specimens*, *A Guide to Museum Pest Control*, etc. There is an annual meeting.

Federation of American Societies for Experimental Biology (FASEB)
    9650 Rockville Pike, Bethesda, MD 20814. (301) 530-7000.

This federation of seven scientific societies, the largest coalition of life sciences societies in the United States (American Physiological Society, American Society for Biochemistry and Molecular Biology, American Society for Pharmacology and Experimental Therapeutics, American Society for Investigative Pathology, American Institute of Nutrition, American Association of Immunologists, American Society for Cell Biology, Biophysical Society, American Association of Anatomists) publishes the *FASEB Journal*, a directory of members, and a newsletter. A Gopher server has been established at FASEB to provide information services to society members and includes announcements of meetings, a member directory of all FASEB societies, the FASEB news letter, pubic policy statements and reports, and job opportunities at FASEB. There is an annual meeting.

Institute of Biology (IOB)
    20 Queensberry Place, London, SW7 2DZ, England.

The purpose of this group is to advance the science and practice of biology. Its publications include the *Biologist*, the *Journal of Biological Education*, and occasional publications.

International Union of Biological Sciences (IUBS)
    51 Blvd. de Montmorency, F-75016 Paris, France.

This is an organization of national societies and international associations and commissions engaged in the study of biological sciences. Their aims are to promote the study of biological sciences; to initiate, facilitate, and coordinate research and other scientific activities; to ensure the discussion and dissemination of the results of cooperative research to promote the organization of international conferences; and to assist in the publication of their reports. There is a triennial general assembly, and the Union also holds an annual executive committee meeting and sponsors scientific meetings.

Society for Experimental Biology and Medicine (SEBM)
    630 W. 168th St., New York, NY 10032.

This society includes 2500 members and 2 regional groups of workers actively engaged in research in experimental biology or experimental medicine. There is an annual meeting with symposium. *Proceedings of the Society* is published 11 times/year and includes an annual membership directory.

Society of Systematic Biologists (SSB)
    National Museum of Natural History, NHB 163, Washington, DC
    20560.

The society represents over 1,600 scientists whose research is aimed at understanding the patterns of life's history and the processes generating them. The society publishes *Systematic Biology*.

Textbooks

Alexander, R. McNeill. *The Human Machine*. New York: Columbia
    University Press, 1992. 176 p. $34.95. ISBN 0231080662.

This book is about how the human body works.

Barnard, Christopher John et al. *Asking Questions in Biology: Design,
    Analysis and Presentation in Practical Work*. New York: John
    Wiley, 1993. 157 p. $20.00. ISBN 0582088542.

Explains important techniques and analytical methods in the context of answering biological questions as clearly as possible. Discussions on formulating hypotheses and predictions, designing critical observations and experiments to test them, choosing appropriate statistical analyses, presenting results and writing reports. Also, see the *CBE Style Manual* under the entry *Scientific Style and Format* (p. 45).

Calladein, C. R. and Horace R. Drew. *Understanding DNA; the Molecule & How It Works*. London: Academic Press, 1992. 220 p. $65.00. ISBN 0121550850. Index.

This book is useful for beginning college students and for informed lay persons who want to know how DNA works at the molecular level.

Cladistics: A Practical Course in Systematics. Edited by P. L. Forey et al. Oxford: Clarendon Press, 1992. 208 p. $52.50. ISBN 0198577672.

Based on a course sponsored by the Systematics Association in Great Britain, this book presents an introduction to the principles of parsimony, methods of tree-building, and tree statistics.

Ethical Dimensions of the Biological Sciences. Edited by Ruth Ellen Bulger et al. New York: Cambridge University Press, 1993. 294 p. ISBN 0521434637. $55.00.

A collection of essays, policy statements, research guidelines, and important works by scientists from many disciplines. Some of the ethics-in-science issues considered: norms of ethical conduct, scientific honesty, ethical standards of laboratory practice, use of human and animal subjects, qualifications for authorship and publication, ethics of learning and teaching, and the relationships of science, industry, and society.

Fletcher, Neville H. *Acoustic Systems in Biology*. New York: Oxford University Press, 1992. 233 p. $65.00. ISBN 0195069404.

Text written for those requiring a brief general survey of the field, a fairly detailed understanding of the concepts and techniques involved, and mathematical arguments and solutions for biological relevance.

Gage, John D. and Paul A. Tyler. *Deep-Sea Biology: A Natural History of Organisms at the Deep-Sea Floor*. New York: Cambridge University Press, 1991. 504 p. $35.95 (paper). ISBN 0521334314.

Introduction to the biology of the deep-sea environment. Species and subject indexes.

Gilbert, Scott F. *Developmental Biology*, 4th ed. Sunderland, MA:
    Sinauer, 1994. 894 p. $57.95. ISBN 0878932496.

A successful developmental text.

Keen, Robert E. and James D. Spain. *Computer Simulation in Biology:
    A Basic Introduction*. New York: Wiley/Liss, 1992. 498 p.
    $39.95 (paper). ISBN 047150971X.

A solid overview concerning simulation of general biological processes, covering
the range from simple differential equation models of growth and decay to physiological control models, models of morphogenesis, and models of epidemics.
There are exercises at the end of each example model.

Keeton, William T. et al. *Biological Science*, 5th ed. New York: Norton,
    1993. 1196 p. ISBN 0393963853. $60.00.

A very successful general biology text.

Kornberg, Arthur and Tania A. Baker. *DNA Replication*, 2nd ed. New
    York: W. H. Freeman and Co., 1992. 931 p. $64.95. ISBN
    0716720035. Indexes.

Descriptions of the individual replication enzymes are followed by the genetic,
physiologic, and structural features of the systems they serve. The goal is to
provide an up-to-date account of DNA replication and metabolism with a strong
biochemical emphasis. Extensive references.

Mayr, Ernst. *The Growth of Biological Thought: Diversity, Evolution,
    and Inheritance*. Cambridge, MA: Belknap Press, 1982. 974 p.
    Bibliography. $45.00. ISBN 0674364457.

A broad survey of biology and its dominant concepts by one of the eminent
players.

Mayr, Ernst. *Toward a New Philosophy of Biology; Observations of an
    Evolutionist*. Cambridge, MA: Belknap Press of Harvard University
    Press, 1988. 564 p. $16.95 (paper). ISBN 0674896653. Index.

Written to strengthen the bridge between biology and philosophy, pointing to the
direction in which a new philosophy of biology will move.

Morphogenesis: An Analysis of the Development of Biological Form.
    Edited by Edward F. Rossomando and Stephen Alexander. New
    York: Dekker, 1992. 449 p. $165.00. ISBN 082478667X.

The aim of this review of several representative systems is to unite work on
morphogenesis and focus on future research efforts.

Panchen, Alec L. *Classification, Evolution, and the Nature of Biology*.

New York: Cambridge University Press, 1992. 403 p. $80.00. ISBN 0521305829.

This book addresses the philosophical and historical relationship between patterns of classification and patterns of phylogeny.

Purves, William K. et al. *Life: The Science of Biology*, 3rd ed. Sunderland, MA: Sinauer, 1992. 1145 p. $59.95. ISBN 0716711763.

Introductory college text presenting biological data and ideas, accompanied by beautiful artwork.

Raven, Peter and George Johnson. *Biology*, 3rd ed. St. Louis, MO: Mosby, 1992. 1,344 p. $56.95. ISBN 0801663725.

A comprehensive account of all aspects and principles of biology.

Rochow, Theodore George and Paul Arthur Tucker. *Introduction to Microscopy by Means of Light, Electrons, X-Rays, or Acoustics*, 2nd ed. New York: Plenum Press, 1994. 436 p. $49.50. ISBN 0306446847.

Information on the cutting edge of microscopies and microscopes, covering advantages and limitations of each kind.

Ruse, Michael. *Philosophy of Biology Today*. Albany, New York: State University of New York Press, 1988. 155 p. $39.50. ISBN 088706910X. Index.

The author views this book as a handbook to the philosophy of biology with the aim of introducing newcomers to the field. There is a 52-page bibliography of references.

Starr, Cecie and Ralph Taggart. *Biology: The Unity and Diversity of Life*, 6th ed. Belmont, CA: Wadsworth Press, 1992. 921 p. $55.95. ISBN 0534165664.

Comprehensive introductory biology text suitable for college level.

Wessells, Norman K. and Janet L. Hopson. *Biology*. New York: Random House, 1988. 1251 p. $49.95. ISBN 0394337328.

Another respected biological textbook.

Wilson, Edward Osborne. *The Diversity of Life*. Cambridge, MA: Belknap Press of Harvard University Press, 1992. (Questions of Science). 424 p. $29.95. ISBN 0674212983.

Stephen Jay Gould reviewed this book on biological diversity conservation as a very successful mixture of details and prophecy.

4

Abstracts and Indexes

Introduction

This chapter can be seen as a companion to Chapter 2, Subject Access to Biological Information. Abstracts and indexes are used to locate articles, proceedings, and occasionally books and book chapters in various subjects. Since the literature of biology is so vast, it should come as no surprise to find that there are many indexes offering access to that literature. This chapter annotates the major indexes and abstracts which cover general science and/or multiple subjects in biology. Those indexes which deal with narrower fields, such as entomology or plant taxonomy, will be covered in the appropriate subject chapter.

Probably the most significant change in access to the biological literature of the past 25 years is the growing number of computerized indexes. These indexes have made locating material on a subject or by a particular author much easier in some ways, and more complicated in others. Indexes are currently available in a number of formats, including paper, online, and on CD-ROM. They are also accessible by a number of methods. While most commercial on-line vendors assume that their users are trained searchers, there are a growing number of options for end-users, the untrained searchers. CD-ROMs and locally mounted databases are the best-known end-user systems. A number of vendors are beginning to utilize the Internet as well. It is now possible to subscribe to a service allowing patrons to use the Internet to search the databases provided by Cambridge Scientific Abstracts and SilverPlatter, for instance.

Most of the major indexes in the sciences are available in computer-readable formats. There are also many databases which have no paper equivalent, or are a combination of several paper indexes. In the following listing of abstracting and indexing tools, the basic bibliographic information is given for the paper version of each title, as well as information on the computer-readable version of the item if it exists. Items which have no paper equivalent are listed in bold print.

The information provided for computer databases includes title changes (if any), format (online or CD-ROM), dates covered, and vendors for both formats. Since some online databases are available from many vendors (at least nine in the United States alone for BIOSIS Previews), listings will be given for only the three most important sci/tech online vendors: CDP, DIALOG, and STN. The number of companies offering the same CD-ROM indexes is less than that of online vendors, so all known CD-ROM vendors are listed for each database. It should be remembered that things change rapidly in the computing world, and the situation for computer databases is far from settled. The information given in this chapter is likely to become outdated rapidly as changes occur on an almost daily basis.

Biological Abstracts and Indexes

Bibliography of Agriculture. v. 1– , 1942– . Phoenix: Oryx. $1,275.
ISSN 0006-1530.

Covers the worldwide literature of agriculture, including journal articles, government documents, reports, and proceedings; from the National Agriculture Library. Available online as AGRICOLA from CDP and DIALOG, 1970 to the present. Also available on CD-ROM from SilverPlatter (1972 to the present) and OCLC (1979 to the present), both updated quarterly.

Bibliography of Bioethics. v. 1– , 1975– . Detroit, MI: Kennedy Institute of Ethics. Annual. $56.25. ISSN 0363-0161.

Annual publication identifying central issues of bioethics. Covers articles, monographs, essays, laws, court decisions, audiovisual materials, and unpublished documents. Each volume provides introduction, list of journals cited, thesaurus, subject entry section, title, and author indexes.

BIOBUSINESS. 1985– . Philadelphia: BIOSIS.

Database indexing approximately 500 technical and business journals, magazines, proceedings, patents, and books covering all areas of the business applications of biological research. Topics covered include agriculture, biotechnology, genetic engineering, medicine, and pollution. Available online on CDP, DIALOG, and STN from 1985 to the present.

BIOETHICS. 1973– . Bethesda, MD: National Library of Medicine.

Database containing references to ethical issues in health care and biomedicine, including topics such as genetic engineering, AIDS, and patient rights. Covers books, book chapters, journal articles, newspaper articles, court decisions, laws, and other documents. Also available on CD-ROM from SilverPlatter as BIO-ETHICSLINE Plus.

Biological Abstracts. v. 1– , 1926– . Philadelphia: BIOSIS. Biweekly. $5,650.00. ISSN 0006-3169.

The most comprehensive biological abstracting service in the English language in the world. Over 9,000 journals reporting original research are scanned. Covers all subjects in biology and biomedicine. Biweekly with six-month cumulations. Abstracts arranged by subject with author, biosystematic, generic, and subject indexes. Five year microfilm cumulations, 1959-1989. Available online as part of BIOSIS Previews from 1969 (DIALOG and STN) or 1970 (CDP) to date, and also from several other vendors. Available on CD-ROM as BA on CD from SilverPlatter, 1985 to date. CD-ROM is updated quarterly. For information on searching *Biological Abstracts*, see Chapter 2, and for a complete listing of publications scanned for *Biological Abstracts*, see *Serial Sources for the BIOSIS Previews Database* (Chapter 3, p. 62).

Biological Abstracts/RRM (Reports, Reviews, Meetings). v. 18– , 1980– . Philadelphia: BIOSIS. Biweekly. $3,715. ISSN 01926985.

Companion to *Biological Abstracts*, cumulated every six months. Worldwide coverage of material not covered in *Biological Abstracts* such as editorials, reports, bibliographies, proceedings, symposia, books, chapters, review journals, translated journals, nomenclature rules, etc. Successor to *Bioresearch Index*. Arranged like *Biological Abstracts* in broad subject categories with five indexes. Available online as part of BIOSIS Previews and on CD-ROM as BA/RRM on CD from SilverPlatter, 1989 to date. The CD-ROM is updated quarterly.

Biological and Agricultural Index. v. 1– , 1916–18– . New York: Wilson. Monthly. Price varies. ISSN 0006-3177.

Appropriate for beginning students and the public. Covers 225 journals and is complementary to *General Science Index* (see p. 77). Alphabetical subject and author index. Available online from CDP and on CD-ROM from Wilson, 1983 to date. The CD-ROM index is updated monthly.

Biology Digest. v. 1– , 1974– . Medford, NJ: Plexus. Monthly. $125.00. ISSN 0095-2958.

This digest covers around 200 biological journals; the level is appropriate for undergraduates and the general public. Unlike *Biological and Agricultural Index*, *Biology Digest* includes abstracts, which are arranged by subject. There are author and key word indexes in each issue which are cumulated annually, as well as a monthly feature article. Available online on the Life Sciences Network.

BIOSIS

See *Biological Abstracts* and *Biological Abstracts/RRM* (both p. 74).

CAB Abstracts. 1973– . Wallingford, UK: CAB International.

This database consists of records from the nearly 50 CABI (Commonwealth Agricultural Bureaux International) abstract journals. Has truly international coverage; important for applied research in plant biology and zoology. Available from 1984 to date online in various sections from CDP, DIALOG, STN and on CD-ROM for the full database; also available on CD-ROM in sections from SilverPlatter from 1939 or 1973, depending on the section.

Chemical Abstracts. v. 1– , 1907– . Columbus, OH: Chemical Abstracts Service. Weekly. $16,800.00. ISSN 0009-2258.

Covers over 18,000 journals, making it the most important English-language index in chemistry. Scans scientific and engineering journals, patents, conference proceedings, reports, and monographs. Essential source for biological topics with chemical facets. Available online from 1967 to date from STN (as CA) and from 1970 to date as CA SEARCH from CDP and DIALOG, and on CD-ROM from Cambridge Scientific Abstracts as CAS 12th Collective Index and CAS 12th Collective Abstracts (covering 1987-1991). STN also offers CAOLD, which consists of *Chemical Abstracts* records 1957–1966. Microfiche and microfilm also available. For a complete list of journals scanned for *Chemical Abstracts*, see annotation for *Chemical Abstracts Service Source Index* in Chapter 3 (p. 61). Searching *Chemical Abstracts* is discussed in Chapter 2.

Conference Papers Index. v. 1– , 1972– . Bethesda, MD: Cambridge Scientific Abstracts. 22/yr. $995.00. ISSN 0162-704X.

Indexes papers presented at scientific conferences; scans final programs, abstracts booklets, and published proceedings. Includes information on publications resulting from meetings. Covers life and physical sciences and engineering. Also available online on DIALOG and STN from 1973. See annotation of the **Life Sciences Collection** (p. 78), for Internet access.

Current Awareness in Biological Sciences. v. 1– , 1954– . Tarrytown, NY: Pergamon. 144/yr. $5,065.00. ISSN 0733-4443.

Consists of 12 print sections of the *Current Advances* series, also available individually. Includes sections on biochemistry, biotechnology, cell and developmental biology, ecology, genetics, neuroscience, physiology, plant biology, others. Covers about 2,700 international journals. Until 1983, called *International Abstracts of Biological Sciences.* Available online from CDP, 1984 to the present.

Current Contents/Agriculture, Biology, and Environmental Sciences. v. 1– , 1970– . Philadelphia: Institute for Scientific Information. Weekly. $460.00. ISSN 0011-3379.

Compilation of tables of contents of over 900 major journals as well as major series. Arranged by subject; there are also title word, subject, and author indexes, as well as publisher address listing in each issue. Subjects covered in *CC/ABES* include agriculture, botany, entomology, ecology, mycology, ornithology, veterinary medicine, and wildlife management. The most current listing of journal contents available; much more timely than most indexes. Available on CD-ROM from ISI (1994 to date); updated weekly. Also available online from DIALOG as part of SCISEARCH (*Science Citation Index*) from 1974 to the present, on CD-ROM from ISI as part of Science Citation Index on CD-ROM from 1986 to the present, and on weekly diskettes, also from ISI.

Current Contents/Life Sciences. v. 1– , 1958– . Philadelphia: Institute for Scientific Information. $460.00. ISSN, 0011-3409.

Companion to *CC/ABES*, covering topics such as biochemistry, biomedical research, biophysics, endocrinology, genetics, immunology, microbiology, neurosciences, and pharmacology. Available in 1,200 and 600 title versions. Available on CD-ROM from ISI (1994 to date); updated weekly. Also available online and on CD-ROM as part of *Science Citation Index* and on diskette (see above).

Dissertation Abstracts International, Section B: Physical Sciences and Engineering. v. 1– , 1938– . Ann Arbor, MI: University Microfilms International, Dissertation Publishing. Monthly. $475. ISSN 0419-4217.

Covers mostly U.S. academic institutions, though many Canadian institutions are also covered, and British and European dissertations are included from 1988 on. Available online from CDP, DIALOG, and STN and on CD-ROM from UMI from 1861 to the present. The CD-ROM is updated annually.

EVENTLINE. 1989– . Amsterdam: Elsevier.

Database listing information on past and future conventions, symposia, exhibitions, and sporting events. Covers all the sciences and business. Includes information on events beyond the year 2000. Available online from DIALOG.

Excerpta Medica. v. 1– , 1947– . Amsterdam: Excerpta Medica Foundation; New York: Elsevier. Freqency varies. Price varies.

Abstracting journals issued in over 50 sections by topic. Entries are in classed arrangement with author and subject indexes for each issue. Annual index cumulations. Abstracts are written for readers using English as a second language so the summaries are usually concise and easy to understand. Although coverage is worldwide, it is not as comprehensive as either *Biological Abstracts* or *Index Medicus*. Suitable for use in basic biological sciences although published for use by medical researchers and/or physicians. The sections most likely of interest to biologists include Section 1: Anatomy, Anthropology, Embryology, and Histology; Section 2: Physiology; Section 3: Endocrinology; Section 4: Microbiology: Bacteriology, Mycology, Parasitology, and Virology; and Section 21: Developmental Biology and Teratology. Available online as EMBASE on CDP and DIALOG from 1974 to the present and on STN from 1980 to the present. Selected medical sections (none of the above sections) are available on CD-ROM from SilverPlatter from 1980 to the present. SilverPlatter also offers the full database from 1984 to the present.

General Science Index. v. 1– , 1978– . New York: Wilson. Monthly. Price varies. ISSN 0162-1963.

Annual cumulation. Suitable for undergraduates and public libraries. Covers 139 magazines and journals in all sciences, including biological. For nonspecialists only, although it does index magazines and journals more completely than most indexes; includes book reviews, editorials and other short reports. General science periodicals such as *Science* and *Nature* are indexed here and not in the companion index, *Biological and Agricultural Index* (see p. 74). Available online on CDP and on CD-ROM from Wilson from 1984 to the present. The CD-ROM is updated annually.

Government Reports Announcements and Index. v. 75– , 1975– . Springfield, VA: U.S. Department of Commerce, National Technical Information Services. Biweekly. $535.00. ISSN 0097-9007.

Lists technical reports available from NTIS, including research supported by federal grants and some state and local governments. Formerly titled *U.S. Government Research and Development Reports* and *Government Reports Announcement*. Available online as NTIS on CDP, DIALOG, and STN from 1964 to the present and on CD-ROM from SilverPlatter from 1983 to the present, updated quarterly, and from DIALOG OnDisk, from 1980 to the present.

Index Medicus. New series. v. 1– , 1960– . Washington, DC: National
    Library of Medicine. Monthly. $284.00. ISSN 0019-3879.

Comprehensive, worldwide indexing service for the biomedical sciences. Indispensable for medically related subjects and for the basic medical sciences.
Covers periodical literature and includes "Bibliography of medical reviews" as
part of its regular monthly issue. Cumulates annually. Subject arrangement
using medical subject headings (MeSH) with author index. Supersedes *Current
List of Medical Literature* (1941–59), *Quarterly Cumulative Index Medicus*
(1927–56), *Quarterly Cumulative Index to Current Medical Literature* (1917–
27), and *Index Medicus, or Quarterly Classified Record of the Current Medical
Literature of the World* (1903–1927). Available online as MEDLINE from
many vendors, including CDP, DIALOG, and STN, all from 1966 to the present;
also on CD-ROM from various vendors, some dating back to 1966 as well. The
Gratedul Med menu-based interface is also available for end-users for modem
or Internet access to the NLM databases. In addition, NLM has made two databases, AIDSLINE and DIRLINE (see annotation, p. 151, in Chapter 7, under
Directory of Biotechnology Information Resources), available for searching
at no cost. Contact NLM at (800) 638-8480 for more information on Grateful
Med or the free databases. For more information on using the MeSH headings,
see Chapter 2 (pp. 15–17).

Index to Scientific and Technical Proceedings. v. 1– , 1978– . Philadel-
    phia: Institute for Scientific Information. Monthly. $1,610.00.
    ISSN 0149-8088.

Appropriate for retrospective and current searches, bibliographic verification, and
acquisition of the published conference literature, although *BA/RRM* covers the
biological sciences more completely. Conferences and individual papers are
indexed. Cumulated semiannually.

Index to Scientific Reviews. v. 1– , 1974– . Philadelphia: Institute for
    Scientific Information. Semiannual. $980.00. ISSN 0360-0661.

Scans journals and books for review articles. Multidisciplinary. Citation,
subject, corporate, and source indexes to the volumes cumulated annually.

INPADOC. 1968– . Vienna: European Patent Office.

Database containing listings of patents from 56 countries. Includes title, inventor, assignee and information on country, patent family, and some legal status
information. Also available on microfiche. Available from DIALOG and STN.

Life Sciences Collection. Bethesda, MD: Cambridge Scientific Abstracts.

This computerized index is the combination of 21 abstracting journals from

Cambridge Scientific Abstracts, including: *Animal Behavior Abstracts, Biochemistry Abstracts* (three sections), *Biotechnology Research Abstracts* (four sections), *Calcified Tissues Abstracts, Chemoreception Abstracts, Entomology Abstracts, Genetics Abstracts, Human Genome Abstracts, Immunology Abstracts, Microbiology Abstracts* (three sections), *Neurosciences Abstracts, Oncogenes and Growth Factors Abstracts, Toxicology Abstracts,* and *Virology and AIDS Abstracts.* The online database is available on DIALOG and STN from 1978 to the present, and the CD-ROM version, from SilverPlatter, is available from 1982 to the present. The individual abstract journals are also searchable via the Internet for an annual subscription fee. The service is made available by Cambridge Scientifc Abstracts.

MEDLINE

See *Index Medicus* (p. 78).

Monthly Catalog of United States Government Publications. v. 1– ,
    1895– . Washington, DC: Government Printing Office. Monthly.
    $229. ISSN 0362-6830.

Essential for locating government documents. List of government publications arranged by department. There are author, title, subject, and series/report indexes in each issue. Complete bibliographic information is supplied for each entry including price and ordering directions. Available online from CDP and DIALOG. Also on CD-ROM from SilverPlatter (as GPO), AutoGraphics (as Government Documents Catalog Service), and Information Access Company (as Government Publications Index on Infotrac), all from 1976 to date.

Official Gazette of the United States Patent and Trademark Office.
    Patents. v. 1– , 1872– . Washington, DC: Government Printing
    Office. Weekly. $375. ISSN 0098-1133.

Listing of patents; includes abstract and sketches. Patentee, classification, and geographical indexes. Available online as CLAIMS/U.S. PATENT ABSTRACTS from DIALOG and STN from 1950 to the present and on CD-ROM as CASSIS/BIB and other sections from the U.S. Patents and Trademark Office from 1977 to date.

PASCAL. 1973– . Paris: CNRS, Centre de Documentation Scientifique
    et Technique.

This database consists of records from the 79 print *PASCAL* (Programme Appliqué à la Sélection et à la Compilation Automatiques de la Literature) indexes, which are the major French indexes. In French and English. International coverage of all science and technical areas, including biology. The print sections of most interest to biologists are: *PASCAL Explore, E 58: Génétique, PASCAL*

Explore, E 61: Microbiologie, PASCAL Explore, E 62: Immunologie, PASCAL Folio, F 52: Biochimie, PASCAL Folio, F 55: Biologie Végétale, PASCAL Folio F 56: Ecologie Animale et Végétale, PASCAL Folio F 43: Anatomie et Physiologie des Vertèbres, and *PASCAL Thema, T 260: Zoologie Fondametale et Appliquée des Invertèbres.* Formerly (to 1984) *Bulletin Signaletique.* Available on-line from DIALOG, 1973 to the present. Also available on CD-ROM from CNRS.

Pollution Abstracts. v. 1– , 1970– . Bethesda, MD: Cambridge Scientific Abstracts. Bimonthly. $885.00. ISSN 0032-3624.

Has annual index. Covers air, water, soil and noise pollution as well as sewage and solid waste management, toxicology, and environmental action. Scans articles and documents. Available online and on CD-ROM as part of the Life Sciences Collection (p. 78).

Psychological Abstracts. v. 1– , 1927– . Washington, DC: American Psychological Association. Monthly. $1307.00. ISSN 0033-2887.

The most important psychology index, covers animal behavior and neurobiology. Author and subject indexes. Also available online as PsycINFO on CDP and DIALOG from 1967 to the present and on CD-ROM from SilverPlatter as PsycLIT, from 1974 to the present.

Referativnyi Zhurnal. v. 1– , 1958– . Moscow: Vsesoyuznyi Institut Nauchno-Tekhnicheskoi Informatsii (VINITI). Monthly. Price varies.

Major, comprehensive abstracting service in Russian; in 64 sections. In subject categories, with annual author and scientific name indexes.

Science Citation Index. v. 1– , 1961– . Philadelphia: Institute for Scientific Information. Bimonthly. $11,730.00. ISSN 0036-827X.

Multidisciplinary index to international science literature. Citation, source, corporate, and subject (keyword) indexes. The citation index groups all articles that have referenced the same earlier work and so provides a different sort of access to the literature than is usually found. Covers about 3,000 journals. Available online as SCISEARCH on DIALOG from 1974 to date and on CD-ROM (as Science Citation Index Compact Disk Edition) from ISI, 1980 to date. The CD-ROM is updated quarterly.

Toxicology Abstracts. v. 1– , 1978– . Bethesda, MD: Cambridge Scientific Abstracts. Monthly. $750.00. ISSN 0140-5365.

Covers all forms of toxicology, including pharmaceuticals, agrochemicals, cosmetics, drug abuse, radiation, and legislation. Includes author and subject indexes. Also available online and on CD-ROM as part of the Life Sciences Collection (see p. 78).

TOXLINE. Bethesda, MD: National Library of Medicine.

This database consists of records from a number of sources dealing with toxicology. Most of the records are from MEDLINE, but a number are from BIOSIS Previews and International Pharmaceutical Abstracts. Also included are files from a number of other sources, including information from various Federal agencies, *Pesticides Abstracts*, the *Poisonous Plants Bibliography*, *Toxicity Bibliography*, and a number of other sources. Dates included vary: some of the files go back to 1950. Sections of the full database are available online from the National Library of Medicine, CDP, and DIALOG, and on CD-ROM from SilverPlatter from the 1980's to the present. Also searchable via Grateful Med (see *Index Medicus*, p. 78).

World Translation Index. v. 1– , 1978– . Delft, Netherlands: International Translations Centre. Monthly. $855.00. ISSN 0259-8264.

Annual cumulation. Formerly (until 1987) *World Transindex*; formed by merger of several other translations indexes. Provides bibliographic and availability information on existing translations of material dealing with science and technology. Available online from DIALOG from 1979 to the present.

Zoological Record. v. 1– , 1864– . Philadelphia: BIOSIS. Annual. $2,400.00. ISSN 0144-3607.

The most comprehensive zoological index in the world. Includes books, proceedings, and over 6,500 periodicals. Covers worldwide literature of the year to which it refers. Exhaustive coverage for systematic zoology. Annual publication means that *Zoological Record* is basically a retrospective tool. Complete bibliographic information provided in author index. Other indexes in each issue are subject, geographic, paleontological, and systematic. Issued in 20 sections relating to a phylum or class of the animal kingdom. Individual sections available separately. List of references used as classification and nomenclature authorities also included. Available online on DIALOG from 1978 to date and on CD-ROM from SilverPlatter (1978 to date). The CD-ROM is updated quarterly. Also available in microform. The *Zoological Record Search Guide* (see Chapter 1) and the *Zoological Record Serial Sources* (Chapter 3) are helpful aids.

Retrospective Tools

Berichte Biochemie und Biologie: Referierendes Organ der Deutschen Botanischen Gesellschaft und der Deutschen Zoologischen Gesellschaft. v. 1–521, 1926–80. New York: Springer-Verlag.

Abstracting periodical in German of the Deutschen Botanische Gesellschaft. Entries are arranged by subject. Cumulative author and subject indexes.

International Catalogue of Scientific Literature, 1st–14th annual issues.
    1901–14. London: Published for the International Council by the
    Royal Society of London.

Outgrowth of the *Catalogue of Scientific Papers* (see below). Covers the years 1901–14. Author and subject catalogue. Divided into sections, for example, section M is devoted to botany, section L to general biology. Journal lists with abbreviated titles are provided.

Royal Society of London. *Catalogue of Scientific Papers, 1800–1900*.
    London, Royal Society of London, 1867–1902.

"Index to the Titles and Dates of Scientific Papers contained in the Transactions of Societies, Journals, and other Periodical Works. . . ." Essential retrospective source. Entries are arranged by author's name. Abbreviations used are explained and are particularly helpful in locating titles of ceased periodicals.

5

Biochemistry and Biophysics

Biochemistry and biophysics have been grouped together in this chapter. Both are integral parts of biology, and their interdisciplinary relationship with basic biological sciences often blurs subject area lines. Frequently, the materials and literature for one discipline will satisfy the demands or questions posed by the other. As an example, consider the similarity between the definitions of biochemistry and *biophysics* from the *McGraw-Hill Dictionary of Scientific and Technical Terms* (1989).

Biochemistry: The study of chemical substances occurring in living organisms and the reactions and methods for identifying these substances.

Biophysics: The hybrid science involving the application of physical principles and methods to study and explain the structures of living organisms and the mechanics of life processes.

There is substantial overlap, also, between biochemistry/biophysics with molecular and cellular biology which have been placed in Chapter 6.

Abstracts and Indexes

Both general and specialized abstracting and indexing tools for biochemistry and biophysics are listed here for the sake of convenience. General biological indexes are annotated in Chapter 4, "Abstracts and Indexes." Most of these resources are available in both print and electronic formats, although online and CD-ROM indexes usually have more restricted backfiles than the in-print counterparts.

Chemoreception Abstracts. v. 1– , 1972– . Bethesda, MD: Cambridge Scientific Abstracts. Quarterly. $495/yr. ISSN 0300-1261.

Covers articles in the areas of chemical senses, including chemotaxis, smell, taste, perfumery, and more. Available on CD-ROM, online, and through the Internet as part of the **Life Sciences Collection** (see Chapter 4, p. 78).

Current Advances in Protein Biochemistry. v. 9– , 1992– . New York: Pergamon. Monthly. $846/yr. ISSN 0741-1618. Continues *Current Advances in Biochemistry*.

One of several *Current Advances* titles from Pergamon. A current awareness service with citations arranged in subject classification; also contains comprehensive listing of review articles.

Current Contents/Physical, Chemical and Earth Sciences. v. 1– , 1961– . Philadelphia: Institute for Scientific Information. Weekly. $460/yr. ISSN 0163-2574.

Provides tables of contents from the major journals in the area. Covers biochemistry and biophysics; see also *Current Contents/Life Sciences* (p. 76) in Chapter 4, "Abstracts and Indexes." Available on diskette. Also available online and on CD-ROM as part of **Science Citation Index**.

Current Physics Index. v. 1– , 1975– . New York: American
   Institute of Physics. Quarterly. $800/yr. ISSN 0098-9819.

"Each quarterly issue contains the approximately 11,000 abstracts of articles
published in the primary journals of one quarter, classified by subject." The
index covers only 75 journals.

General Physics Advance Abstracts. v. 1– , 1985– . New York:
   American Institute of Physics. Semimonthly. $570/yr. ISSN
   0749-4823.

"Contains early drafts of the abstracts of papers accepted for publication in the
journals of AIP and its member societies." *Physical Review Abstracts* (see below) offers the same service for journals of the American Physical Society. Preprints and advance warning of forthcoming publications are more important for
physicists than biologists, so there are no equivalent abstracting services for
biology.

Physical Review Abstracts. v. 1– , 1970– . New York: American
   Institute of Physics. Semimonthly. $320/yr. ISSN 0048-4024.

Offers the same service as *General Physics Advance Abstracts* for the publications of the American Physical Society, including *Physical Review, Physical
Review Letters*, and *Reviews of Modern Physics*.

Physics Abstracts. v. 1– , 19– . London: Institution of Electrical
   Engineers. Semimonthly. $2835/yr. ISBN 0036-8091. Continues
   Science Abstracts, Part A: Physics Abstracts.

The most important abstracting service in physics. In subject order, with author
index. Also available on CD-ROM as part of **INSPEC** and online (also as
INSPEC) from CDP, DIALOG, and STN.

Toxicology Abstracts. v. 1– , 1978– . Bethesda, MD: Cambridge
   Scientific Abstracts. Monthly. $795/yr. ISSN 0140-5365.

Covers all areas of toxicology. Available on CD-ROM, online, and through the
Internet as part of the **Life Sciences Collection** (see Chapter 4, p. 78).

See also *Biological Abstracts* (p.74), *Biology Digest* (p. 75), *Chemical Abstracts*
(p. 75), and *Current Contents/Life Sciences* (p. 76) in Chapter 4, "Abstracts and
Indexes."

Dictionaries, Encyclopedias, and Nomenclature

Biochemical Nomenclature and Related Documents; A Compendium, 2nd ed.
See the annotation (p. 60) in Chapter 3, "General Sources."

Compendium of Chemical Terminology: IUPAC Recommendations.
  Compiled by Victor Gold et al. Oxford: Blackwell, 1987. 456
  p. ISBN 063201751.

Alphabetical listing of terms recommended by the International Union of Pure and Applied Chemistry.

Concise Dictionary of Chemistry, new ed. New York: Oxford University Press, 1990. 336 p. $9.95 (paper). ISBN 192861107.

Chemical terms including biochemistry and a limited selection from biophysical chemistry.

Concise Dictionary of Physics, new ed. New York: Oxford University Press, 1990. 320 p. $9.95 (paper). ISBN 0192861115.

Derived from *Concise Science Dictionary* by the same publisher, includes all entries relating to physics. See *Concise Dictionary of Biology* (Oxford University Press, 1985) for more biophysical terms.

Dictionary of Chemical Names and Synonyms. Edited by Philip
  Howard and Michael Neal. Boca Raton, FL: Lewis/CRC Press, 1992.
  1,800 p. $150.00. ISBN 0873713966.

Finding-aid for approximately 20,000 chemicals by names and synonyms, Chemical Abstracts Services registry numbers, chemical formulae, and structure.

Dictionary of Natural Products. New York: Chapman & Hall, 1993.
  7 v. $4,995.00. ISBN 0412466201. Updated annually. Also
  available in CD-ROM, updated every six months: ISSN 0966-2146.

An authoritative, comprehensive source of information for 100,000 natural products and related compounds containing chemical, structural and bibliographic data. There are extensive indexes, including a species index, and a type-of-compound index. Chapman & Hall also publishes other major chemical dictionaries that are in print or on CD-ROM: *Dictionary of Organic Compounds, Dictionary of Inorganic Compounds, Dictionary of Alkaloids, Dictionary of Analytical Reagents, Dictionary of Steroids*.

Dictionary of the Physical Sciences: Terms, Formulas, Data, edited
  by Cesare Emiliani. New York: Oxford University Press, 1987.
  365 p. $16.00. ISBN 0195036514.

Five thousand definitions and 70 tables explaining physical, chemical, geological, and cosmological terms.

Dictionary of Physics, 3rd ed. Edited by H. J. Gray and Alan Isaacs.
  London: Longman, 1991. 636 p. ISBN 0582037972.

Comprehensive.

Encyclopedia of Modern Physics. Edited by Robert A. Meyers. San
　　Diego: Academic Press, 1990. 773 p. $75.00. ISBN
　　0122266927.

Survey of the most rapidly advancing fields of theoretical and applied physics,
chosen for scientists in physics or the biological sciences who have basic knowledge of chemistry, physics, and calculus.

Encyclopedia of Physics, 2nd ed. Edited by Rita G. Lerner and
　　George L. Trigg. New York: VCH, 1991. 1,408 p. $95.00.
　　ISBN 0895737523.

Alphabetical arrangement with references for further reading.

Encyclopedia of Physics, 3rd ed. Edited by Robert M. Besancon.
　　New York: Van Nostrand Reinhold, 1990. 1,378 p. $54.95.
　　ISBN 0442005229.

Articles are arranged alphabetically and include the role of physics in medicine
and biology. This encyclopedia is a condensation of 34 textbooks chosen to
comprehensively detail theory and applications of recent advances in physics.
All articles include a glossary of terms with concise definitions, first-princi-
pal–based treatment of the subject, and a bibliography.

*Enzyme Nomenclature 1992: Recommendations of the Nomenclature
　　Committee of the International Union of Biochemistry and
　　Molecular Biology on the Nomenclature and Classification of
　　Enzymes*

See the annotation (p. 61) in the Chapter 3, "General Sources."

Facts on File Dictionary of Chemistry, rev. expanded ed. Edited by
　　John Daintith. New York: Facts on File, 1990. $24.95. ISBN
　　0816023670.

2,500 entries explaining the most important and commonly used chemical terms.

Facts on File Dictionary of Physics, rev. ed. by John Daintith. New
　　York: Facts on File, 1988. 235 p. $24.95. ISBN 0816018685.

Updated by including new terms in solid-state physics and quantum physics and
containing more than 2,000 definitions appropriate at the student level.

Glick, David M. *Glossary of Biochemistry and Molecular Biology*.
　　New York: Raven Press, 1990. 194 p. ISBN 0881675636.

*Longman Chemistry Handbook: The Fundamentals of Chemistry Ex-
　　plained and Illustrated*. Edited by Arthur Godman. Harlow,
　　Essex, UK: Longman, 1992. 256 p. ISBN 0582088100.

This handbook is appropriate at the high school or college undergraduate student level. It explains nearly 1,500 terms and concepts, many of them illustrated in color. The book is arranged by broad topic and there is a comprehensive index.

McGraw-Hill Encyclopedia of Chemistry, 2nd ed. Sybil P. Parker, editor-in-chief. New York: McGraw-Hill, 1993. 1,236 p. ISBN 00704-54558.

Although biochemistry as such is not included, this comprehensive encyclopedia is valuable for its overview of theoretical chemistry. The 2nd edition is taken from the 7th edition of the *McGraw-Hill Encyclopedia of Science and Technology*.

McGraw-Hill Encyclopedia of Physics. Sybil P. Parker, editor-in-chief. New York: McGraw-Hill, 1993. 1,624 p. ISBN 0070514003.

Condensed from the 7th edition of *McGraw-Hill Encyclopedia of Science and Technology*. All important developments in classical and modern physics including the most recent advances in theoretical and experimental research.

Physics in Medicine and Biology Encyclopedia. Tarrytown, NY: Pergamon, 1986. 2 v. $420. ISBN 0080264972.

Comprehensive and convenient source for hospital physicists, medical technicians, and clinicians. Aimed at the novice, there are classified and alphabetic lists of articles, and indexes by subject and author. Each article includes a glossary of terms and a bibliography.

Stenesh, J. *Dictionary of Biochemistry and Molecular Biology*, 2nd ed. New York: Wiley, 1989. 525 p. $74.95. ISBN 0471840890.

16,000 entries, many of them from allied sciences of chemistry, biophysics, genetics, microbiology, etc.

Directories

Chemical Research Faculties: An International Directory. Edited by Gisella Linder Pollock. Washington, DC: American Chemical Society, 1988. $159.95. 559 p.

International directory listing institutions and universities granting advanced degrees in chemistry, chemical engineering, biochemistry, and medical/pharmaceutical chemistry. There are 960 departments listed including 10,000 faculty in countries worldwide, the United States and Canada.

Chemical Sciences Graduate School Finder. Washington, DC: American Chemical Society. 1993–1994, 620 p. $44.95. ISSN 1058-1227. Available on CD-ROM with *Chemical Research Faculties* and *Directory of Graduate Research*, $180.00.

List of graduate schools with complete information on admissions, fields of study (including biochemistry), and a geographic index.

College Chemistry Faculties, 9th ed. By Cornelia A. Talmadge. Washington, DC: American Chemical Society, 1993. 208 p. $84.95. ISBN 0841226639.

Comprehensive directory of chemistry, biochemistry, and chemical engineering teachers in universities, two- and four-year colleges in the United States and Canada. Available on CD-ROM with *Directory of Graduate Research* and *Chemical Sciences Graduate School Finder*, $180.00.

Directory of Graduate Research. Washington, DC: American Chemical Society, 1991. 1,440 p. (Academic price $159.95). ISSN 0193-5011. Also available on CD-ROM.

Lists faculties, academic research, publications, Doctoral and Master's theses in departments or divisions of chemistry, chemical engineering, biochemistry, medical/pharmaceutical chemistry, clinical chemistry, and polymer sciences in the United States and Canada. Available on CD-ROM with *College Chemistry Faculties* and *Chemical Sciences Graduate School Finder*, $180.00.

Directory of Physics and Astronomy Staff 1993–1994. New York: American Institute of Physics, 1993. Biennial. 502 p. $60.00 (paper). ISBN 1563962578.

Names, addresses, and phone numbers for physicists and astronomers in North America, with additional sections for academic institutions and research and development organizations. Foreign organizations and member societies of the American Institute of Physics comprise separate sections.

Graduate Programs in Physics, Astronomy and Related Fields. New York: American Institute of Physics, 1993–1994. 988 p. Annual. $45.00 (paper). ISBN 156396239X.

Information on graduate programs in North America. Includes general information for specific universities, number of faculty, admission and graduate requirements, enrollment. Faculty and research specialties are listed.

Regulated Chemicals Directory 1994, rev. and updated. New York: Chapman & Hall, 1994. 1,600 p. $299.00 (paper). ISBN 0412052814.

This is a compilation of information on hazards and regulations for over 10,000 chemical substances from all relevant federal sources, e.g., OSHA, EPA, NIOSH, DOT, RCRA, CERCLA/SARA, NTP, IARC. Data includes: 22,000 synonyms; CAS numbers and generic names; workplace exposure limits; carcinogenicity, target organ and reproductive effects; environmental exposure levels; transportation, fire, and reactivity hazards.

Guides to Internet Resources

The American Institute of Physics (AIP) operates a Gopher server that is especially relevant to this chapter. This server is the "client" program to navigation of the Internet, and it is also home for some local files, directories, and other links to the physical sciences, including access to physics FTP archives, physics education news, physics news update, and **PINET**, the Physics Information Network. **PINET** is a full-service interactive network of searchable databases, communications services, news, announcements, and job listings created specifically for the physics and astronomy community. For more information or to register, telnet to pinet.aip.org.

The American Chemical Society (ACS) Gopher system contains items of interest to chemists, including information about the ACS books catalog, information numbers, other chemistry Gophers, Internet chemistry resources, ACS publications, membership, and special reports. The most recent version of the directory to Internet chemistry resources may be found either on the Clearinghouse for Chemical Information Instructional Materials sites listed in the Teaching Resources section, or the Indiana University Chemistry Library Gopher. Headings in the Aug. 20, 1994 revision to resources on the Internet include book catalogs, chemical information source guides, databases, document delivery, e-mail servers, listservs, newsgroups, FTP resources, gophers, guides to Internet resources, periodicals and conference proceedings, software, teaching resources, and World Wide Web resources.

There are also a number of pertinent discussion groups, including the USENET groups bionet.prof-society.biophysics, bionet.structural-nmr, sci. chem, sci.chem.labware, sci.med.physics, and sci.physics.

Guides to the Literature

Guide to Sources of History of Solid State Physics. Edited by J. Warnow-Blewett and J. Teichman. Woodbury, NY: AIP Press, 1992. 156 p. $19.00 (paper). ISBN 1563960680.

A compilation of historical source materials documenting the field of solid-state physics. It includes over 400 papers, records, and oral history interviews.

Information Sources in Physics, 2nd ed. Edited by Dennis F. Shaw.
    London: Butterworths, 1985. 456 p. $60.00. ISBN 0408014741.

Different authors have contributed to the following chapters: science libraries, abstracting and indexing tools, various subjects (particle physics, acoustics, etc.), grey literature, patent literature, and important physics journals according to INSPEC. There is very little information concerning biophysics.

Maizell, Robert E. *How to Find Chemical Information: A Guide for
    Practicing Chemists, Teachers and Students*, 2nd ed. New
    York: Wiley, 1987. 496 p. $59.95. ISBN 0471867675.

Useful discussions of current awareness sources, patents, Chemical Abstracts Services, government documents, searching the chemical literature, major reference books, and business information sources for chemists.

Wiggins, Gary. *Chemical Information Sources*
See the annotation (p. 46) in Chapter 3, "General Sources."

Handbooks

This section includes both print and electronic formats. Rather than duplicate the discussion on computer-based and electronic databases that appears in Chapter 6, "Molecular and Cellular Biology," we refer the reader to the fully annotated Databases section in that chapter.

There is a great deal of overlap between biochemistry and other biological disciplines. For example, databanks for protein and enzyme sequences are annotated in Chapter 6, "Molecular and Cellular Biology," and gene sequence databases are discussed in Chapter 7, "Genetics."

ACS Style Guide: A Manual for Authors and Editors. Washington,
    DC: American Chemical Society, 1986. 264 p. $24.95. ISBN
    0841209170.

Covers American Chemical Society style conventions, plus newly emerging technology and concepts. There are chapters on writing scientific papers, grammar, style, usage, illustrations and tables, copyright, submission for machine-readable manuscripts, chemical literature, oral presentations, ethical guidelines, manuscript typing hints, etc.

AIP Style Manual, prepared under the direction of the AIP Publication
    Board, 4th ed. New York: American Institute of Physics,
    1990. 64 p. $10.00 (paper). ISBN 088318642X. Index.

"For guidance in writing, editing, and preparing physics manuscripts for publication"—from the cover.

American Institute of Physics Handbook, 3rd ed. New York: McGraw-Hill, 1972. 2,368 p. $159.00. ISBN 007001485X.

Authoritative reference material, data tables, graphs, bibliographies arranged by broad topic. Updated by *Physicist's Desk Reference*, annotated on p. 94.

Ayad, Shirley et al. *The Extracellular Matrix FactsBook*. San Diego, CA: Academic Press, 1994. (FactsBook series). 163 p. $42.00. ISBN 0120689103.

Basic information and catalog of the essential properties for over 40 entries for the diverse group of macromolecules that form the extracellular matrix. Data includes molecular structure, isolation, primary structure, database accession numbers, structural and functional sites, gene structure, and key references.

Biochemistry Labfax. Edited by J. A. A. Chambers and David Rickwood. San Diego: Academic Press, 1993. 356 p. $59.95 (spiral bound). ISBN 0121673405.

Designed for practicing scientists, this reference contains information on radioisotopes, enzymes, proteins, nucleic acids, lipids, and selected techniques such as chromatography, electrophoresis, etc. Databases, suppliers, chemical and mathematical formulae, and references are provided.

CRC Handbook of Chemistry and Physics.

See the annotation (p. 48) in Chapter 3, "General Sources."

Data for Biochemical Research, 3rd ed. Edited by R. M. Dawson, et al. New York: Oxford University Press, 1986. 580 p. $55.00 (paper). ISBN 0198553587. Index.

This laboratory handbook makes no attempt to be comprehensive and was compiled to be of use to a wide range of readers. The book consists of tables and references in 25 sections providing data on amino acids, vitamins, steroids, carotenoids, biochemical reagents, biochemical procedures, definitions, and atomic weights, to name a few examples. References are included to enable readers to find the relevant literature.

Enzyme Handbook. Edited by D. Schomburg and M. Salzmann. New York: Springer-Verlag. v. 1– , 1990– . Each volume priced separately. v. 6: *Oxidoreductases*, 1993. $182.00. ISBN 3540564357.

An impressive compendium covering 3,000 enzymes for the practicing enzymologist. Entries will be updated as necessary in print; the entire work is stored in a databank with plans for it to be computer searchable.

Journal of Physical and Chemical Reference Data. v. 1– , 1972– .
New York: American Chemical Society and the American Institute of physics for the National Institute of Standards and Technology. Bimonthly. $510/yr. ISSN 0047-2689.

Data compilation and reviews produced under the National Standard Reference Data System provides reliable, up-to-date reference data for atomic and molecular science, chemical kinetics, spectroscopy, thermodynamics, transport phenomena, crystallography, materials science, etc.

Kanare, Howard M. *Writing the Laboratory Notebook.* Washington, DC: American Chemical Society, 1985. 150 p. $14.95 (paper). ISBN 0841209332.

Information on how to keep a proper and permanent laboratory notebook, including creating records, writing with clarity, and examples of notebook entries and the electronic notebook.

Lange's Handbook of Chemistry, 14th ed. Edited by John A. Dean. New York: McGraw-Hill, 1992. $82.50. ISBN 0070161941. Index.

One-volume source of factual information for chemists to satisfy general information needs. This excellent time saver provides data on organic compounds, conversion tables, mathematics, inorganic chemistry, properties of atoms, physical properties, thermodynamic properties, spectroscopy, electrolytes, electromotive forces, chemical equilibrium, physicochemical properties, polymers, and practical laboratory information.

Maynard, John T. and Howard M. Peters. *Understanding Chemical Patents: A Guide for the Inventor,* 2nd ed. Washington, DC: American Chemical Society, 1991. 183 p. $39.95. ISBN 0841219974.

Description of the patent system from understanding the terminology, to working with patent attorneys, agents, and technical liaison personnel.

Merck Index: An Encyclopedia of Chemicals and Drugs.

See the annotation (p. 37) in Chapter 3, "General Sources."

The Physicist's Companion. Edited by E. Richard Cohen. Woodbury, NY: AIP Press, 1994. 200 p. $13.95 (paper). ISBN 1563961431.

This book contains material from the first chapter of *A Physicist's Desk Reference* (see p. 94), updated and supplemented by additional new data. References to more complete sources of data are included.

A Physicist's Desk Reference: The Second Edition of Physics Vade Mecum, 2nd ed. Herbert L. Anderson, editor-in-chief. New York: American Institute of Physics, 1989. 356 p. $70.00. ISBN 0883186292. Index.

Arranged by topic, this handbook "presents the essence of concepts and numerical data contained primarily in archival journals in which physicists and astronomers publish." The first section reports physical constants, followed by acoustics, astronomy and astrophysics, etc. There is a section for medical physics.

Pigott, Rod and Christine Power. *The Adhesion Molecule FactsBook*. San Diego, CA: Academic Press, 1993. (FactsBook series). 190 p. $42.00 (paper). ISBN 0125551800.

Concise information on structure, function, and biology of cell adhesion molecules in an accessible format, including key references, and PIR, SWISSPROT, GenBank/EMBL accession numbers. Arranged by most commonly used name.

Practical Handbook of Biochemistry and Molecular Biology. Edited by Gerald D. Fasman. Boca Raton, FL: CRC Press, 1989. 601 p. $59.95. ISBN 0849337054.

Material is derived and updated from the multi-volume *CRC Handbook of Biochemistry and Molecular Biology*, 1975–1977. Physical/chemical data for proteins, nucleic acids, lipids, etc.

Rapid Guide to Hazardous Chemicals in the Workplace. Edited by N. Irving Sax and Richard J. Lewis, Sr. New York: Van Nostrand Reinhold, 1986. 236 p. $19.95. ISBN 0442282206.

Quick reference to 700 of the most frequently encountered hazardous materials. Each entry includes identifying information, standards and recommendations, toxic and hazardous material reviews, and physical properties. There are appendices and cross-references to Chemical Abstracts registry numbers, Registry of Toxic Effects of Chemical Substrates numbers, Department of Transportation hazard codes. If you need more information, consult *Hazardous Chemicals on File* (New York: Facts on File, 1988. 3 Vols. $250/set. ISBN 0816022135 (set); and *Hazardous Chemicals on File 1989 Update*. $50.00. ISBN 0816022127.

Shugar, Gershon J. and John A. Dean. *The Chemist's Ready Reference Handbook*. New York: McGraw-Hill, 1989. unpaged. $79.80. ISBN 0070571783.

Covers real-world questions and problems in the chemical laboratory. A compilation of basic data, procedures, precautions, and trouble-shooting hints. A companion to *Chemical Technicians' Ready Reference Handbook*, 3rd ed. (McGraw-Hill, 1990, 889 p.), which was written for all lab personnel, this volume is appropriate for high school and graduate students who need "every single step" for conventional lab procedures. There are sections on mathematics, statistics, safety, first aid, and a glossary, to name a few of the important chapters.

Watson, Steve and Steve Arkinstall. *The G-Protein Linked Receptor FactsBook*. San Diego, CA: Academic Press, 1994. (Facts-Book series). 427 p. $29.95. ISBN 0127384405.

Similar to the other titles in the *FactsBook* series, this compilation provides a catalog of over 50 entries on all members of the seven-transmembrane family of cell surface receptors and their associated G-proteins and effectors, including acetylcholine, adrenaline, dopamine, glutamine, 5-HT, G-proteins, phospholipase, adenylyl cyclase. There is information on structure, molecular weights and glycosylation sites; distribution; receptors; pharmacology; effector pathways; amino acid sequence; PIR, SWISSPROT, and EMBL/GenBank accession numbers; gene structure and organization; and key references.

Histories

Biographical Dictionary of Scientists: Physicists. General editor, David Abbott. New York: Peter Bedrick Books, 1984. 212 p. $28.00. ISBN 0911745793. Glossary. Index.

Historical introduction to physics. Alphabetical biographical entries of several paragraph's length include important physicist's dates and major works.

Brock, William H. *The Norton History of Chemistry*. New York: Norton, 1993. (History of Science). 768 p. $35.00. ISBN 0393035360.

For the general reader as well as the chemistry student, this history surveys chemistry from its early beginnings through the 20th century. Each chapter is devoted to a significant development in chemical history.

Brush, Stephen G., ed. *Resources for the History of Physics: I. Guide to Books and Audiovisual Materials. II. Guide to Original Works of Historical Importance and Their Translations into Other Languages*. Ann Arbor, MI: Books on Demand, originally published 1972. 176 p. $48.00. ISBN 031710599X.

Resources for physics teachers, arranged by topic or format. Entries are evaluated for age group with subject and institution codes.

Brush, Stephen G. and Lanfranco Belloni. *The History of Modern Physics: An International Bibliography*. New York: Garland, 1983. 334 p. ISBN 0824091175.

Guidance for novices, including specific information about the subject in brief annotations. "Modern physics" is defined as beginning with the discovery of x-rays in 1895.

Chemistry of Life: Eight Lectures on the History of Biochemistry. Edited, with an introduction, by J. Needham. New York: Cambridge University Press, 1970. $49.95. ISBN 0521073790.

This collection is part of a series of lectures given by Cambridge University biochemists under the aegis of the history of Science Committee and Department. Topics addressed by important figures in the field include photosynthesis, biological oxidations, microbiology, neurology, animal hormones, vitamins, foundations of modern biochemistry, and pioneers of biochemistry in the 19th century.

Contrasts in Scientific Style; Research Groups in the Chemical and Biochemical Sciences. Philadelphia: American Philosophical Society, 1990. (Memoirs series, Vol. 191). 473 p. $40.00. ISBN 0871691914. Index. Bibliography.

Sequel to *Molecules and Life* (p. 97), discussing the emergence of large research groups during the 19th century.

Fruton, Joseph S. *A Bio-Bibliography for the History of the Biochemical Sciences since 1800.* Philadelphia: American Philosophical Society, 1982. $20.00. ISBN 0871699834. *A Supplement to a Bio-Bibliography for the History of the Biochemical Sciences Since 1800.* Philadelphia: American Philosophical Society, 1986. 262 p. $15.00. ISBN 087169980X.

Data on the lives and work of people who participated in providing chemical explanations of biological phenomena. Each of the entries includes reference to biographical or bibliographical reference works or citations of books and articles in serial publications.

Heilbron, J. L. and Bruce R. Wheaton. *Literature on the History of Physics in the 20th Century.* Berkeley, CA: Office for History of Science and Technology, University of California, 1981. (Berkeley papers in history of science, 5). 485 p. ISBN 0918102057.

Survey of sources gleaned from *Isis, Dictionary of Scientific Biography*, Poggendorff's *Handworterbuch*, etc., for 20th century physicists. Entries are coded for category, classifier (person, institution, field), and subject breakdown. *An Inventory of Published Letters . . .* is a companion volume; see the annotation under the first author, Bruce R. Wheaton (p. 98).

History of Biochemistry. Section 6 of the series *Comprehensive Biochemistry.* Edited by M. Florkin and E. H. Stotz.

Vol. 37: *Selected Topics in the History of Biochemistry: Personal Recollections*, 1990. 396 p. $114.25. ISBN 0444812164. See the annotation under series title in the Textbooks and Treatises section (p. 112).

History of Physics. Edited by Spencer R. Weart and Melba Phillips. Woodbury, NY: AIP Press, 1985. 375 p. $39.00 (paper). ISBN 0883184680.

Readings from *Physics Today*. Also, see *The Life and Times of Modern Physics* (see below).

Home, R. W. *The History of Classical Physics: A Selected, Annotated Bibliography*. New York: Garland, 1984. 324 p. Index.

Classical physics embraces the years 1700–1900. This annotated bibliography is arranged in broad topics for general bibliographical works, individual biographies, collected works, general histories, 18th century physics, physics in transition, 19th century physics.

The Life and Times of Modern Physics; History of Physics II. Edited by Melba Phillips. Woodbury, NY: AIP Press, 1992. 300 p. $40.00. ISBN 0883188465.

Recent articles from *Physics Today* chronicles the people and events shaping modern science and society. Includes profiles, personal memoirs, and histories of important institutions and organizations.

Molecules and Life; Historical Essays on the Interplay of Chemistry and Biology. New York: Wiley-Interscience, 1972. 579 p. References: pages 505–554. ISBN 0471284483.

Covers the period from 1800-1950 discussing "Ferments to Enzymes," "Nature of Proteins," "Nuclein to the Double Helix," "Intracellular Respiration," "Pathways of Biochemical Change."

Nobel Prize Winners: Chemistry. Edited by Frank N. Magill. Pasadena, CA: Salem Press, 1990. 3 Vols. ISBN 089356561X (set).

The three volumes cover the years 1901–1989. There are descriptions of the laureate's life, work and scientific career, speeches and critical commentary, most significant experiments, theories, and publications. Comprehensive index of nationality, important people, key terms, theories, labs, institutions, and area of concentration. History and overview of the prize in chemistry, and a portrait of each laureate are included.

Nobel Prize Winners: Physics. Edited by Frank N. Magill. Pasadena, CA: Salem Press, 1989. 3 Vols. ISBN 0893565571 (set).

Arranged in an identical manner to *Nobel Prize Winners: Chemistry.*

The Physical Review: The First Hundred Years; A Selection of Seminal Papers and Commentaries. Edited by H. Henry Stroke. Woodbury, NY: AIP Press, 1994. 1,200 p. $75.00. ISBN 1563961881. CD-ROM and book packaged together; CD-ROM contains over 6,000 additional pages of text.

This set contains select seminal papers published in the first 100 years of *The Physical Review*, including commentaries from many eminent physicists.

Slater, Edward Charles. *Biochimica et Biophysica Acta: The Story of a Biochemical Journal.* New York: Elsevier, 1986. 122 p. ISBN 0444807691.

Interesting history of one of the most important biochemical/biophysical journals in the world.

Solomon, A. K. "A Short History of the Foundation of the International Union for Pure and Applied Biophysics," *Quarterly Review of Biophysics* 1: 107–124 (1968).

Teich, Mikulas, with Dorothy M. Needham. *A Documentary History of Biochemistry 1770–1940.* Leicester, England: Leicester University Press, 1992. 579 p. ISBN 071851341X. Bibliography.

Selected collection of reprints, over half of them translated into English for the first time, on the evolution of the study of the chemistry of life into modern biochemistry.

Wheaton, Bruce R. and J. L. Heilbron. *An Inventory of Published Letters to and from Physicists, 1900–1950.* Berkeley, CA: Office for History of Science and Technology, University of California, Berkeley, 1982. Microform. (Berkeley papers in history of science, 6). ISSN 0145-0379.

Contains quotations from the correspondence of almost 6,000 physicists, active from 1895 to 1955, who appear in printed books and articles in *Literature on the History of Physics in the 20th Century* listed under Heilbron and Wheaton (p. 96).

Methods and Techniques

Consult the Handbooks section (p. 91) for practical laboratory information, and don't forget to check out the Methods section in Chapter 6, "Molecular Biology."

Alexander, Renee R. and Joan M. Griffiths. *Basic Biochemical Methods*, 2nd ed. New York: Wiley, 1993. 353 p. $39.95. Index. ISBN 0471561533.

Text intended for a biochemistry laboratory course at the advanced undergraduate/beginning graduate level for students in the biological sciences. Designed to introduce methods used in isolation and quantification of various cell fractions or compounds having biological significance. Appendices provide information on equipment, supplies, and preparation of reagents, and problem sets.

Current Chemical Reactions. v. 1– , 1979– . Philadelphia: Institute for Scientific Information. Monthly. $935/yr. ISSN 0163-6278.

Guide to revised and newly modified reactions and syntheses reported in 100 current source journals. Each entry includes complete bibliographic information, description of the reaction, flow chart, and notice of explosive reactions. Author, journal, corporate, and permuted subject indexes are provided.

Freifelder, David Michael. *Physical Biochemistry: Applications to Biochemistry and Molecular Biology*, 2nd ed. San Francisco: W. H. Freeman, 1982. 761 p. ISBN 071673152. Bibliographies, index, and answers to problems.

Techniques are arranged in sections reviewing direct observation, general laboratory methods, separation and identification, hydrodynamic methods, spectroscopic methods, and miscellaneous techniques.

Graham, Richard C. *Data Analysis for the Chemical Sciences: A Guide to Statistical Techniques.* New York: VCH, 1993. 536 p. ISBN 1560810483. Index.

Statistical methods; bibliographical references. (See p. 112 for full annotation.)

Journal of Biochemical and Biophysical Methods. v. 1– , 1979– . Amsterdam: Elsevier Science. Monthly. $460.00/yr. ISSN 0165-022X.

Deals with all methodological aspects of biochemistry, biophysics, and molecular biology. Accepts original papers and short notes; includes book reviews.

Laboratory Techniques in Biochemistry and Molecular Biology. v. 1– , 1969– . General editor: P. C. van der Vliet. Amsterdam: Elsevier. Irregular. Price varies. ISSN 0075-7535.

Technique review series. Vol. 23: *Monoclonal Antibody and Immunosensor Technology; The Production and Application of Rodent and Human Monoclonal Antibodies*, by Ailsa M. Campbell, 1991. Vol. 24: *Hybridization with Nucleic Acid Probes: Part I and II*, by P. Tijssen. 2 v., 1993. $267.50 (set). ISBN 04444898840 (v. 1).

Meier, Peter C. and Richard E. Zund. *Statistical Methods in Analytical Chemistry*. New York: Wiley, 1993. (Chemical Analysis, v. 123). 321 p. plus 1 computer disk. $59.95.00. ISBN 0471584541.

This volume is part of a series of monographs on analytical chemistry and its applications. The authors intend to convey, in a very practical manner, an appreciation of the importance of statistics to analytical chemistry. The use of computers and computerized graphics is included.

Methods in Enzymology. v. 1– , 1955– . New York: Academic Press.

See the annotation for this and its companion *Methods* (p. 57) in Chapter 3, "General Sources."

Methods in Protein Sequence Analysis. Edited by Kazutomo Imahori and Fumio Sakiyama. New York: Plenum, 1993. 310 p. ISBN 0306444887. Bibliography. Index.

"Proceedings of the Ninth International Conference on Methods in Protein Sequence Analysis, Sept. 1992. Up-to-date information about micropreparation and microsequencing, mass spectroscopy, post-translational modifications, prediction and database analysis, and protein structures of special interest.

Methods of Biochemical Analysis. v. 1– , 1954– . New York: Wiley. Irregular. $75. ISSN 0076-6941.

Each volume deals with biochemical methods and techniques used in different areas of science. Vol. 37: *Bioanalytical Instrumentation*. Edited by C. H. Suelter, 1994. 240 p. $89.95.

Modern Physical Methods in Biochemistry. Edited by A. Neuberger and L. L. M. van Deenen. New York: Elsevier, 1985–1988. (New Comprehensive Biochemistry, vol. 11 A-B). ISBN 0444806490.

Chapters on protein crystallography, NMR spectroscopy, electron spin resonance, mass spectroscopy. Part B extends the range of spectroscopic techniques to fluorescence and Raman spectroscopy.

Patterns in Protein Sequence and Structure. Edited by William R. Taylor. (Springer Series in Biophysics, Vol. 7). New York: Springer-Verlag, 1992. 262 p. ISBN 0387540431.

This volumes describes the latest techniques to interpret protein sequence data using pure theoretical sequence analysis, detailed structural analysis, experimental approaches, all with the common tactic of computer analysis.

Practical Approach Series. New York: Oxford University Press.

This series provides nuts and bolts experimental protocols relevant to a specific area of biology as well as an overview of general methods and background information about the topic. A recent volume: *Protein Blotting: A Practical Approach*, 1994, 242 p., $45.00, ISBN 0199634386.

Principles and Techniques of Practical Biochemistry, 4th ed. Edited
    by Keith Wilson and John M. Walker. New York: Cambridge
    University Press, 1994. 586 p. $99.95. ISBN 0521428092.

Updated edition of *A Biologist's Guide to Principles and Techniques of Practical Biochemistry* published in 1986. Contents of the 4th edition include general principles of biochemical investigations, and techniques for immunochemistry, molecular biology, proteins and enzymes, radioisotopes, centrifugation, spectroscopy, electrophoresis, chromatography, and electrochemistry.

Periodicals

Journals are selected that are key to the dissemination of original research in the areas of biochemistry and biophysics. Increasingly, journals are available in both in-print and electronic formats, e.g., *Biochemistry,* published by the American Chemical Society, is available on CD-ROM. The transition from print to electronic availability is a gradual and steady development, and one that should not be overlooked when searching for current journals.

Analytical Biochemistry. v. 1– , 1960– . San Diego: Academic
    Press. Monthly. $1,208/yr. ISSN 0003-2697.

Publishes original research on methods and methodology of interest to the biological sciences and all fields that impinge on biochemical investigations.

Applied Biochemistry and Biotechnology.

This is a journal of contemporary biotechnology that is annotated (p. 159) in Chapter 7, "Genetics."

Archives of Biochemistry and Biophysics. v. 1 – , 1942– . San
    Diego: Academic Press. Monthly. $1,224/yr. ISSN 0003-
    9861.

An international journal dedicated to the dissemination of fundamental knowledge in all areas of biochemistry and biophysics.

Biochemical and Biophysical Research Communications. v. 1– ,
    1959– . San Diego: Academic Press. Semimonthly. $996/yr.
    ISSN 0006-291X.

Devoted to the rapid dissemination of timely and significant observations in the diverse fields of modern experimental biology.

Biochemical Journal. v. 1– , 1906– . Colchester, England: Portland Press. Semimonthly. $1,495/yr. ISSN 0264-6021.

Journal published on behalf of the Biochemical Society, London. Includes high-quality research papers from all fields of biochemistry discussing new results, new interpretations, or new experimental methods.

Biochemical Society Transactions. v. 1– , 1973– . Colchester, England: Portland Press. Quarterly. $225/yr. ISSN 0300-5127.

Contains lectures, colloquia, and communications presented at Biochemical Society meetings. Submitted research papers are not published.

Biochemistry (U.S.). v. 1– , 1962– . Washington, DC: American Chemical Society. Weekly. $1,186/yr. ISSN 0006-2960. Also available on CD-ROM.

Publishes experimental results on all areas of biochemistry, "Perspectives in Biochemistry," and concise reviews on topics of timely interest.

Biochemistry and Cell Biology. v. 1– , 1986– . Ottawa, Canada: National Research Council of Canada. Monthly. $237/yr. ISSN 0829-8211. Continues *Canadian Biochemistry and Cell Biology.*

Selected by the Canadian Biochemical Society and the Canadian Society for Cellular and Molecular Biology as their recommended medium for the publication of scientific papers.

Biochimica et Biophysica Acta. International Journal of Biochemistry and Biophysics published in 9 sections: Bioenergetics, Biomembranes, General Subjects, Protein Structure and Molecular Enzymology, Lipids and Lipid Metabolism, Gene Structure and Expression, Molecular Cell Research, Molecular Basis of Disease, and Review Sections. v. 1– , 1947– . Amsterdam: Elsevier Science. Frequency varies. $6,900/yr for all sections. ISSN varies by section.

Bioelectromagnetics. v. 1– , 1980– . New York: Wiley. Bimonthly. $335/yr. ISSN 0197-8462.

Published for the Bioelectromagnetics Society and devoted to research on biological systems as they are influenced by natural or manufactured electric and/or magnetic fields at frequencies from DC to visible light.

Biophysical Chemistry. v. 1– , 1973– . Amsterdam: Elsevier Science. Monthly. $672/yr. ISSN 0301-4622.

"International journal devoted to the physics and chemistry of biological phenomena."

Biophysical Journal. v. 1– , 1960– . New York: Rockefeller University Press. Monthly. $650/yr. ISSN 0006-3495.

Edited by the Biophysical Society in cooperation with the Division of Biological Physics of the American Physical Society, this publication is a leading journal for original research in molecular and cellular biophysics. It features full-length original research papers, brief communications on topics of special interest, reviews of important topics, brief discussions of recent work, overviews of curricular design, book and software reviews, calendar, and employment opportunities.

Cell Biophysics: An International Journal. v. 1– , 1979– Totowa, NJ: Humana Press. Bimonthly. $190/yr. ISSN 0163-4992.

Major thrust of the journal is the presentation of original analytical and quantitative studies of the properties, structures, and interactions of cells, tissues, and their constituents.

Chemico-Biological Interactions; A Journal of Molecular and Biochemical Toxicology. v. 1– , 1969– . Ireland: Elsevier. Monthly. $800/yr. ISSN 0009-2797.

Devoted to the mechanisms by which exogenous chemicals produce changes in biological systems.

European Biophysics Journal. v. 11– , 1984– . New York: Springer-Verlag. Bimonthly. $522/yr. ISSN 0175-7571.

Continues *Biophysics of Structure and Function.* Publishes papers reporting research in the study of biological phenomena by using physical methods and concepts.

European Journal of Biochemistry. v. 1– , 1967– . New York: Springer-Verlag. Frequency varies. $2,196/yr. ISSN 0014-2956.

Published for the Federation of European Biochemical Societies. This journal covers research on molecular biology, nucleic acids, biochemistry, and biophysics.

FASEB Journal: Official Publication of the Federation of American Societies for Experimental Biology. v. 1– , 1987– . Bethesda, MD: Federation of American Societies for Experimental Biology. Monthly. $250/yr. ISSN 0892-6638.

Continues *Federation Proceedings*. Designed to report on rapidly changing developments in biological sciences, publishing original research, state-of-the-art reviews, a book list, news, calendar, and employment opportunities. The American Society for Biochemistry and Molecular Biology, and the Biophysical Society are both members of this federation.

FEBS Letters. v. 1– , 1968– . Amsterdam: Elsevier Science on the behalf of the Federation of European Biochemical Societies. Weekly. $3,192/yr. ISSN 0014-5793.

"An international journal for the rapid publication of short reports in biochemistry, biophysics and molecular cell biology."

Journal of Biochemistry (Tokyo). v. 1– , 1922– . Tokyo: Japanese Biochemical Society. Monthly. $204/yr. ISSN 0021-924X.

In English. Publishes original research and communications requiring prompt publication.

Journal of Bioenergetics and Biomembranes. v. 1– , 1970– . New York: Plenum Press. Bimonthly. $260/yr. ISSN 0145-479X.

Devoted to publication of original research that contributes to fundamental knowledge in the areas of bioenergetics, membranes, and transport.

Journal of Biological Chemistry. v. 1– , 1905– . Baltimore, MD: Williams & Wilkins for the American Society for Biochemistry and Molecular Biology. Weekly. $710/yr. ISSN 0021-9258.

A leading journal that publishes papers on a broad range of topics of interest to biochemists.

Journal of Cellular Biochemistry. v. 18- , 1982– . New York: Wiley-Liss. Monthly. $1,255/yr. ISSN 07730-2312.

Continues *Journal of Supramolecular Structure and Cellular Biochemistry*. Publishes original research on complex cellular, pathologic, clinical, or animal model systems studied by molecular biological, biochemical, quantitative ultrastructural or immunological approaches. There is an annual supplement containing abstracts of the *Keystone Symposia on Molecular & Cellular Biology*.

Journal of Lipid Research. v. 1– , 1959– . Bethesda, MD: Federation of American Societies for Experimental Biology. Monthly. $252/yr. ISSN 0022-2275.

Promotes basic research in the lipid field. Invited reviews.

Journal of Membrane Biology. v. 1– , 1969– . New York: Spring-
    er Verlag. Semimonthly. $1,053/yr. ISSN 0022-2631.

An international journal for studies on the structure, function and genesis of
biomembranes.

Journal of Photochemistry and Photobiology. A, Chemistry. v.
    40– , 1987– . Switzerland: Elsevier. Semimonthly. $1,986/yr.
    ISSN 1010-6030.

Continues, in part, *Journal of Photochemistry* and is concerned with quantitative
or qualitative aspects of photochemistry, including papers on applied photochemistry.

Journal of Photochemistry and Photobiology. B, Biology. v. 1– ,
    1987– . Switzerland: Elsevier. Quarterly. $1,241/yr. ISSN 1011-
    1344.

Official journal of the European Society for Photobiology. Accepts research
papers on light and its interactions with the processes of life; also includes news
on technological developments, forthcoming events, and conference reports.

Photochemistry and Photobiology. v. 1– , 1962– . Lawrence, KS:
    Allen Press for the American Society for Photobiology. Month-
    ly. $525/yr. ISSN 0031-8655.

Official organ of the American Society for Photobiology. Publishes papers on
all aspects of photochemistry and photobiology although it is concerned
primarily with articles having current or foreseeable biological relevance.

Physiological Chemistry and Physics and Medical NMR. v. 15– ,
    1983– . Wake Forest, NC: Pacific Press. Quarterly. $80/yr.
    ISSN 0748-6642.

Continues *Physiological Chemistry and Physics*. Original research on biochemistry, biophysics, magnetic resonance imaging, and nuclear magnetic resonance.

Protein Engineering. v. 1– , 1986– . New York: Oxford University
    Press. Monthly. $275/yr. ISSN 0269-2139.

Application of specific theoretical and experimental disciplines promoting understanding of the structural and biochemical basis of protein function. The 1993
supplement reports the proceedings of the Miami Bio/Technology Winter Symposium: *Advances in Gene Technology: Protein Engineering and Beyond*.

Proteins: Structure, Function, and Genetics. v. 1– , 1987– . New
    York: Wiley. Monthly. $576/yr. ISSN 0887-3585.

Concentrates on advances in all areas of biochemistry: structure, function,
genetics, computation, and design.

Radiation and Environmental Biophysics. v. 11– , 1974– . New
    York: Springer-Verlag. Quarterly. $462/yr. ISSN 0301-634X.

Continues *Biophysik.* Devoted to fundamentals and applications of biophysics of ionizing and non-ionizing radiation; biological effects of physical factors such as temperature, pressure, gravitational forces, electricity and magnetism; biophysical aspects of environmental and space influence.

Reviews of the Literature

Advances in Biophysics. v. 1– , 1970– . Limerick, Ireland: Elsevier
    Scientific and Japan Scientific Societies Press for the Biophysical
    Society of Japan. Annual. $117/yr. ISSN 0065-227X.

Annual survey of biophysical research in English.

Annual Review of Biochemistry. v. 1– , 1932– . Palo Alto, CA:
    Annual Reviews. Annual. Vol. 62, $46.00. ISSN 0066-4154.

The leading review serial in biochemistry. The table of contents includes citations to related articles in other *Annual Reviews*, e.g., *Biophysics and Biophysical Structure, Cell Biology, Genetics, Immunology, Medicine, Microbiology, Neuroscience, Nutrition, Physiology, Plant Physiology and Molecular Biology.*

Annual Review of Biophysics and Biomolecular Structure. v. 21– ,
    1992– . Palo Alto, CA: Annual Reviews. Annual. Vol. 22,
    $59.00. ISSN 0883-9182.

Continues *Annual Review of Biophysics and Bioengineering.* This series reviewing biophysics began in 1972 and has held an important place in the science literature ever since. Like the *Annual Review of Biochemistry, Biophysics* includes references to related articles in other *Annual Reviews*.

Comments on Molecular and Cellular Biophysics. v. 1– , 1980– .
    New York: Gordon and Breach. $484/yr. ISSN 0143-8123.
    Part A of *Comments on Modern Biology.*

An interdisciplinary journal devoted to critical analysis of contemporary research with focus on physical and quantitative investigations in molecular and cellular biosciences. The purpose is to facilitate the dissemination and analysis of emerging ideas.

Critical Reviews in Biochemistry and Molecular Biology. v. 1– ,
    1972– . Boca Raton, FL: CRC. Bimonthly. $320/yr. ISSN
    1040-9238. Formerly: *CRC Critical Reviews in Biochemistry
    and Molecular Biology.*

Includes critical surveys of specific topics of current interest selected on the advice of the Editorial Board.

Current Topics in Bioenergetics. v. 1– , 1966– . San Diego, CA: Academic Press. Irregular. Price varies. ISSN 0070-2129. Volume 16 was published in 1991, $115.00.

A refereed review series.

Essays in Biochemistry. v. 1– , 1965– . London: Portland Press. Annual. $25.00. ISSN 0071-1365.

Covers rapidly developing areas of biochemistry and molecular and cellular biology that are of particular interest to students and their teachers.

Life Chemistry Reports. v. 1– , 1982– . New York: Gordon and Breach. Irregular. $404/vol. ISSN 0278-6281.

Overview of advances in the fields of biochemistry molecular biology, pharmacology, and medicine. This review journal publishes the latest developments in chemistry as they relate to the life sciences.

Membrance Protein Structure: Experimental Approaches. Edited by Stephen H. White. New York: Oxford University Press, 1994. (American Physiological Society, Methods in Physiology Series). 395 p. $65.00. ISBN 0195071123.

A critical examiniation of membrane protein structure and new structural methods.

New Comprehensive Biochemistry. Edited by A. Neuberger and L. L. M. van Deenen. v. 1– , 1981– . Amsterdam: Elsevier. Price varies. ISSN 0167-7306.

Review series, each volume has a distinctive title. Vol. 21, *Molecular Aspects of Transport Proteins*, 1992.

Progress in Biophysics and Molecular Biology. v. 1– , 1950- . Tarrytown, NY: Pergamon. Bimonthly. $448/yr. ISSN 0079-6107.

International review journal covering the ground between the physical and biological sciences.

Quarterly Reviews of Biophysics. v. 1– , 1968– . New York: Cambridge University Press. Quarterly. $226/yr. ISSN 0033-5835.

Official journal of the International Union for Pure and Applied Biophysics. This publication provides a forum for general and specialized communication between biophysicists working in different areas.

Reviews of Physiology, Biochemistry and Pharmacology. v. 1– ,
    1902– . New York: Springer-Verlag. Irregular. $211/yr.
    ISSN 0303-4240.

Survey of the literature including special issues on selected topics: Vol. 121,
1992: *Signal Transduction*.

Structural Biology: The State of the Art. Proceedings of the Eighth
    Conversation in the Discipline Biomolecular Stereodynamics
    held at the State University of New York at Albany, June
    22–26, 1993. Edited by R. H. Sarma and M. H. Sarma. Sche-
    nectady, NY: Adenine, 1994. 2 v. $255.00 (set). ISBN
    094003042X (set).

A review of biological structure, dynamics, interactions and expression.

Theoretical Biochemistry & Molecular Biophysics. Edited by David L.
    Beveridge and Richard Lavery. Schenectady, NY: Adenine
    Press, 1991. 2 vol. $180.00 (set). ISBN 0940030284 (set).

Current state of research. Volume 1: *DNA*, volume 2: *Proteins*. The survey is
theoretical and computational, discussing mathematical models, numerical
calculations, and computer simulation.

Trends in the Biochemical Sciences. v. 1– , 1976– . Amsterdam:
    Elsevier. Monthly. $412/yr. ISSN 0376-5067.

Reviews of the literature plus research news, perspectives, techniques, letters to
the editor, book reviews, job trends, calendar. Other *Trends* of interest include
*Biotechnology, Endocrinology and Metabolism, Genetics, Neurosciences, Pharm-
acological Sciences*.

Societies

For complete details, consult the most recent *Encyclopedia of Associations*, in
print, CD-ROM, or online.

American Chemical Society (ACS)
        1155 16th St. NW, Washington, DC 20036.

Founded: 1876. 142,00 members. Scientific and educational society of chemists
and chemical engineers. Conducts studies, surveys, special programs; monitors
legislation; presents courses, career guidance, etc. Computerized services include
a data base containing the full text of 22 ACS journals, 10 Royal Society of
Chemistry journals, and 5 polymer journals; Chemical Abstracts Service Online;
and STN International, a scientific and technical information network.
Publications: 22 journals including *Analytical Chemistry, Biochemistry,*

Chemical Abstracts, Chemical and Engineering News, Journal of Agricultural and Food Chemistry, Journal of Chemical Education, Journal of the American Chemical Society, Journal of Physical and Chemical Reference Data, etc. Semiannual meetings.

American Institute of Chemists (AIC)
7315 Wisconsin Avenue, Bethesda, MD 20814.

Founded: 1923. 4,000 members. Chemists and chemical engineers. Promotes advancement of chemical professions in the United States; protects public welfare by establishing and enforcing high practice standards; represents professional interests of chemists and chemical engineers. Publications: annual *Professional Directory*; *The Chemist*. Annual meeting.

American Institute of Physics (AIP)
500 Sunnyside Boulevard, Woodbury, NY 11797-2999.

Founded: 1931. Eleven national member societies, including 75,000 members. Promotes the advancement and diffusion of the knowledge of physics and its application to human welfare. Presents science writing awards, prizes for research and service to physics; compiles statistics; provides placement service; maintains biographical archives; operates computerized information services such as **PINET**. Publications: AIP and its member societies publish more than one quarter of the world's research literature in physics and astronomy. Meetings and conferences are scheduled and conducted by individual societies.

American Society for Biochemistry and Molecular Biology (ASBMB).
9650 Rockville Pike, Bethesda, MD 20814. Charles C. Hancock, Executive Officer.

Founded: 1906, formerly American Society of Biological Chemists. 8,500 members. Biochemists and molecular biologists who have conducted and published original investigations in biological chemistry and/or molecular biology. Publications: *Journal of Biological Chemistry*. Annual meeting.

Biochemical Society (BS)
7 Warwick Court, High Holborn, London WC1R 5DP, England.

Founded: 1911. 5,500 members. Objectives are to promote the biochemistry field and to provide a forum for information exchange and discussion of various aspects of teaching and research in the field of biochemistry. Publications: *Biochemical Journal, Bioscience Reports, Biochemical Society Transactions, Essays in Biochemistry*, etc. Five scientific meetings per year in the United Kingdom.

Biophysical Society (BPS)
>   c/o Emily M. Gray, Biophysical Society Office, 9650 Rockville
>   Pike, Rm. 512, Bethesda, MD 20814.

Founded: 1957. 4,000 members. Biophysicists, physical biochemists, and physical and biological scientists interested in the application of physical laws and techniques to the analysis of biological or living phenomena. Publications: *Annual Meeting Abstracts, Annual Review of Biophysics and Biophysical Chemistry, Biophysical Journal, Biophysical Society Directory,* and *Biophysical Society Newsletter.* Annual meeting.

Federation of European Biochemical Societies (FEBS)
>   Dept. of Biochemistry, Jozef Stefan Institute, Jamova 39, SLO-
>   61000 Ljubljana, Slovenia. Prof. Vito Turk, General Secretary.

Founded: 1964. 39,000 members. Purpose is to further research and education in the field of biochemistry and to disseminate research findings. Publications: *European Journal of Biochemistry* and *FEBS Letters.* Annual Congress.

International Union of Biochemistry and Molecular Biology (IUBMB)
>   c/o Dr. Horst Kleinkauf, Technical University of Berlin, Frank-
>   linstrasse 29, W-1000 Berlin 10, Germany.

Founded: 1955. 51 members. National academies, research councils, or biochemical societies; associate members are regional bodies; special members are organizations representing individual biochemists. Promotes international cooperation in the research, discussion, and publication of matters relating to biochemistry. Seeks to: standardize methods, nomenclature, and symbols used in biochemistry; contribute to the advancement of biochemistry; promote high standards; aid biochemists in developing countries. Publications: *Biochemical Education, Biochemical Nomenclature and Related Documents, Biochemistry International, BioFactors, Journal of Biotechnology and Applied Biochemistry, Trends in Biochemical Sciences.*

International Union of Pure and Applied Biophysics (IUPAB)
>   Institute of Molecular Biology and Biophysics, ETH-Hongger-
>   berg, CH-8093 Zurich, Switzerland.

Founded: 1961. National committees appointed by academies and research councils. Purposes are to organize international cooperation in biophysics and to promote communication between the various branches of biophysics and allied subjects; to encourage within each country cooperation between the societies that represent the interests of biophysics; to contribute to the advancement of biophysics in all aspects. Publications: *Quarterly Reviews of Biophysics.* Triennial International Congress of Biophysics.

Pan American Association of Biochemical Societies (PAABS)
c/o Dr. Marino Martinez-Carrion, University of Missouri—
Kansas City, School of Basic Life Sciences, 109 Biological Sciences
Building, Kansas City, MO 64110.
Founded: 1969. 14 members. Societies of professional biochemists in the
Americas and culturally related European Countries. Promotes the sciences of
biochemistry by disseminating information and encouraging contacts among its
members. Cooperates with other organizations having similar objectives.
Publications: *Symposium Proceedings*. Triennial meeting (1995).

Textbooks and Treatises

Adams, Roger L. P. et al. *The Biochemistry of the Nucleic Acids*,
11th ed. New York: Chapman & Hall, 1992. 704 p. $95.00.
ISBN 0412460300.
Written for the researcher and the student who need information on DNA structure, replication and repair, gene expression and its control, and protein
synthesis.

Alpen, Edward L. *Radiation Biophysics*. Englewood Cliffs, NJ:
Prentice Hall, 1990. (Prentice Hall Advanced Reference Series.
Physical and Life Sciences Series; Prentice Hall Biophysics and
Bioengineering Series). 392 p. ISBN 0137504802.
Text useful for advanced undergraduates and graduate students, and as a desk-
top reference for working scientists.

Armstrong, Frank Bradley. *Biochemistry*, 3rd ed. New York: Oxford
University Press, 1989. 675 p. $49.95. ISBN 0195053567.

Bergethon, P. R. and E. R. Simons. *Biophysical Chemistry: Molecules to Membranes*. New York: Springer-Verlag, 1990. 340
p. $59.00. ISBN 0387970533.
Addresses the fundamental physical chemistry of solutions and membranes
appropriate for beginning graduate students in biochemistry and physiology.

Bloomfield, Molly M. *Chemistry and the Living Organism*, 5th ed.
New York: Wiley, 1992. 749 p. $40.00. ISBN 0471512923.
Introduction to the basic principles of general, organic, and biological
chemistry.

Branden, Carl and John Tooze. *Introduction to Protein Structure*.
    New York: Garland Pubs., 1992. 302 p. $49.95. ISBN
    0815303440.

Biological processes and functions are explained in terms of the structural analysis of proteins, and the chemistry and physics of the macromolecule.

Comprehensive Biochemistry. Edited by M. Florkin and E. H. Stotz.
    Vol. 1– , 1962– . Amsterdam: Elsevier. Price varies.

"An advanced treatise in biochemistry which assembles the principal areas of the subject in a single set of books." Vol. 37, *A History of Biochemistry*, 1990.

Graham, Richard C. *Data Analysis for the Chemical Sciences; A
    Guide to Statistical Techniques*. New York: VCH, 1993. 536
    p. ISBN 1560810483.

This text provides statistical and data analysis tools, introducing univariate, bivariate, and multivariate statistical techniques. The book includes problems and solutions, statistical tables, lists of software for statistical analysis, and exercises.

Holme, David J. and Hazel Peck. *Analytical Biochemistry*, 2nd ed.
    New York: Halstead Press, 1993. 520 p. $49.95 (paper).
    ISBN 0470220457.

The principles of analytical biochemistry are presented in three sections: 1) analytical techniques for spectroscopy, chromatography, etc.; 2) enzymes, antibodies, and radioisotopes; 3) methods of analysis of major groups of biologically important compounds, carbohydrates, amino acids, proteins, lipids, and nucleic acids.

Lipid Biochemistry: An Introduction, 4th ed. Edited by M. I. Guss
    and J. L. Harwood. New York: Chapman & Hall, 1991. 406
    p. $89.95. ISBN 0412266105.

Useful lipid biochemistry text.

New Era of Bioenergetics. Edited by Yasuo Mukohata. San Diego:
    Academic Press, 1991. 308 p. $64.95. ISBN 0125098545.

Overview of bioenergetics for students and those new to the field.

Physical Forces and the Mammalian Cell. Edited by John A. Frangos. San Diego: Academic Press, 1993. 400 p. $105.00.
    ISBN 0122653300.

Principles of Biochemistry, 2nd ed. By Albert L. Lehninger et al.
    New York: Worth Pubs., 1993. 1,013 p. $65.95. ISBN
    0879015004.

This textbook is an excellent resource. It is divided into sections describing foundations of biochemistry, each major class of biomolecules, bioenergetics and metabolism, and information pathways, the conversion of information contained in DNA to cellular macromolecules. The last chapter addresses experimental methods; all techniques are indexed.

Stryer, Lubert. *Biochemistry*, 3rd ed. New York: W. H. Freeman, 1988. 1089 p. $57.95. ISBN 071671843X.

An established, well-respected textbook.

Sybesma, Christian. *Biophysics, An Introduction*. Dordrecht; Boston: Kluwer Academic, 1989. 320 p. $131.00. ISBN 0792300297.

A revised edition of the 1977 publication *An Introduction to Biophysics*.

Voet, Donald and Judith G. Voet. *Biochemistry*. New York: Wiley, 1990. 1,223 p. $62.95. ISBN 0471617695. 1993 supplement: 102 p. ISBN 0471303585.

Kept up to date by supplements.

Vol'kenshtein, Mikhail Vladimirovich. *Biophysics*, revised and translated from the 1983 Russian edition. UK: Collets, 1983. 640 p. $85.00. ISBN 0317465813.

Yeargers, Edward K. *Basic Biophysics for Biology*. Boca Raton: CRC Press, 1992. 202 p. Index. $39.95. ISBN 0849344247.

Introductory text covering thermodynamics and molecular structure.

Bibliography

McGraw-Hill Dictionary of Scientific and Technical Terms, 4th ed. Sybil P. Parker, Editor-in-Chief. New York: McGraw-Hill, 1989. 2088 p. ISBN 0070452709.

Stankus, Tony. *Making Sense of Journals in the Life Sciences: From Specialty Origins to Contemporary Assortment*. New York: Haworth Press, 1992. 278 p. ISBN 1560241810. (Monographic supplement #8 to the *Serials Librarian*, ISSN 0897-8409).

6

Molecular and Cellular Biology

There is a natural affinity between this chapter and the previous one, which discusses reference materials for biochemistry and biophysics. In fact, the definition for molecular biology taken from the *McGraw-Hill Dictionary of Scientific and Technical Terms* (1989) corroborates this viewpoint by defining molecular biology as "that part of biology which attempts to interpret biological events in terms of the physico-chemical properties of molecules in a cell." Given this very substantial overlap, it is essential for the reader to review the sources annotated in Chapter 5, as well as Chapters 4, "Abstracts and Indexes," and 7, "Genetics," for a more complete picture of the literature of molecular biology.

Abstracts and Indexes

See also the specialized physics indexes annotated in Chapter 5, "Biochemistry and Biophysics."

Current Advances in Cell and Developmental Biology. v. 1– ,
      1984– . Oxford, England: Pergamon Press. Monthly.
      $950/yr. ISSN 0741-1626.
Specialized current awareness service for cell and developmental biology; one of a series of similar Pergamon titles.

Current Advances in Genetics and Molecular Biology. v. 1– ,
      1984– . Oxford, England: Pergamon Press. Monthly.
      $998/yr. ISSN 0741-1642.
Current awareness service for genetics and molecular biology.

Nucleic Acids Abstracts. v. 1– , 1970– . Bethesda, MD: Cambridge
      Scientific Abstracts. Monthly. $815/yr. ISSN 1070-2466.
Covers RNA, DNA, and enzyme research. Available on CD-ROM, online, and through the Internet as part of the **Life Sciences Collection** (see Chapter 4, p. 78).

See also *Biological Abstracts* (p. 74), *Biology Digest* (p. 75), *Chemical Abstracts* (p. 75), *Current Contents/Life Sciences* (p. 76), and *Index Medicus* (p. 78) in Chapter 4, "Abstracts and Indexes."

Databases (*Also, see* Handbooks, p. 120)

The proliferation of information concerning biochemistry, genetics, and molecular and cell biology is astounding. Numerous computer-based databases provide access to this material, and in many cases, may be the only entry to this vast mountain of facts in the fastest moving fields of biology. Major databanks are annotated in this section; one should also consult Chapters 5, "Biochemistry and Biophysics," and 7, "Genetics," for descriptions of other relevant databases. Guides to accessing electronic databanks are listed under Guides to Internet Resources, annotated on p. 119.

Entrez: Sequences and *Entrez: References*. Bethesda, MD: National Center for Biotechnology Information, National Library of Medicine, National Institutes of Health. Available on compact disc (GenInfo Compact Library Series) and the Internet.

This database, which is updated continuously, offers access to nucleotide and protein sequences from *GenBank, PIR*, and *SWISS-PROT* (see entries for these titles under *LiMB*) and to **MEDLINE** citations for the papers in which the sequences were published. In the compact disc version, each section is on a separate disc.

LiMB (Listing of Molecular Biology Databases). Los Alamos National Laboratory, Los Alamos, NM. Available on the Internet via Gopher.

The goal of LiMB is to provide the scientific community with a comprehensive overview of databases relevant to molecular biology and related data sets. LiMB contains a brief listing of all the databases, a data dictionary describing the meanings of the fields in a LiMB entry, the full listing of the databases, and a cross-index of the database and the data types they contain. Some of the major molecular biology databases/databanks are listed below; also consult the genetics and biochemistry chapters for relevant information, and browse LiMB for a complete, up-to-date listing.

DCT (Drosophila Codon Tables). Codon usage for Drosophila, nucleotide sequences, cross-references, base composition.

DDBJ (DNA Data Bank of Japan). Nucleotide sequences.

EMBL (EMBL Nucleotide Sequence Database). Collects, organizes, documents, and makes freely available the body of known nucleotide sequence data.

ENZYME (The ENZYME Data Bank). Contains EC number, recommended and alternative names, catalytic activity, cofactors, and cross-references to SWISS-PROT for each type of characterized enzyme for which an EC number has been designated.

GDB (Genome Data Base). Provides a genetic mapping and disease database to support the mapping and sequencing of the human genome.

GENBANK (GenBank Genetic Sequence Data Bank). U.S. government-sponsored, internationally available collection of all reported nucleotide sequences, catalogued and annotated with functional physical and administrative context.

HDB (Hybridoma Data Bank: A Data Bank on Immunoclones). Collects and maintains information on hybridomas, cloned cell lines, and monoclonal antibodies.

MOUSE (List of Mouse DNA Clones and Probes). Booklet containing list of mouse probes and clones and a map showing the location of the genes in the mouse for which probes and clones are listed.

PDB (Protein Data Bank). Comprehensive coverage of bibliographic, atomic coordinate and crystallographic structure factor data for biological macromolecules.

PIR (National Biomedical Research Foundation Protein Identification Resource and Protein Sequence Database). Data on completed sequence proteins, amino-terminal sequences, and bibliographic citations for amino acid sequences.

SIGSCAN (Signal Scan). Developed to aid the molecular biologist in determining what eukaryotic transcription factor elements, and other significant elements, may exist in a DNA sequence.

SWISSPROT (SWISS_PROT Protein Sequence Data Bank). Lists all protein sequences and related data.

Sillince, John A. A. and Maria Sillince. Databases for Structure–Function Studies in Molecular Biology. *Science & Technology Libraries* 14(1): 37–56, 1993.

An appendix lists some of the main search and analysis programs.

Sillince, John A. A. and Maria Sillince. *Molecular Databases for Protein Sequences and Structure Studies; An Introduction*. New York: Springer-Verlag, 1991. 236 p. $69.00. ISBN 3540543325.

This volume discusses the structure and function of proteins and nucleic acids, how molecular data is represented and registered in software, and how data is manipulated; presents a state-of-the-art review of existing databanks for biochemistry and molecular science.

Sillince, Maria and John A. A. Sillince. Sequence and Structure
    Databanks in Molecular Biology: The Reasons for Integration.
    Journal of Documentation 49(1): 1–28, March 1993.

Dictionaries and Encyclopedias

Dictionary of Cell Biology, 2nd ed. Edited by J. M. Lackie and J. A.
    T. Dowe. San Diego: Academic Press, 1995. 380 p. $45.00.
    ISBN 0124325624.

This new edition contains over 5,000 entries; over 20 tables on amino acids, antibiotic mode of action, types of light microscopy, sugars, vitamins, etc., have been included among the dictionary entries.

Encyclopedia of Molecular Biology. Sir John Kendrew, Editor-in-Chief.
    Cambridge, MA: Blackwell Scientific, 1994. 1,152 p. $149.95.
    ISBN 0632021829.

Over 6,000 definitions covering molecular biology, from the structure of genes to the intricacies of the cell, the development of organisms and the molecular analysis and therapeutics of human disease. Two hundred main articles, enhanced by 806 illustrations including an 8-page full-color section, are written by 300 of the leading authorities in molecular and cell biology.

Glick, David M. *Glossary of Biochemistry and Molecular Biology*.
    New York: Raven Press, 1990. 194 p. ISBN 0881675636.

See the annotation (p. 87) in the Chapter 5, "Biochemistry."

Maclean, Norman. *Dictionary of Genetics & Cell Biology*. New York: New
    York University Press, 1987. 422 p. ISBN 0814754384.

Contains short dictionary of terms encountered in cell biology and genetics. Appendices include common and Latin names of key organisms, chromosome numbers of select species, the DNA content of haploid genomes, and a brief guide to the classification of living organisms.

Schaeffer, Warren I. Usage of Vertebrate, Invertebrate and Plant
    Cell, Tissue and Organ Culture Terminology. *In Vitro* 20(1):
    19–24, January 1984.

Terms to describe phenomena of cell, tissue and organ culture used for plants and animals.

Singleton, P. and D. Sainsbury. *Dictionary of Microbiology and
    Molecular Biology*, 2nd ed. Chichester, England: Wiley, 1993.
    1,032 p. $35.00 (paper). ISBN 0471940526.

Up-to-date coverage of the terminology of molecular biology and microbiology. References are included to numerous papers, review articles, and monographs in microbiology and its allied subjects.

Stenesh, J. *Dictionary of Biochemistry and Molecular Biology,*
    2nd ed. New York: Wiley, 1989. 525 p. $74.95. ISBN
    0471840890.
See the annotation (p. 88) in Chapter 5, "Biochemistry."

Guides to Internet Resources

"Guide to Molecular Biology Databases," compiled by Damian Hayden, is available on the Internet via the University of Michigan Gopher for the Clearinghouse for Subject-Oriented Internet Resource Guides, Guides to the Sciences. In the compiler's words, "This guide provides specific information regarding matters of access to biological sequence information," including a comprehensive list of molecular biology resources arranged in two parts: 1) Guide to Using Molecular Biology Resources, explaining methods of sequence similarity searching, using Internet searching tools, and CD-ROM products; 2) Guide to Internet Molecular Biology Resources, listing Internet Molecular Biology Resources with Universal Resource Locators and short annotations.

There are various other subject guides available on World Wide Web (WWW) Resources for Biology and Medicine that are relevant; for example, under "Biosciences," see "Biochemistry and Molecular Biology." Also included are "AI," a database of molecular biologists working in the field of AI: literature reference to algorithms related to molecular biology; and periodical references to journals in the area of molecular biology, just to name a few promising categories of information.

Also, see the review article by Robert Harper, "Access to DNA and Protein Databases on the Internet," in *Current Opinion in Biotechnology* 5(1): 4–18, 1994.

There are also a large number of molecular and cellular biology discussion groups on the Internet. For instance, in the bionet hierarchy of USENET, there are several pertinent groups such as bionet.cellbiol, bionet.molbio.ageing, bionet.molbio.embldatabank, bionet.molbio.methds-reagnts, and bionet.molbio. yeasts.

Guides to the Literature

Refer to Chapter 3, "General Sources," for a listing of guides that include molecular biology as part of a broader guide to the biological sciences.

Handbooks (*Also, see* Databases, p. 115)

ATCC Catalogue of Recombinant DNA Materials, 3rd ed. Rockville, MD: American Type Culture Collection, 1994. 150 p. Free of charge in the United States. Electronic versions available online through Internet and ATCC Online.

The catalog/directory lists materials needed by molecular biologists, including thoroughly cross-referenced indexes or hosts, libraries, vectors, and clones. ATCC catalogs and databases are also available on the Internet, on diskette, and on CD-ROM.

Cell Biology: A Laboratory Handbook. Edited by Julio E. Celis. San Diego: Academic Press, 1994. 3 v. $120.00 (comb bound). ISBN 0121647145.

Contains over 200 articles dealing with a wide variety of protocols in cell biology.

The Cytokine Handbook, 2nd ed. Edited by Angus Thomson. San Diego, CA: Academic Press, 1994. 615 p. $95.00. ISBN 0126896615.

Comprehensive source of information of the hemopoietic growth factors and the interleukin and interferon families of molecules, cytokine by cytokine.

Cross, Patricia C. and K. Lynne Mercer. *Cell and Tissue Ultrastructure: A Functional Perspective*. New York: W. H. Freeman, 1993. 420 p. $39.95. ISBN 0716770334.

This collection of electron micrographs provides a complete up-to-date introduction to cell ultrastructure, with emphasis on cell anatomy and physiology. The volume is arranged with the right-hand pages presenting micrographs, or portraits of cells from major organs in every organ system. The left-hand pages describe the basic structure and function of the cell in the micrograph on the opposite page. Accompanying line drawings and diagrams place the micrographs in context with surrounding tissue. Also, see *The Cell* by Don W. Fawcett (below), and *Cell and Tissue Ultrastructure* (under Cross and Mercer, p. 140).

Enzymes of Molecular Biology. Edited by Michael M. Burrell. Totowa, NJ: Humana Press, 1993. 384 p. $59.50. ISBN 0896032345.

Background information, size, and structure for each of 24 enzymes used as tools in molecular biology, along with parameters for use, source, and application. Practical procedures and protocols are included.

Fawcett, Don W. *The Cell*, 2nd ed. Philadelphia: Saunders, 1981. 928 p. $55.85. ISBN 0721635849.

An extremely useful atlas of cytology, with micrographs of cell ultrastructure and accompanying text.

Flow Cytometry: New Developments. Edited by A. Jacquemin-
Sablon. New York: Springer-Verlag, 1993. (NATO ASI series:
subseries H: cell biology. v. 67). $200.00. ISBN 3540546065.

Practical applications of flow cytometry in specific biological systems, ranging
from cell biology to chromosome analysis. Major areas addressed are cell
activation, membrane-ligand interactions and the form and function of nuclear
components.

GUIDEBOOK SERIES: New York: Oxford University Press, published
in association with Sambrook and Tooze Scientific Publishers.

These handbooks draw together information from widely scattered literature and
are designed to provide a convenient reference source.

Guidebook to the Cytoskeletal and Motor Proteins. Edited by
Thomas Kreis and Ronald Vale. 276 p. $70.00. ISBN
0198599323.

Summarizes the essential features of these proteins.

Guidebook to the Extracellular Matrix and Adhesion Proteins. Edited
by Thomas Kreis and Ronald Vale. 176 p. $65. ISBN
019859934X.

Up-to-date information on proteins involved in cell adhesion, the
cytoskeleton, and the extracellular matrix.

Forthcoming books in the series are *Guidebook to the Homeobox Genes* by
Denis Duboule; *Guidebook to the Oncogenes,* by Brad Ozanne et al.; *Guidebook
to the Secretory Pathway,* by Tom Stevens et al; and *Guidebook to the Small
GTPases,* by M. Zerial and L. Huber.

LABFAX Series

The following three volumes are part of the LABFAX series published by the
Oxford, England–based Bios Scientific, distributed in the United States by
Academic Press. The aim of the series is to provide ready access to critical data
on materials and methods that are routinely used by scientists. Each volume is
limited to a narrow topic so that the compendia of information is focused for
convenient and rapid access to data and protocols. For example, the *Cell Biology LABFAX* covers formulation of salt solutions, subcellular fractionation
techniques and characteristics of electron microscopy, DNA content of cells,
membrane composition, properties and receptors, metabolic inhibitors, cyclic
nucleotides, steroid and peptide hormones, growth regulators, and cell cycle
oncogenes. Information may be presented in tables or in text and/or diagrams.

Cell Biology LABFAX. Edited by G. B. Dealtry and D. Rickwood.
San Diego: Academic Press, 1992. 254 p. $49.95. ISBN
012207890X.

Cell Culture LABFAX. Edited by Michael Butler and Maureen
     Dawson. San Diego: Academic Press, 1992. 247 p. $49.95.
     ISBN 0121480607.

Molecular Biology LABFAX. Edited Terence Austen Brown. San
     Diego: Academic Press, 1991. 322 p. $49.95. ISBN
     0121375102.

Macromolecular Structures. Philadelphia, PA: Current Biology,
     1991– . Annual. 1991–1993: $810.00/set; single volume:
     $385.00.

Atomic structures of biological macromolecules reported since 1990. Includes
X-ray, NMR, fiber and electron crystallography structures, and comprehensive
bibliography.

Practical Handbook of Biochemistry and Molecular Biology. Edited
     by Gerald D. Fasman. Boca Raton, FL: CRC Press, 1989. 601
     p. ISBN 0849337054.

Material is derived and updated from the multi-volume *CRC Handbook of Biochemistry and Molecular Biology.* For more information see the annotation (p.
94) in the Chapter 5, "Biochemistry and Biophysics."

Shapiro, Howard M. *Practical Flow Cytometry*, 3rd ed. New York: Wiley-
     Liss, 1994. 588 p. $79.95. ISNB 0471303763.

Histories

Edsall, J. T. and D. Bearman. "Survey of sources for the history of
     biochemistry and molecular biology." *Federation Proceedings*
     36(8): 2069–2073, July, 1977.

This paper describes the effort to collect and preserve sources for the study of
the history of biochemistry and molecular biology. Thirteen histories are cited
the references. The "Survey of sources . . . [is] the first major undertaking in
this field, for the biological sciences, in the 20th century."

Judson, Horace. *Eighth Day of Creation*. New York: Simon &
     Schuster Trade, 1980. $15.95 (paper). ISBN 0671254103.

Historical account of molecular biology.

Olby, R. C. *The Path to the Double Helix*. Seattle: University of
     Washington Press, 1974.

Although this book is out of print and may be hard to find, it retains its value
by presenting a comprehensive account of the origins of molecular biology.

Phage and the Origins of Molecular Biology. Edited by John Cairns,
Gunther S. Stent, and James D. Watson. Cold Springs Harbor,
NY: Cold Spring Harbor Laboratory of Quantitative Biology,
1992. 366 p. $35.00. ISBN 0879694076. Bibliographies.

Published on the occasion of the sixtieth birthday of Max Delbruck. Expanded
edition of the 1966 collection of 35 essays by pioneers of molecular biology.

Schrodinger, Erwin. *What is Life? with Mind & Matter & Autobiographical Sketches*. New York: Cambridge University Press,
1992. 200 p. $9.95 (paper). ISBN 0521427088.

The reviewer Adam S. Wilkins, writing in *Bioessays* (vol. 15(11): 767–769,
Nov. 1993) commented that Schrodinger's book was a "milestone in conceptual
development of modern biology, and specifically of molecular biology."

Watson, James D. *Double Helix: Being a Personal Account of the
Discovery of the Structure of DNA*. New York: Macmillan reprint of
the 1969 ed. 144 p. $11.95 (paper). ISBN 0689706022.

This account is also available edited by Gunther S. Stent as *Double Helix: A
Norton Critical Edition*. New York: Norton, 1980, $8.95 (paper), ISBN
0393950751.

Methods and Techniques

Also, refer to Chapters 5, "Biochemistry and Biophysics," and 7, "Genetics."

Basic Methods in Molecular Biology, 2nd ed. Edited by Leonard G. Davis,
W. Michael Juehl, James F. Battey. Norwalk, CT: Appleton &
Lange, 1994. 776 p. $49.95 (spiral). ISBN 0838506429.

A step-by-step guide for molecular biology lab procedures: a cookbook format.
There is a great deal of useful information including preparation of stock
solutions, time required, special equipment, reagents, methods, and commentary.

The Basics. Oxford University Press, 1994– .

This major new series provides a range of new laboratory guides for a wide
range of techniques. The first three volumes:

DNA Sequencing: The Basics. 1994. 128 p. $15.00. ISBN
0199634211.

Nucleic Acid Blotting: The Basics. 1994. 100 p. $15.00. ISBN
0199634467.

Somatic Cell Hybrids: The Basics. 1994. 100 p. $15.00. ISBN
0199634432.

Cell and Tissue Culture: Laboratory Procedures. Edited by A. Doyle,
     J. B. Griffiths, and D. G. Newell. New York: Wiley, 1994. 2
     v. $575.00 (looseleaf). ISBN 0471928526. Includes 1994
     core volumes and 4 updates of 100 pages each.

Comprehensive collection of cell and tissue culture *in vitro* techniques.

Cell Biology: A Laboratory Handbook. Edited by Julio Celis. San
     Diego, CA: Academic Press, 1994. 3 v. $120.00 (comb-
     bound set). ISBN 0121647145.

The aim is to assemble in one place the most important methods in cell biology
which have withstood the test of time—in other words, classic methods for cell
biology, including over 200 articles by hundreds of world-renowned scientists.
Vol. 1: *Tissue Culture and Associated Techniques*; Vol. 2: *Microscopy Techniques*; Vol. 3: *Transfer of Macromolecules and Small Molecules.*

Copeland, Robert A. *Methods for Protein Analysis; A Practical Guide
     for Laboratory Protocols.* New York: Chapman & Hall, 1993.
     224 p. $49.95. ISBN 0412037416.

A concise summary of the methods most relevant to the generalist bench
scientist who works with proteins and is without specialized equipment or
expertise.

Current Protocols in Molecular Biology. Edited by Frederick M.
     Ausubel, et al. New York: Wiley, 1993. 2 v. $415.00.
     Update service $170/yr. ISBN 047150338X. Also available
     on CD-ROM, $450.00.

Designed for maximum ease of use in a laboratory setting, covering a wide
range of techniques in molecular biology. Protocols, background information,
and a guide to the choice of methods are provided, often with extensive troubleshooting guides. This review of molecular biology techniques may be updated,
in the looseleaf format, by supplements, corrections, and improved protocols.

DNA Probes, 2nd ed. Edited by George H. Keller and Mark M.
     Manak. New York: Stockton Press, 1993. 659 p. $90.00.
     ISBN 1561591025.

Useful reference on the development and use of nucleic acid hybridization.
Synthesis, labelling, and detection of nucleic probes, sample preparation, and
specific applications.

DNA Sequencing Protocols. Edited by Hugh G. Griffin and Annette
     M. Griffin. Totowa, NJ: Humana Press, 1993. 405 p.
     $59.50. ISBN 0896032485.

Comprehensive guide to over 38 DNA-sequencing methods and techniques. Includes cycle sequencing, sequencing PCR products, sequencing lambola and cosmids, multiplex sequencing, direct blotting electrophoresis, sequencing by chemiluminescence, and automated sequencing.

Farrell, Robert E., Jr. *RNA Methodologies; A Laboratory Guide for Isolation and Characterization*. San Diego: Academic Press, 1993. 317 p. $49.95. ISBN 0122497007.

The isolation of chemically stable RNA is central to molecular biology. This text presents mammalian RNA isolation strategies and a collection of protocols for basic research, as well as for more sophisticated procedures. Includes flowcharts, tables, and graphs to aid explanation and learning.

Flow Cytometry, Part A, Part B, 2nd ed. San Diego: Academic Press, 1994. (*Methods in Cell Biology*. v. 41, 42). 2 v. $59.95 (comb-bound), each volume. ISBN 0122030516 (v. 1).

This authoritative manual presents a comprehensive array of methods applicable to chromosome analysis, plant biology, marine biology, fluorescence, *in situ* hybridization, etc.

Flow Cytometry: A Practical Approach, 2nd ed. Edited by M. G. Ormerod. New York: IRL Press, 1994. (Practical Approach series, 142). 282 p. $50.00. ISBN 01996342629.

Introduction and guide for operation of a flow cytometer or a fluorescence-activated cell sorter. Topics covered: fluorescence technology, DNA analysis and the measurement of cell proliferation, chromosome analysis and sorting, membrane potential and pH, and other applications to cell biology.

Freshney, R. Ian. *Culture of Animal Cells*, 3rd ed. New York: Wiley, 1994. 500 p. $69.95. ISBN 0471589667.

Updated methods manual suitable for class work, with detailed instruction for proven methods. Includes resources for specialized equipment and supplies.

Genetic Manipulation: Techniques and Applications. Edited by J. M. Grange, et al. Oxford, England and Boston: Blackwell Scientific, 1991.

Fully annotated (p. 157) in Chapter 7, "Genetics."

Introduction to Molecular Cloning Techniques. Edited by G. Lucotte and F. Baneyx. New York: VCH, 1993. 320 p. $50.00. ISBN 1560816139.

Concise summary of the techniques commonly used in research and development of molecular biology and genetic engineering. Focusing entirely on the most widely used host, *E. coli*, the book provides descriptions of cloning vectors and essential recombinant DNA methodologies. The book can also function as a text, although the reader is expected to have a firm background in biochemistry, including laboratory experience.

Journal of Tissue Culture Methods. v. 1– 1978– . Columbia, MD: Tissue Culture Association. Quarterly. $80/yr. ISSN 0271-8057.

Formerly *TCA Manual*. Lists specific directions for cell, tissue, and organ culture procedures.

Likhtenshtein, Gertz Ilich. *Biophysical Labeling Methods in Molecular Biology*. New York: Cambridge University Press, 1993. 305 p. $54.95. ISBN 0521431328.

This volume covers all aspects of biophysical labeling methods including the theoretical basis, the experimental techniques, and how to interpret the resulting data. Results are critically discussed along with projections for the future.

Martin, Bernice M. *Tissue Culture Techniques; An Introduction*. Secaucus, NJ: Birkhauser, 1994. 247 p. $95.00. ISBN 0817637184.

Designed to teach the fundamentals of tissue culture techniques to a wide audience, emphasizing safety and sterility. There are troubleshooting tips and problem sets for the undergraduate student.

Methods in Cell Biology. v. 1– , 1964– . San Diego: Academic Press. Irregular. Price varies. ISSN 0091-679X.

Vol. 44: *Drosophila melanogaster: Practical Uses in Cell and Molecular Biology*. Edited by Lawrence S. B. Goldstein and Eric A. Fyrberg. 732 p. $65. A compendium of short technical chapters designed to provide state-of-the-art methods used by cell biologists.

Methods in Enzymology, vol. 216: *Recombinant DNA*. San Diego: Academic Press, 1993. 631 p. $80.00. ISBN 012182117X.

Included isolation, synthesis and detection of DNA and RNA; enzymes and methods for cleaving and manipulating DNA; reporter genes and vectors for cloning genes. This is a valuable series that should not be overlooked.

Methods in Molecular and Cellular Biology. v. 1– , 1989– . New York: Wiley/Liss. Bimonthly. $140/yr. ISSN 0898-7750.

An international journal devoted to the rapid publication of original manuscripts describing new methods and methodology, improvements of common protocols, troubleshooting guides, and all-new simplified protocols related to molecular biology. The aim is to publish methods of general interest performed at the molecular level for characterization of macromolecules, molecular cloning, expression of cloned sequences, interactions between nucleic acids and proteins, mutagenesis procedures, formation of antigen–antibody complexes, computerized programs, cell and organ culture.

Methods in Molecular Biology. Edited by John M. Walker. v. 1– , 1984– . Clifton, NJ: Humana Press. Price and frequency vary.

Practical, hands-on laboratory protocols for a wide range of basic and advanced techniques in all areas of experimental biology and medicine. Vol. 28: *Protocols for Nucleic Acid Analysis by Nonradioactive Probes*, 1994, 280 p., $49.50, ISBN 089603254X; Vol. 29: *Chromosome Analysis Protocols*, 1994, 528 p., $69.50, ISBN 0896032434; Vol. 30: *DNA–Protein Interactions; Principles and Protocols*, 1994, 444 p., $64.50, ISBN 0896032566.

Methods in Nucleic Acids Research. Edited by Jim D. Karam et al. Boca Raton: CRC Press, 1991. 403 p. $228.00. ISBN 0849353114.

This manual on basic techniques also includes principles and background information on the evolution of new technology in the field. The volume incorporates a survey of approaches, and each chapter provides a brief literature review for principles and experimental details with the hope that the volume will be useful, practical, and educational.

Molecular Structures in Biology. Edited by R. Diamond et al. New York: Oxford University Press, 1993. 326 p. $60.00. ISBN 0198547714.

This volume includes 10 articles that survey techniques used in structure determination, on modelling studies, and on specific groups of biomacromolecules. There is a CD-ROM version of this book, with the same title, which integrates text, data, and computer graphics, offering a powerful electronic reference research, and teaching tool.

Non-Invasive Techniques in Cell Biology. Edited by J. Kevin Foskett and Sergio Grinstein. New York: Wiley-Liss, 1990. (Modern cell biology series, v. 9). 423 p. ISBN 0471568090.

Review of the approaches and techniques for probing a variety of dynamic cellular processes with a minimum of structural or chemical invasion.

PCR Methods and Applications. v. 1– , 1990– . Cold Spring
    Harbor, NY: Cold Spring Harbor Laboratory Press. Bimonthly.
    $276/yr. ISSN 1054-9803.

A top-quality, timely source of research information on PCR and other
amplification techniques.

PCR Protocols: A Guide to Methods and Applications. Edited by
    Michael A. Innis et al. San Diego: Academic Press, 1990. 482
    p. $49.95 (paper). ISBN 0123721814.

Basic instruction in PCR methods which have been tested repeatedly in the
various authors' laboratories. This book can also serve as a resource on novel
variations and applications, and is divided into five sections: Basic methodology,
Research applications, Genetics and evolution, Diagnostics and forensics, and
Instrumentation and supplies. Refer, also, to *PCR* in the Practical Approach
series (p. 129).

PCR Protocols; Current Methods and Applications. Edited by Bruce
    A. White. Totowa, NJ: Humana Press, 1993. 408 p. $49.50.
    ISBN 0896032442.

Up-to-date, comprehensive collection of over 35 detailed laboratory procedures
for the use of polymerase chain reactions in a wide range of applications.

PCR Technology: Current Innovations. Edited by Hugh G. Griffin and
    Annette M. Griffin. Boca Raton, FL: CRC, 1994. 370 p.
    $49.95 (spiral bound). ISBN 0849386748.

A selection of the most widely used applications including the generation and
detection of genetic mutations, diagnosis of clinical disease, detection of food-
borne pathogens, and determination of genetic relatedness of plant and animal
species.

Physical Forces and the Mammalian Cell. Edited by John A. Fran-
    gos. San Diego: Academic Press, 1993. 400 p. $105.00.
    ISBN 0122653300. Index.

Review of current status, guidelines, and techniques for examining the
intersection between biophysics and cell biology. Chapters survey effects of
physical forces on mammalian cells, and effects of various physical forces such
as mechanical strain, blood flow, gravity, and the like.

The Polymerase Chain Reaction. Edited by K. Mullis, R. Gibbs, and
    F. Ferre. Basel: Birkhaeusr, 1994. 600 p. $79.00. ISBN
    3764336072.

Co-edited by the inventor (Mullis) of PCR, this texts examines the most up-to-date methodological protocols, including new techniques and enhanced methods for the novice and the experienced PCR user.

Practical Approach Series. New York: Oxford University Press.

This laboratory series aims to provide tried and tested protocols for all areas, along with expert guidance and reference materials. Some recent volumes:

> No. 72 and 81: *Essential Molecular Biology*. v. 1–2. 1991. $75.00 (spiral-bound set of 2). ISBN 019963114X.

The essential techniques in molecular biology.

> No. 77: *PCR*. 274 p. $45.00 (spiral-bound). ISBN 019963226X.

An acclaimed best seller!

> No. 104: *Animal Cell Culture; A Practical Approach*, 2nd ed. Oxford, England and New York: Oxford university Press, 1992. 348 p. $45.00. ISBN 019963212X.

An update of the first edition, the emphasis remains on presenting techniques in a readily accessible form. Detailed protocols are provided for research and professional readers.

> No. 137: *The Cell Cycle*, 1994. $45.00. ISBN 0199633959.

All the main biochemical, genetic, and cellular manipulations required for cell cycle investigations in popular experimental systems: yeasts and fungi, amphibian and invertebrate eggs, insects, and vertebrate cells.

> No. 140: *Protein Blotting: A Practical Approach*. Edited by B. S. Dunbar, 1994. $45.00. ISBN 0199634386.

Written by experts, this book details all the key techniques used for protein blotting, including basic equipment, principles, methods for sample preparation, detection methods, and important applications.

> Unnumbered: *Basic Cell Culture: A Practical Approach*, edited by J. M. Davis, 1994. $50.00. ISBN 0199634343.

A guide for the newcomer covering all areas of cell culture, including animal and human.

Sambrook, Joseph, E. F. Fritsch, and T. Maniatis. *Molecular Cloning: A Laboratory Manual*, 2nd ed. Cold Spring Harbor, NY: Cold Spring Harbor Laboratory, 1989. 3 v.

Originally written as a collection of laboratory protocols used at Cold Spring Harbor for a course on molecular cloning, and used throughout molecular biology by researchers, this set contains much basic material. The second edition has more advanced material, making it a resource for the experienced cloner as well as a guidebook. Updated by *Current Protocols in Molecular Biology.*

YAC Libraries: A User's Guide. Edited by David L. Nelson and Bernard Brownstein. New York: Oxford University Press, 1993. 240 p. $39.95.

Techniques for the development of yeast artificial chromosome (YAC) vectors used for cloning long stretches of DNA. Useful for all researchers interested in cloning and cloning techniques.

Zyskind, Judith W. and Sanford I. Bernstein. *Recombinant DNA Laboratory Manual,* rev. ed. San Diego: Academic Press, 1992. 224 p. $29.95. ISBN 0127844015.

Key features of this edition include a new chapter on PCR and new protocols, random primed labeling of DNA, rapid plasmid miniprep, and double-stranded DNA sequencing.

Nomenclature

Rather than duplicate entries, we refer the reader to the Dictionaries, Encyclopedias, and Nomenclature section (p. 85) in Chapter 5, "Biochemistry and Biophysics."

Periodicals

In addition to the leading biological journals (*Journal of Biological Chemistry, Nature, Proceedings of the National Academy of Sciences,* and *Science* annotated elsewhere in the book) refer to the periodicals listed below and in Chapters 5, "Biochemistry and Biophysics," and 7, "Genetics." All of these journals have been chosen because of their recognized importance and impact.

Biology of the Cell. v. 40– , 1981– . Paris: Elsevier. Under the auspices of the European Cell Biology Organization. Continues *Biologie Cellulaire.* Monthly. $475/yr. ISSN 0248-4900.

Original work concerning the structure and function of cells, organelles and macromolecules. Areas of interest include all aspects of cellular and molecular biology in the context of developmental biology, genetics, immunology, physiology and virology.

Cell. v. 1– , 1974– . Cambridge, MA: Cell Press. Semi-monthly. $325/yr. ISSN 0092-8674.

A leading journal in the field.

Cell and Tissue Research. v. 1– , 1924– . New York: Springer-
Verlag. Monthly. $2,638/yr. ISSN 0302-766X.

Vertebrate and invertebrate structural biology and functional microanatomy, emphasizing neurocytology, neuroendocrinology, endocrinology, reproductive biology, morphogenesis, immune cells and systems, immunocytology, and molecular cell structure.

Cellular and Molecular Biology. v. 22– , 1977– . Paris: C. M. B.
Association. 8 issues/yr. $300/yr. ISSN 0145-5680.

Original articles in cellular and molecular biology and enzymology, applied to human tissues as well as to animals and plants. All papers demonstrating the link between structure and biochemical, biophysical, or physiological function are considered.

Cytometry. v. 1– , 1980– . New York: Wiley-Liss. The Journal of
the International Society for Analytical Cytology. 9 times/yr.
$396/yr. ISSN 0196-4763.

All aspects of analytical cytology, which is defined broadly as characterization and measurement of cells and cellular constituents for biological, diagnostic, and therapeutic purposes. Covers cytochemistry, cytophysics, cell biology, molecular biology, statistics, instrumentation, clinical laboratory practice, and other relevant subjects.

DNA and Cell Biology. v. 9- , 1990– . New York: Mary Ann Liebert. Continues *DNA*. Monthly. $390/yr. ISSN 1044-5498.

Publishes papers, short communications, reviews, laboratory methods, and editorials on any subject dealing with eukaryotic or prokaryotic gene structure, organization, expression, or evolution.

Differentiation: Ontogeny, Neoplasia and Differentiation Therapy. v. 1– ,
1973– . Berlin: Germany. Monthly. $1,031/yr. ISSN 0301-4681.

The Journal of the International Society of Differentiation. Multidisciplinary journal devoted to the problems of biological diversification. Its ambition is to cover biological differentiation and evolution from the sub-cellular level to species differentiation.

EMBO Journal. v. 1– , 1982– . New York: Oxford University
Press. Published for the European Molecular Biology Organization. Biweekly. $695/yr. ISSN 0261-4189.

Rapid publication of full-length papers describing original research of general rather than specialist interest in molecular biology and related areas.

European Journal of Cell Biology. v. 19- , 1979- . Stuttgart, Germany: Wissenschaftliche Verlagsgesellschaft mbh. Continues *Cytobiologie.* Bimonthly. $840/yr. ISSN 0171-9335.

Under the auspices of the European Cell Biology Organization, this journal publishes papers on the structure, function, and macromolecular organization of cells and cell components. Aspects of cellular dynamics, differentiation, biochemistry, immunology, and molecular biology in relation to structural data are preferred fields.

Experimental Cell Research. v. 1- , 1950- . San Diego: Academic Press. Monthly. $1380/yr. ISSN 0014-4827.

The chief purpose of this journal is to promote the understanding of cell biology by publishing experimental studies on the general organization and activity of cells. The scope includes all aspects of cell biology.

Gene. v. 1- , 1976- . New York: Elsevier. Semi-monthly. $3,235.50/yr. ISSN 0378-1119.

"An international journal focusing on gene cloning and gene structure and function."

In Vitro Cellular & Developmental Biology—Animal. v. 27A- , 1991- . Columbia, MD: Tissue Culture Association, Inc. Continues *In Vitro.* 13 issues/yr. $180/yr. ISSN 0883-8364.

Devoted to the advancement and dissemination of basic and applied knowledge concerning the in vitro cultivation of cells, tissues, organs, or tumors from multicellular animals.

In Vitro Cellular & Developmental Biology—Plant. v. 27P- , 1991- . Columbia, MD: Tissue Culture Association, Inc. Split from the *Animal* section. Quarterly. $80/yr. ISSN 1054-5476.

Devoted to original papers in plant cellular and developmental biology, with emphasis on the developmental, molecular and cellular biology of cells, tissues, and organs.

Journal of Cell Biology. v. 1- , 1955- . New York: Rockefeller University Press. Semimonthly. $420/yr. ISSN 0021-9525. Edited in cooperation with the American Society for Cell Biology.

Reports substantial and original findings on the structure and function of cells, organelles, and macromolecules. Includes mini-reviews and commentaries offering a personalized perspective or synthesis of information on a topic of interest to the general readership.

Journal of Cell Science. v. 1– , 1966– . Cambridge, England: Company of Biologists Ltd. Monthly. $1,050/yr. ISSN 0021-9533.

Critical work over the full range of cell biology. Scientific excellence is the single most important criterion for acceptance.

Journal of Cellular Biochemistry. v. 18– , 1982– . New York: Wiley-Liss. Continues *Journal of Supramolecular Structure and Cellular Biochemistry*. Monthly. $1,255, including supplements. ISSN 0730-2312.

Description of original research in which complex cellular, pathologic, clinical, or animal model systems are studied by molecular biological, biochemical, quantitative ultrastructural or immunological approaches.

Journal of Molecular Biology. v. 1– , 1959– . San Diego: Academic Press. Semimonthly. $1,860/yr. ISSN 0022-2836.

Studies of living organisms or their components at the molecular level. Suitable subject areas: proteins, nucleic acids, genes, viruses and bacteriophages, cells.

Molecular and Cellular Biology. v. 1– , 1984– . Washington, DC: American Society for Microbiology. Monthly. $361/yr. ISSN 0270-7306.

Devoted to the advancement and dissemination of fundamental knowledge concerning the molecular biology of eukaryotic cells of both microbial and higher organisms.

Molecular and General Genetics. v. 1– , 1908– . New York: Springer-Verlag. 13 issues/yr. $2,194/yr. ISSN 0026-8925.

Publishes in all areas of general and molecular genetics (developmental genetics, somatic cell genetics, and genetic engineering) irrespective of the organism.

Natural Structural Biology. v. 1– , 1994– . New York: Nature Publishing. Monthly. $495/yr. ISSN 1072-8368.

Published by the same company as *Nature*, this journal contains many of the same features, including editorials, news, book reviews, and full-length articles. The coverage is limited to molecular and structural biology, however.

Nucleic Acids Research. v. 1– , 1974– . New York: IRL Press at Oxford University Press. Semimonthly. $1,100/yr. ISSN 0305-1048.

Rapid publication for papers on physical, chemical, biochemical and biological aspects of nucleic acids, and proteins involved in nucleic acid metabolism and/or interactions.

Reviews of the Literature

Advances in Enzyme Regulation, v. 1– , 1963– . Oxford, England:
Pergamon Press. Annual. $350/yr. ISSN 0065-2571.

Original research papers and critical overviews of metabolic regulation in normal and cancer cells at the level of molecular biology, enzyme and metabolic regulation, and in clinical investigations.

Advances in Enzymology and Related Areas of Molecular Biology. v.
1– , 1941– . New York: Wiley. Irregular. Price varies. ISSN
0065-258X.

Formerly *Advances in Enzymology and Related Subjects of Biochemistry.* Review of literature in designated areas. Example of an article in v. 67 (1993): "Development of enzyme-based methods for DNA sequence analysis and their applications in the genome projects," by Ray Wu, p. 431–468.

Advances in Molecular and Cell Biology. v. 4– , 1992– . Greenwich, CT: JAI Press. Irregular. Price varies. ISSN 0898-8455.
Continues *Advances in Cell Biology*, 3 v. 1987–1990.

Publication reviewing research in molecular and cell biology, each volume has a distinctive title. Vol. 8: *Organelles in Vivo*, 1994. 191 p. $90.25.

Advances in Structural Biology. v. 1– , 1992– . Greenwich, CT:
JAI Press. Annual. $90/yr. ISSN 1064-6000.

"A research annual" for cytology and molecular structure.

Animal Cell Technology: Basic and Applied Aspects; Proceeding of
the Fourth Annual Meeting of the Japanese Association for
Animal Cell Technology, November, 1991. Dordrecht: Kluwer
Academic, 1992. 600 p. $150.00. ISBN 079231882X.

New data on animal cell technology, with emphasis given to basic characterization of cell lines. Merits of different cell culture systems are examined and investigations into the factors influencing cell growth and productivity are presented.

Annual Review of Cell Biology. v. 1– , 1985– . Palo Alto, CA:
Annual Reviews, Inc. Annual. $46/yr. ISSN 0743-4634.

Reflects current state of scientific research in cellular and molecular biology. Besides review articles in this very active and intense area, the volumes also lists other reviews of interest to cell biologists from other *Annual Review* publications.

BioEssays: Advances in Molecular, Cellular and Developmental Biology. v. 1– , 1984– . Cambridge, UK: Company of Biologists, for the International Council of Scientific Unions. Monthly. $295/yr. ISSN 0265-9247.

Review articles, features, book reviews, forthcoming events.

Cell Mechanics and Cellular Engineering. Edited by Van C. Mow et al. New York: Springer-Verlag, 1994. 564 p. $54.00. ISBN 0387943072.

A collection of articles reviewing the relationship between mechanistic forces, engineering principles and methods, and cell functions in living tissues and organisms.

Cold Spring Harbor Symposia on Quantitative Biology, 1933– . Cold Spring Harbor, NY: Cold Spring Harbor Laboratory. Price varies. ISSN 0091-7451.

Distinguished annual proceedings series, each volume reviewing a particular topic in detail. This is an important reference for molecular biologists as well as biochemists. Vol. 56: *The Cell Cycle*, 1992, $210.00, ISBN 0879690615; Vol. 57: *The Cell Surface*, 1993, $210.00, ISBN 08796990631.

Current Communication in Cell and Molecular Biology. Cold Spring Harbor, NY: Cold Spring Harbor Laboratory.

Irregular book series.

Apoptosis II: The Molecular Regulation of Cell Death. Edited by L. David Tomei and Frederick O. Cope. (v. 8 of the series) 300 p. $65.00. ISBN 0879693959.

Current Opinion in Cell Biology. v. 1– , 1989– . Philadelphia, PA: Current Science. Bimonthly. $390/yr. ISSN 0955-0674.

Reviews all advances; evaluation of key references; comprehensive listing of papers. Beginning late in 1995, this journal will be available electronically through the OCLC Electronic Journals Online services.

Current Opinion in Structural Biology. v. 1– , 1991– . Philadelphia, PA: Current Science. Bimonthly. $390/yr. ISSN 0959-440X.

Similar in format and content to other *Current Opinion* journals, reviewing advances, and providing a bibliography of current world literature. Available electronically in 1995 through the OCLC Electronic Journals Online services.

International Review of Cytology: A Survey of Cell Biology. v. 1– ,
 1952– . San Diego: Academic Press. Irregular. $925/yr,
 including supplements. ISSN 0074-7696.

A leading series presenting current advances and comprehensive reviews in cell biology, both plant and animal. Vol. 137: *Molecular Biology of Receptors and Transporters*, parts A,B, and C. 1993, each part approximately $65.00 if sold separately.

NATO ASI Series H: Cell Biology. New York: Springer-Verlag.

Series covers current advances in molecular biology, biochemistry, genetics, and microbiology.

> *Protein Synthesis and Targeting in Yeast.* v. 71. 1993. $198.00
> ISBN 0387565213.

Nucleic Acids and Molecular Biology. v. 1– , 1987– . New York:
 Springer-Verlag. Price and frequency vary.

Review of nucleic acids and their structure, function, and interaction with proteins. Vol. 8: 1994, $169.00, ISBN 0387574859.

The Photosynthetic Reaction Center. Edited by Johann Deisenhofer
 and James R. Norris. San Diego, CA: Academic Press, 1993.
 2 v. $258.00. ISBN 0122086619 (Vol. 1).

Summary of the work of Nobel laureates for a model of photosynthetic reaction centers, including structural analysis and functional properties using tools of molecular genetics.

Progress in Molecular and Subcellular Biology. v. 1– , 1969– .
 New York: Springer-Verlag. Irregular. Price varies.

Original articles review current advances at the frontier of life sciences at the molecular level. Topics in v. 12 (1992, ISBN 038753900X) include synthesis of small nuclear RNAs, DNA-activated protein kinase, interactions of water and proteins in cellular functions, heat-shock protein synthesis, and the cytoskeleton during early development.

Progress in Nucleic Acid Research and Molecular Biology. v. 1– ,
 1963– . San Diego: Academic Press. Irregular. Price varies.
 ISSN 0079-6603.

A forum for discussion of new discoveries, approaches, and ideas in molecular biology, including contributions from leaders in their fields. Vol. 48: 1994, 363 p. $95.00.

Results and Problems in Cell Differentiation. v. 1– , 1968– . New York: Springer-Verlag. Irregular. Price varies.

A series of topological volumes in developmental biology. Vol. 19: *Structure, Cellular Synthesis and Assembly of Biopolymers*. 1992, $198.00, ISBN 0387555498. The reader is introduced to the biological systems using biopolymers and the highly differentiated cells responsible for their synthesis.

Seminars in Cell Biology. v. 1– , 1990– . San Diego, CA: Academic Press. Bimonthly. $185/yr. ISSN 1043-4682.

Each issue, edited by an international authority, is devoted to a topical, important subject in cell biology. Recent issues review RNA editing, protein tyrosine phosphatases, cell motility, photoregulation of gene expression, and signal transduction in neuronal cells.

Trends in Cell Biology. v. 1– , 1993– . Cambridge, UK: Elsevier Trends Journals. Monthly. $514/yr. ISSN 0962-8924.

Covers all aspects of current research in cell biology.

Societies

Associations and societies relevant to molecular and cellular biology are included here with complete information, unless they have been annotated elsewhere. Societies whose activities are fully annotated in Chapters 3 or 5 ("General Sources" or "Biochemistry and Biophysics") are only listed. For current information about societies, membership directories, and announcements, consult the Internet via Gopher or World Wide Web.

American Society for Biochemistry and Molecular Biology (ASBMB)

Annotated (p. 109) in Chapter 5, "Biochemistry and Biophysics."

American Society for Cell Biology (ASCB)
9650 Rockville Pike, Bethesda, MD 20814.

Founded: 1960. 7,100 members. Includes scientists with educational or research experience in cell biology or an allied field. Placement service is offered. Publications: *Cell Regulation*; *Journal of Cell Biology*; *Methods in Cell Biology*; *Newsletter*. There is an annual convention with symposium and exhibits.

European Cell Biology Organization (ECBO)
University of Milan, Via Vanvitelli 32, I-20129 Milan, Italy.

Founded: 1975. There are European national cell biology societies in 18 countries. They promote the study of cell biology and its applications, and encourage cooperation among individual cell biologists and their respective national societies. The biennial congress is in 1994 at Prague.

European Molecular Biology Organization (EMBO)
   Postfach 10 22 40, 69012 Heidelberg, Germany. Dr. John Tooze, Exec. Sec.

Founded: 1963. 750 member organizations. Promotes the advancement of molecular biology in Europe and neighboring countries, administers programs funded by the European Molecular Biology Conference consisting of fellowships and courses. The organization holds courses and workshops, and presents an annual award. Publications: *EMBO Journal*. Periodic general assemblies and an annual symposium.

Federation of American Societies for Experimental Biology (FASEB)
Annotated (p. 67) in Chapter 3, "General Sources."

Federation of European Biochemical Societies (FEBS)
Annotated (p. 110) in Chapter 5, "Biochemistry and Biophysics."

Institute of Biology (IOB)
Annotated (p. 68) in Chapter 3, "General Sources."

International Cell Research Organization (ICRO)
   Organisation Internationale de Recherche sur la Cellule, Maison de
   l'UNESCO, 7, place de Fontenoy, F-75700 Paris, France. Dr. G.
   Cohen, Exec. Sec.

Founded: 1962. 400 members. Scientists, researchers, and laboratories in 50 countries encourage and facilitate the exchange of information on basic cell biology research. The organization organizes international training courses in microbiology, biotechnology, and cell and molecular biology; it also compiles statistics. There is an annual meeting.

International Federation of Cell Biology (IFCB)
   Federation Internationale de Biologie Cellulaire, University of Toronto,
   Dept. of Zoology, Ramsay Wright Zoological Laboratories, 25 Harbor
   St., Toronto, ON, Canada M5S 1A1. Dr. Arthur M. Zimmerman,
   Sec. Gen.

Founded: 1972. 21 member organizations. National and regional associations of cell biologists promoting international cooperation among scientists working in cell biology and related fields, and contributing to the advancement of cell biology in all of its branches. The organization acts as a coordinating body that initiates special studies and encourages research in subjects outside the normal scope of national societies, such as the problem of scientific communication. Conducts seminars and holds a quadrennial international congress.

International Union of Biochemistry and Molecular Biology (IUBMB)
Annotated (p. 110) in Chapter 5, "Biochemistry and Biophysics."

International Union of Biological Sciences (IUBS)
Annotated (p. 68) in Chapter 3, "General Sources."

Society for Experimental Biology and Medicine (SEBM)
Annotated (p. 68) in Chapter 3, "General Sources."

Society for In Vitro Biology
    8815 Centre Park Drive, Suite 210, Columbia, MD 21045.

Founded: 1946. 2,000 members. Formerly the Tissue Culture Association. For all scientists interested in the use of cells, tissues, and organs in vitro. Publications: *In Vitro Cellular and Molecular Biology—Animal* and *In Vitro Cellular and Molecular Biology—Plant*.

Women in Cell Biology (WICB)
    c/o Dr. Mary Lou King, University of Miami, Rm. 124, Dept. of Anatomy and Cell Biology, 1600 N. W. 10th Ave., Miami, FL 33101.

800 members. Sponsored by the American Society of Cell Biology, this organization serves as a forum for the discussion of various women's issues. Publications: *How to Get a Job*; *How to Keep a Job*; *Alternate Careers in Cell Biology*; and list of female members of ASCB. Annual convention is held in conjunction with the ASCB in November or December.

Textbooks

Alberts, Bruce et al. *Molecular Biology of the Cell*, 3rd ed. New York: Garland, 1994. 1,294 p. $59.95. ISBN 0815316194. Index.
Landmark textbook has all-star team of authors.

Baserga, R. *Cell Growth and Division: A Practical Approach*. New York: Oxford University Press, 1989. 210 p. $38.00. ISBN 0199630275.

Brachet, Jean. *Molecular Cytology*. San Diego: Academic, 1985–86. Vol. 1: *The Cell Cycle*, $49.00. ISBN 0121233707. Vol. 2: *Cell Interactions*, $55.00. ISBN 0121233715.

Calladine, C. R. and Horace R. Drew. *Understanding DNA: The Molecule and How It Works*. San Diego: Academic Press, 1992. 220 p. $65.00. ISBN 0121550850. Bibliography Index.

Cell Biology. Molecular and Cell Biochemistry. Edited by Chris A. Smith and E. J. Wood. New York: Chapman & Hall, 1992. 366 p. $31.00 (paper). ISBN 041240740X.

Introduces undergraduate students to the cell.

Cell Physiology Source Book. Edited by Nicholas Sperelakis. San Diego: Academic Press, 1995. 742 p. $99.00. ISBN 0126569703.

Designed as a resource for graduate students in sub-disciplines other than cell biology.

Cross, Patricia C. and K. Lynn Mercer. *Cell and Tissue Ultrastructure; A Functional Perspective*. Oxford, England: W. H. Freeman, 1993. ISBN 0716770334.

Introduction to the cell ultrastructure of all tissues and organs including text, diagrams, and micrographs. Also, see the atlases by Patricia C. Cross (p. 120) and the Don W. Fawcett (p. 120).

Darnell, James E. et al. *Molecular Cell Biology*, 2nd ed. Salt Lake City, UT: Freeman, 1990. 1,105 p. $57.95. ISBN 0716719819.

DeRobertis, E. D. and E. M. DeRobertis. *Cell and Molecular Biology*, 8th ed. Philadelphia, PA: Lea and Febiger, 1987. $45.00. ISBN 0812110129.

Essentials of Molecular Biology, 2nd ed. By David Freifelder; edited by George M. Malacinski. Boston: Jones and Bartlett, 1993. 478 p. $41.25. ISBN 0867201371.

Introductory text for first- and second-year undergraduates.

Frank-Kamenetskii, Maxim D. *Unraveling DNA*. Deerfield Beach, FL: VCH, 1993. 216 p. $24.95. ISBN 1560816171.

A concise, highly readable introduction to the structure, properties, and functions of the DNA molecule for both general readers and scientists.

Goodsell, David S. *The Machinery of Life*. New York: Springer-Verlag, 1993. 140 p. $29.00. ISBN 0387978461.

Presents scale-model images of life including macromolecules, substrates and large assemblies. Excellent drawings explain the mechanisms of life. Written for the non-specialist but also of use to the biochemist. Divided into three sections: molecules and life, molecules into cells, and cells in health and disease.

Lauffenburger, Douglas A. and Jennifer J. Linderman. *Receptors: Models for Binding, Trafficking, and Signalling*. New York: Oxford University Press, 1993. 365 p. $69.95. ISBN 0195064666.

Bridges the gap between chemical engineering and cell biology with math modelling approach to quantitative experiments for enhanced understanding of cell phenomenon. The book includes the entire spectrum of receptor processes.

Loewy, Ariel et al. *Cell Structure and Function: An Integrated Approach*, 3rd ed. Philadelphia: Saunders, 1991. 947 p. $56.00. ISBN 0030474396.

Comprehensive and lavishly illustrated cell biology text for undergraduate students.

Molecular Biology: An Illustrated Introduction from Cells to Atoms, 2nd ed. Oxford, England: Blackwell, 1993. $10.00 (paper). ISBN 0632024658.

New editions of *From Cells to Atoms* presents essential elements of molecular and cellular biology suitable for a wide audience.

Molecular Biology and Biotechnology. Edited by Chris A. Smith and E. J. Wood. New York: Chapman & Hall, 1991.

Refer to the complete annotation (p. 174) in Chapter 7, "Genetics."

Murray, Andrew Wood and Tim Hunt. *The Cell Cycle: An Introduction*. New York: Freeman, 1993. 251 p. $45.00. ISBN 071677044X. Bibliography. Index.

Describes origin of different experimental approaches to the cell cycle and traces contributions to current knowledge of cell growth and division. The book concentrates on a small number of organisms with critical experiments of genes that affect the cell cycle. There is an alphabetical index to these genes including their CDC entry (cell cycle mutants) to their GenBank reference.

Principles of Molecular Recognition. Edited by A. D. Buckingham et al. New York: Chapman & Hall, 1993. 224 p. $49.95. ISBN 075140120.

How molecules of all sizes interact.

Smith, Chris A. and E. J. Wood. *Cell Biology. Molecular and Cell Biochemistry*. New York: Chapman & Hall, 1992. 366 p. $31.00 (paper). ISBN 041240740X.

Ideal for a one-semester course for beginning biology students.

Thomas, Lewis. *Lives of a Cell: Notes of a Biology Watcher*. Bantam
    Classics reprint, 1984. 192 p. $4.95. ISBN 0553275801.

A biology classic.

Weber, Gregorio. *Protein Interactions*. New York: Chapman & Hall,
    1992. 293 p. $60.00. ISBN 0412030314.

Complete conceptual and quantitative description of proteins and their
associations.

Widnell, Christopher. *Essentials of Cell Biology*. Baltimore, MD:
    Williams & Wilkins, 1990. $33.95. ISBN 06830905518.

Bibliography

McGraw-Hill Dictionary of Scientific and Technical Terms, 4th ed. Sybil P. Parker,
Editor-in-Chief. New York: McGraw-Hill, 1989. 2088 p. ISBN 0070452709.

7

Genetics

This chapter deals with the related fields of genetics, genetic engineering, and biotechnology. **Genetics** is "the science that is concerned with the study of biological inheritance," while **genetic engineering** is "the intentional production of new genes and alteration of genomes by the substitution or addition of new genetic material," and **biotechnology** is "the use of advanced genetic techniques to construct novel microbial and plant strains and obtain site-directed mutants to improve the quantity or quality of products" (*McGraw-Hill Dictionary of Scientific and Technical Terms*, 1989). Traditional animal and plant breeding is excluded from this chapter, as are the chemical engineering–related aspects of biotechnology.

Abstracts, Bibliographies, and Indexes

Agricultural and Environmental Biotechnology. v. 1– , 1993– .
Bethesda, MD: Cambridge Scientific Abstracts. Bimonthly.
$195.00. ISSN 1063-1151.

Covers biotechnology in food science, agriculture, and the environment; plant genome studies are also covered, as well as topics such as bioremediation, transgenic plants and animals, and water treatment. Available online as part of the Life Sciences Collection from DIALOG and STN and on CD-ROM from Silver-Platter.

ASFA Marine Biotechnology BioEngineering. v. 1– , 1989– . Bethesda,
MD: Cambridge Scientific Abstracts. Quarterly. $185.00. ISSN
1043-8971.

Covers marine biotechnology, including aquaculture, biofouling, chemical products, etc. Available online as part of the Life Sciences Collection from DIALOG and STN and on CD-ROM from SilverPlatter. Also searchable via the Internet (see annotation fo the **Life Sciences Collection**, p. 78.)

Biotechnology Citation Index. v. 1– , 19 – . Philadelphia: Institute for
Scientific Information. Bimonthly.

A subsection of *Science Citation Index* dealing with biotechnology. Available only on CD-ROM. Also searchable via the Internet (see above).

Current Advances in Genetics and Molecular Biology. v. 1– , 1984– .
New York: Pergamon. Monthly. $998. ISSN 0741-1642.

Current awareness service for genetics and molecular biology.

Current Biotechnology. v. 1– , 1983– . Cambridge, England: Royal
Society of Chemistry. Monthly. $657.00. ISBN 0264-3391.

Formerly *Current Biotechnology Abstracts*. Covers all aspects of biotechnology, including news items, patents, general information such as forthcoming events, and article citations. About 200 journals and newsletters are scanned. A list of source items is listed with each issue. Available through DIALOG as Current Biotechnology Abstracts.

Derwent Biotechnology Abstracts. v. 1– , 1982– . London: Derwent
  Publications. Bimonthly. ISSN 0262-5318.

Scans over 1,300 journals as well as proceedings and patents in all areas of biotechnology. Available online through DIALOG and on CD-ROM from SilverPlatter.

Excerpta Medica. Section 22: Human Genetics. v. 1– , 1963– .
  Amsterdam: Excerpta Medica. $1178.00 10 issues per year. ISSN
  0014-4266.

Continues *Human Genetics Abstracts.* Available on CD-ROM as part of the complete Excerpta Medica Library Service from SilverPlatter.

Genetics Abstracts. v. 1– , 1968– . Bethesda, MD: Cambridge
  Scientific Abstracts. Monthly. $835. ISSN 0016-674X.

Covers all areas of genetics, including biotechnology and basic genetic research. Available online as part of the Life Sciences Collection from DIALOG and STN and on CD-ROM from SilverPlatter. Also searchable via the Internet (see annotation under the **Life Sciences Collection**, p. 78).

Human Genome Abstracts. v. 1– , 1990– . Bethesda, MD:
  Cambridge Scientific Abstracts. Bimonthly. $195.00. ISSN
  1045-4470.

Covers all aspects of human genome projects, basic and applied. Available online as part of the Life Sciences Collection from DIALOG and STN and on CD-ROM from SilverPlatter. (See above for Internet access.)

Medical and Pharmaceutical Biotechnology Abstracts. v. 1– , 1993– .
  Bethesda, MD: Cambridge Scientific Abstracts. Bimonthly.
  $195.00. ISSN 1063-1178.

Covers genetically engineered drug delivery systems, including bioreactors, vaccines, and cell culture. Available online as part of the Life Sciences Collection from DIALOG and STN and on CD-ROM from SilverPlatter. (See above for Internet access.)

Plant Genetic Resources Abstracts. v. 1– , 1992– . Wallingford, UK:
  CAB International. Quarterly. $238.00. ISSN 0966-0100.

Covers plant genetic resources for plants of economic importance, including scientific, legal, economic, and agricultural aspects. (Also available as PlantGeneCD.)

See also **BioBusiness** (p. 74), *Biological Abstracts* (p. 74), *Biological Abstracts/RRM* (p. 74), *Chemical Abstracts* (p. 75), *Current Contents/Life Sciences* (p. 76), and *Index Medicus* (p. 78), described in Chapter 4, "Abstracts and Indexes."

Databases

The databases discussed here are genetics databases which provide access to gene sequences. Each time a researcher sequences a gene, he or she is expected not only to publish the sequence in a research journal, but also to submit the sequence to a sequence database. Sometimes journals will not accept articles until the sequences have appeared in a database, which is one of the few cases in which journals will accept data previously published elsewhere. There are many databases currently available; only the most important are listed here. Most of the sequence databases are available for searching at no charge through the Internet. For more information, see *Nucleic Acids Research* Database issue (1994), Corteau (1991), or Fuchs and Cameron (1991), all in the Bibliography (p. 177).

DDBJ (DNA Data Bank of Japan). Mishima, Japan: National Institute of Genetics.

Collaborates with GenBank and EMBL to collect nucleotide sequences. Available on magnetic tape, on CD-ROM as part of **Entrez**, or through the Internet (e-mail, FTP, and Gopher) at the National Institute of Genetics in Mishima.

EMBL (European Molecular Biology Laboratory). 1980– . Heidelberg, Germany: EMBL Data Library.

A nucleotide sequence database, created in collaboration with GenBank and DDBJ. Available on CD-ROM and through the Internet (e-mail, FTP, and Gopher) at Heidelberg.

Entrez. 1992– . Bethesda, MD: National Center for Biotechnology Information. Updated irregularly.

This CD-ROM database is a molecular sequence retrieval system which provides access to nucleotide and protein sequences from a number of databases, including GenBank, DDBJ, EMBL, PIR, SWISS-PROT and others (for the latter two, see p. 117 in Chapter 6, "Molecular and Cellular Biology)." Entrez is on two disks, one with sequences and one with MEDLINE references to the articles in which the sequences were published.

GenBank. 1982– . Bethesda, MD: National Center for Biotechnology Information.

NIH's database of all known nucleotide and protein sequences. Available on magnetic tape, CD-ROM, online from STN, and through the Internet (e-mail, FTP, and Gopher) at NCBI in Bethesda.

Dictionaries and Encyclopedias

Babel, W., M. Hagemann, and W. Hohne. *Dictionary of Biotechnology: English-German*. New York: Elsevier, 1989. 113 p. ISBN 0444989005.

Translates over 7,000 terms in English to German; includes an appendix listing some biotechnologically important micro-organisms.

Bains, W. *Biotechnology from A to Z* New York: Oxford University Press, 1993. 358 p. $19.95 (paper). ISBN 0199633347 (paper).

An "extended glossary" of over 280 terms in biotechnology. Designed to provide an introduction to the concepts discussed, rather than to simply define them.

Biotechnology: A Multi-Volume Comprehensive Treatise, 2nd, completely rev. ed. H.-J. Rehm and G. Reed, eds. New York: VCH, 1991– . $365.00 per volume. ISBN 1560816023 (set).

This massive work covers all areas in biotechnology. Twelve volumes are planned, with volumes covering topics including biochemical and biological fundamentals, products of primary and secondary metabolism, biotransformations, legal and social issues and many others.

Coombs, J. *Dictionary of Biotechnology*, 2nd ed. New York: Stockton Press, 1992. 364 p. $90.00. ISBN 1561590746.

Covers over 3,000 definitions in all aspects of biotechnology. Includes illustrations.

King, Robert C. and William D. Stansfield. *A Dictionary of Genetics*, 4th ed. New York: Oxford University Press, 1990. 406 p. $39.95. ISBN 0195063708.

Includes non-genetic terms often encountered in genetics literature as well as more strictly genetics terms. Also has appendices on the classification of organisms, major domesticated species, a chronology and index of major events and geneticists, and a list of genetics periodicals.

Kirk-Othmer Encyclopedia of Chemical Technology. Jacqueline I. Kroschwitz, executive ed. New York: John Wiley and Sons, 1993– . 27 v. $275 per v. ISBN (set) 0471527041.

Includes articles of interest to biotechnology, including biosensors, genetic engineering, and biotechnology.

Moo-Young, Murray. *Comprehensive Biotechnology: The Principles,*
    Applications and Regulations of Biotechnology in Industry,
    Agriculture and Medicine. New York: Pergamon Press, 1985. 4 v.
    $1,940.00 (set). ISBN 008026204X (set). v. 1: Principles of Bio-
    technology—Scientific Fundamentals. v. 2: Principles of Biotechnol-
    ogy—Engineering Considerations. v. 3: The Practice of Biotechnol-
    ogy—Current Commodity Products. v. 4: The Practice of Biotech-
    nology—Specialty Products and Service Activities.

This is "intended to be the standard reference work in the field." It attempts to
draw together information from the entire range of fields which make up
biotechnology.

Oliver, Stephen G. and John M. Ward. *A Dictionary of Genetic Engineer-*
    ing. New York: Cambridge University Press, 1985. 153 p.
    $29.95. ISBN 0521260809.

Illustrated; includes over 500 terms used in genetic engineering. Also includes
list of restriction enzymes and genetic maps of *Escherichia coli* and *Bacillus
subtilis.*

Rieger, R., A. Michaelis, and M. M. Green. *Glossary of Genetics: Classi-*
    cal and Molecular, 5th ed. New York: Springer-Verlag, 1991. 553
    p. $39.00 (paper). ISBN 0387520546.

Some definitions consist of a short essay, though most are more brief. Includes
genetic engineering terms.

Schmid, Rolf and Saburo Fukui. *Dictionary of Biotechnology in English-*
    Japanese-German. New York: Springer-Verlag, 1986. 1324 p.
    ISBN 038715566X.

Includes over 6,000 terms in three main sections, one translating English terms
into Japanese and German, one translating Japanese terms into English and
German, and one translating German terms into English.

Steinberg, Mark L. and Sharon D. Cosloy. *Facts on File Dictionary of*
    Biotechnology and Genetic Engineering. New York: Facts on File,
    1994. 197 p. $27.95. ISBN 0816012504.

Comprehensive dictionary intended primarily for undergraduates and laypeople.

Walker, John M. and Michael Cox. *The Language of Biotechnology: A*
    Dictionary of Terms. Washington, DC: ACS Professional Reference
    Book, 1988. 255 p. $49.95, $29.95 (paper). ISBN 0841214891,
    0841214905 (paper).

Attempts to "define routinely used specialized language in the various areas of
biotechnology." Includes illustrations.

Directories

BioScan. v. 1– , 1986– . Phoenix, AZ: Oryx Press. Annual directory
with bimonthly updates. $795.00. ISSN 0887-6207.

Directory of biotechnology companies. Also available as an online database
(updated monthly) and on diskette (updated bimonthly, $1,200.00).

Biotech Buyers' Guide. v. 1– , 1990– . Washington, DC: American
Chemical Society. Annual. $50.00.

Includes instruments, equipment, and biologicals with a listing of suppliers.

Biotechnology Research Directory: 4000 Faculty Profiles. Prepared by the
Biotechnology Information Division of the North Carolina
Biotechnology Center. Washington, DC: Bureau of National Affairs,
1991. 662 p. $125.00. ISBN 1558712267.

A directory of individuals doing biotechnology research, arranged by state.
Provides the address, research focus, organisms studied, and techniques used for
the academics listed. An expertise index is also included.

Coombs, J. and Y. R. Alston. *The Biotechnology Directory: Products,
Companies, Research and Organizations*, 11th ed. New York:
Stockton Press, 1995. 645 p. $265.00. ISBN 1561591114.

Profiles over 10,000 commercial and non-commercial biotechnology organizations. Includes information sources, international societies, university
departments, government institutes, and a buyer's guide to products and services.

Dibner, Mark D. *Biotechnology Guide USA: Companies, Data and Analysis*, 2nd ed. New York: Stockton Press, 1991. 652 p. $199.00.
ISBN 1561590150.

Lists companies, along with areas of interest such as biosensors, and company
data such as revenue, numbers of patents received, numbers and types of em-
ployees, etc. More analytical than most directories.

Federal Biotechnology Information Resources Directory. Washington, DC:
OMEC International, 1987. 151 p. $95.00 (paper). ISBN
0931283035.

Lists resources such as databases available at various federal agencies; examples
include GenBank, *Energy from Biomass Bulletin*, and registries.

Federal Biotechnology Programs Directory. Washington, DC: OMEC
International, 1987. 162 p. $95.00 (paper). ISBN 0931283027.

Companion to the above directory; lists and describes programs by federal
agency, including contact names and addresses.

Guide to Scientific Products, Instruments and Services. Washington, DC:
American Association for the Advancement of Science. $20.00.

Annual supplement to *Science*, and free to subscribers. Lists products and
suppliers, as well as services such as DNA sequencing. Each issue of *Science*
also includes a listing of products and services.

National Biotech Register. v. 1– , 1992– . Wilmington, MA: Barry.
$35.00.

Consists of three sections, including about 1,300 brief company profiles, a
subject listing of companies by research area, and a listing of trademarks.

*State-by-State Biotechnology Directory: Centers, Companies, and
Contacts*, 2nd ed. Prepared by the Biotechnology Information
Division of the North Carolina Biotechnology Center. Washington,
DC: Bureau of National Affairs, 1991. 175 p. $79.00. ISBN
1558712283.

Lists contact persons in state governments, regulatory agencies, and research
centers along with addresses and telephone numbers. Also provides a list of
biotechnology companies in each state.

Guides to Internet Resources

The Internet is particularly useful for geneticists, due to the large number of
genome databases which are accessible through the Internet (see the Databases
section, p. 146). The IUBio Archives Gopher at Indiana University is a good
starting place to find genetics resources, particularly molecular genetics. There
are also a number of BIOSCI/bionet discussion groups relating to genetics and
biotechnology, including bionet.drosophila, bionet.genome.*, several bionet.
molbio groups including bionet.molbio.genbank, bionet.molbio.gene-linkage,
embnet.general, and sci.bio.technology. In addition, many genome projects are
accessible through either Gopher or the World Wide Web, including genomes
for the dog, *C. elegans*, forest trees, *Arabidopsis*, zebrafish, the mouse, maize,
Mycoplasma capricolum, *Saccharomyces*, and *E. coli*, just to name a few.

Guides to the Literature

Alston, Yvonne and James Coombs. *Biosciences: Information Sources
and Services*. New York: Stockton Press, 1992. 407 p. $99.00.
ISBN 1561590657.

Covers a wide variety of information sources in biotechnology, including organizations, publications, services (software, consultants, etc), and indexes. Ex-

panded version of Coombs and Alston's *The Biotechnology Directory* (see p. 149 or below).

Biotechnology Information Sources: North and South America. Compiled and edited by Barbara A. Rapp. Medford, NJ: Published by Learned Information for the International Council for Scientific and Technical Information, 1994. 144 p. $32.50 (paper). ISBN 0938734814 (paper).

Includes a wide variety of information sources available in North and South America, including the usual print sources and Internet-accessible resources.

Coombs, J. and Y. R. Alston. *The Biotechnology Directory: Products, Companies, Research and Organizations*.

See p. 149 for full citation. Includes sections on biotechnology information sources.

Crafts-Lighty, A. *Information Sources in Biotechnology*, 2nd ed. New York: Stockton Press, 1986. 403 p. $130.00. ISBN 0943818184.

Covers the literature of biotechnology in detail, including books, conferences, journals, trade information, abstracts, databases, patents, and market surveys.

Directory of Biotechnology Information Resources (DBIR). National Library of Medicine. Updated monthly.

Database containing directory information on a wide range of biotechnology resources. Includes print publications, electronic media, sequence databases, agencies and their reports, and much more. Very similar in format and coverage to the other guides to the literature covered in this section. Searchable as part of DIRLINE via NLM/MEDLARS and Grateful Med (see annotation under *Index Medicus*, p. 78).

Handbooks

Animal Cell Biotechnology. Edited by R. E. Spier and J. B. Griffiths. v. 1– , 1985– . Orlando, FL: Academic Press. v. 5 (1992). $129.00. ISBN 012657555X.

Intended to provide both state-of-the-art reviews of animal cell biotechnology for experts, and methods for culturing animal cells.

Ashburner, Michael. *Drosophila: A Laboratory Handbook*. Cold Spring Harbor, NY: Cold Spring Harbor Press, 1989. 1331 p. $180.00. ISBN 0879693215.

A summary of *Drosophila* biology, including chromosomes, taxonomy, and developmental and molecular biology. Designed for rapid reference.

Atkinson, Bernard and Ferda Mavituna. *Biochemical Engineering and Biotechnology Handbook*. New York: Stockton, 1991. 1271 p. $265.00. ISBN 1561590126.

Includes information on a wide variety of biotechnological topics, including properties of important microorganisms, microbial metabolism, product information, reactors, downstream processing, and so on.

Beckmann, Jacques S. and Thomas C. Osborn. *Plant Genomes: Methods for Genetic and Physical Mapping*. Dordrecht: Kluwer Academic, 1992. 250 p. ISBN 0792316304.

This manual provides methods for mapping plant genomes, including Southern blot and pulsed-field gel electrophoresis. Several of the chapters were originally published in the *Plant Molecular Biology Manual*, Supplement 5.

Brown, T. A., ed. *Molecular Biology Labfax*. San Diego, CA: Academic Press, 1991. 322 p. $49.95. ISBN 0121375102.

For a complete annotation, see p. 122 in Chapter 6, "Molecular and Celular Biology." Includes data on genomes and genes of interest to geneticists.

King, Robert C. *Handbook of Genetics*. New York: Plenum, 1974–1976. 5 v. (v. 1: Bacteria, Bacteriophages, and Fungi; v. 2: Plants, Plant Viruses and Protists; v. 3: Invertebrates of Genetic Interest; v. 4: Vertebrates of Genetic Interest; v. 5: Molecular Genetics).

While some of the information in these volumes has become dated, there is still no single source to replace them.

Kitani, Osamu and Carl W. Hall, eds. *Biomass Handbook*. New York: Gordon and Breach, 1989. 963 p. ISBN 2881242693.

Covers all forms of biomass—plant and animal—and includes information on biomass production, conversion, and utilization. Also includes information on biotechnology for biomass production and utilization, and statistics and properties of biomass.

Lindsley, Dan L. and Georgianna G. Zimm. *The Genome of* Drosophila melanogaster. San Diego, CA: Academic Press, 1992. 1,133 p. $79.00. ISBN 0124509908.

A compendium of information about the genetics and chromosomes of the fruit fly. An update of Lindsley and Grell's *Genetic Variations of Drosophila melanogaster*, published in 1968. Also includes chromosome maps, which are available separately ($19.95, ISBN 0124509916).

McKusick, Victor A. *Mendelian Inheritance in Man: Catalogs of Autosomal Dominant, Autosomal Recessive, and X-Linked Phenotypes*, 11th ed. Baltimore: Johns Hopkins University Press, 1992. 2 v. 2560 p. $165.00. ISBN 0801844118.

This handbook catalogs hereditary diseases in humans; it is useful for genetic counseling. All known genetic disorders are described, along with their loci, if known, and references to the literature. The catalog is also available through the Internet as OMIM (Online Mendelian Inheritance in Man) at The Johns Hopkins University.

Morgan, S. J. and D. C. Darling. *Animal Cell Culture*. Oxford, UK: BIOS Scientific Publishers, 1993. ISBN 1872748163.

This manual serves as an introduction to animal cell culture, including the basic methodology, equipment, media preparation, various techniques, and a brief listing of suppliers and manuals for further reading.

O'Brien, Stephen J., ed. *Genetic Maps: Locus Maps of Complex Genomes*. 6th ed. Cold Spring Harbor, NY: Cold Spring Harbor Press, 1993. $175.00. ISBN 0879694149.

Provides information on the genetic organization of different species; the only comprehensive source for genetic maps. Available as a comprehensive reference volume or as six individual paperback volumes (viruses, bacteria, lower eukaryotes, nonhuman vertebrates, humans, and plants).

Sukatsch, Dieter A. and Alexander Dziengel. *Biotechnology: A Handbook of Practical Formulae*. New York: Wiley, 1987. 160 p. (paper). ISBN 0470207299.

A "summary of current nomenclature, definitions and associated equations currently in use in the field of biotechnology." The handbook has separate chapters on microbiology, biochemistry, physical chemistry, biochemical engineering, and an appendix providing physical constants, symbols, coefficients, and other useful data.

Histories

Bud, Robert. *The Uses of Life: A History of Biotechnology*. Cambridge, England: Cambridge University Press, 1993. 299 p. $49.95. ISBN 0521382408.

Discusses the history of biotechnology, from the ancient "zymotechnology" (fermentation technology), through chemical engineering and the current genetic engineering. Emphasis on European personalities and institutions.

Carlson, Elof Axel. *The Gene: A Critical History*. Ames, IA: Iowa State University Press, 1989. 301 p. ISBN 0813814065. Reprint of 1966 ed.

A history of the gene concept, arranged by themes rather than a timeline.

Cook-Deegan, Robert. *The Gene Wars: Science, Politics, and the Human Genome*. New York: Norton, 1993. 416 p. $25.00. ISBN 0393035727.

A history of the early years of the Human Genome Project, covering events and personalities both in the United States and abroad, written by an author who was witness to much of the events described. The first four chapters offer a discussion of the science involved in the Human Genome Project at a level suitable for the general public.

Dunn, L. C. *A Short History of Genetics: The Development of Some Main Lines of Thought, 1864–1939*. New York: McGraw-Hill, 1965. 261 p.

Covers the period of "classical" genetics, from Mendel to the development of the gene theory.

The Dynamic Genome: Barbara McClintock's Ideas in the Century of Genetics. Edited by Nina Fedoroff and David Botstein. Plainview, NY: Cold Spring Harbor Laboratory Press, 1992. 422 p. $65.00. ISBN 08769422X.

A festschrift honoring Barbara McClintock (winner of the 1983 Nobel Prize for Physiology or Medicine) on her 90th birthday. Contains reprints of six of McClintock's papers on maize genetics, plus chapters discussing various phases of McClintock's career.

The Early Days of Yeast Genetics. Edited by Michael H. Hall and Patrick Linder. Plainview, NY: Cold Spring Harbor Laboratory Press, 1993. 477 p. $75.00. ISBN 0879693789.

Personal remniscences of the beginning of the study of yeast genetics.

Harwood, Jonathan. *Styles of Scientific Thought: The German Genetics Community 1900–1933*. Chicago: University of Chicago Press, 1993. 423 p. $65.00. ISBN 0226318818.

Discusses the German geneticists of the early part of the twentieth century, who were more interested in development and evolution than in the gene concept studied by American geneticists.

Jacob, Francois. *The Logic of Life: A History of Genetics*. New York: Viking Penguin, 1989. ISBN 0140552421.

A history of genetics written by one of the great geneticists of the 1950's.

Keller, Evelyn F. *A Feeling for the Organism: The Life and Work of Barbara McClintock*. New York: W. H. Freeman, 1983. 235 p. $20.95, $14.95 (paper). ISBN 0716714337, 071671504X (paper).

A biography of one of the most important twentieth-century geneticists and winner of the 1983 Nobel Prize for Physiology or Medicine, who discovered transposable elements in corn.

Kohler, Robert E. *Lords of the Fly: Drosophila Genetics and the Experimental Life*. Chicago: University of Chicago Press, 1994. 321 p. $45.00, $17.95 (paper). ISBN 0226450627, 0226450635 (paper).

"This book is about the material culture and way of life of experimental scientists. It is also about a particular and familiar community of experimental biology, the *Drosophila* geneticists, and their no less familiar co-worker, the fruit fly."

Medvedev, Zhores A. *The Rise and Fall of T. D. Lysenko*. New York: Columbia University Press, 1969. 284 p.

Presents a first-hand account of this "bizarre chapter" in the history of Soviet genetics from 1937 to 1964.

Sarkar, Sahotra, ed. *The Founders of Evolutionary Genetics: A Centenary Reappraisal*. Boston: Kluwer Academic, 1992. 300 p. ISBN 0792317777.

Discusses the role of four geneticists, R. A. Fischer, J. B. S. Haldane, H. J. Muller, and S. Wright in founding what the editor calls "evolutionary genetics." Several chapters were written by students or co-workers of the subjects.

Sturtevant, A. H. *A History of Genetics*. New York: Harper and Row, 1965. 165 p.

Covers much the same time period as Dunn (p. 154) with the addition of material up to the mid-50's.

Tiley, N. A. *Discovering DNA: Meditations on Genetics and a History of the Science*. New York: Van Nostrand Reinhold, 1983. 288 p. ISBN 044226204.

A philosophical discussion of the history of genetics, from ancient times to genetic engineering, including ethical aspects. Has appendices containing the full text of the original Watson-Crick articles and other important works.

Watson, James D. *Double Helix: Being a Personal Account of the*

Discovery of the Structure of DNA. New York: Atheneum, 1980. 226 p. $11.95 (paper). ISBN 0689706022.

The classic account of the race to discover the secrets of DNA.

Methods

Anand, R. *Techniques for the Analysis of Complex Genomes*. San Diego, CA: Academic Press, 1992. 256 p. $39.95. ISBN 0120576201.

A spiral-bound lab manual describing genomic mapping by pulsed-field gel electrophoresis, cloning of DNA using yeast artificial chromosomes, and other methods.

Ashburner, Michael. Drosophila: *A Laboratory Manual*. Cold Spring Harbor, NY: Cold Spring Harbor Press, 1989. 434 p. $60.00 (comb bound). ISBN 0879693223.

Provides 131 protocols for working with *Drosophila* genetics, including protocols dealing with chromosomes, molecular biology, tissue culture, developmental biology, and transformations.

Biotechnology: Applications and Research. Paul N. Chcremisinoff and Robert P. Ouellette, eds. Lancaster, PA: Technomic Publishing Co., 1985. 699 p. $49.00. ISBN 877623910.

Intended as a reference for practitioners and students.

Demain, Arnold L. and Nadine A. Solomon. *Manual of Industrial Microbiology and Biotechnology*. Washington, DC: American Society for Microbiology, 1986. 466 p. $55.00, $43.00 (paper). ISBN 0914826727, 0914826735 (paper).

The purpose of this manual is to "bring together in one place the biological and engineering methodology required to develop a successful industrial process from the isolation of the culture to the isolation of the product."

Donis-Keller, H. *Human Gene Mapping Techniques*. Oxford, England: W. H. Freeman, 1993. 250 p. $52.50. ISBN 0716770121.

A practical guide to gene-mapping techniques, including sources for reagents and equipment and an extensive bibliography.

Edwards, Clive, ed. *Monitoring Genetically Manipulated Microorganisms in the Environment*. New York: John Wiley and Sons, 1993. 216 p. $64.95. ISBN 0471937959.

Covers techniques for monitoring or detecting genetically engineered microbes.

Freshney, R. Ian. *Culture of Animal Cells*, 3rd ed. New York: Wiley-Liss, 1993. 496 p. $69.95. ISBN 0471589667.

For both novices and experienced researchers. Classic work covering all aspects of cell culture, from setting up a laboratory to advanced techniques.

Genetic Manipulation: Techniques and Applications. J. M. Grange, A. Fox, and N. L. Morgan, eds. Boston: Blackwell Scientific Publications, 1991. (Society for Applied Bacteriology Technical Series, 28). 401 p. ISBN 0632029269.

Provides techniques for genetic manipulation, as well as applications such as DNA fingerprinting and the identification of pathogenic bacteria. Both biotechnology and genetic engineering techniques are included.

Genome Analysis. Kay E. Davies and Shirley M. Tilghman, series editors. v. 1– , 1990– . Plainview, NY: Cold Spring Harbor Laboratory Press. Price varies. ISSN 1050-8430.

"*Genome Analysis* is a series of short, single-theme books that review the data, methods, and ideas emerging from the study of genetic information in humans and other species." Volumes to date include *Genetic and Physical Mapping*, *Gene Expression and its Control*, *Genes and Phenotypes*, *Strategies for Physical Mapping*, and *Regional Physical Mapping*.

Glick, Bernard R. and John E. Thompson. *Methods in Plant Molecular Biology and Biotechnology*. Boca Raton, FL: CRC Press, 1993. 360 p. ISBN 0849351642.

Provides techniques for plant biotechnology, including methods for using recombinant DNA technology, isolation and characterization of plant DNA, and many other techniques.

Guide to Human Genome Computing. Edited by Martin J. Bishop. San Diego: Academic Press, 1994. 350 p. $39.50 (paper). ISBN 0121020509.

Covers basic computing, related genome databases, and other topics in genome computing. An extension of a training course run by the UK Human Genome Mapping Project.

Innis, Michael A., David H. Gelfand, John J. Sninsky, and Thomas J. White. *PCR Protocols: A Guide to Methods and Applications*. San Diego, CA: Academic Press, 1990. 482 p. $89.95, $49.95 (comb bound). ISBN 0123721806, 0123721814 (comb bound).

Includes protocols for the PCR (polymerase chain reaction) technique for DNA analysis.

Keller, George and Mark Manak. *DNA Probes*, 2nd Ed. New York: Stockton Press, 1993. 659 p. $90.00. ISBN 0333573846.

"*DNA Probes* is a one-stop techniques manual covering DNA probes for scientists, technicians and managers in all fields."

Methods in Enzymology. v. 1– , 1955– . San Diego, CA: Academic Press. Also available: CD-ROM version of eighteen select *Methods in Enzymology* volumes, including *Recombinant DNA* parts A-I, *DNA Structures*, parts A and B, in addition to a cumulative index for volumes 1-244. (Academic Press, $999.00, ISBN 0120001004).

For full annotation, see p. 57 in Chapter 3, "General Sources." Includes many volumes of methods relating to genetics and biotechnology, such as v. 183, *Molecular Evolution: Computer Analysis of Protein and Nucleic Acid Sequences* and v. 211 and 212, *DNA Structures, Parts A and B.*

Methods in Molecular Biology. v. 1– , 19– . Totowa, NJ: Humana Press. Price varies.

For full annotation (p. 127), see Chapter 6, "Molecular and Cellular Biology." Several of these volumes refer to molecular genetics and biotechnology, such as v. 23, *DNA Sequencing Protocols*, or v. 24–25, *Computer Analysis of Sequence Data.*

Methods in Molecular Genetics. v. 1– , 1993– . San Diego, CA: Academic Press. Price varies.

"A new series, *Methods in Molecular Genetics*, provides practical experimental procedures for use in the laboratory." Volumes to date include *Gene and Chromosome Analysis, Parts A, B, and C; Molecular Microbiology*; and *Molecular Virology.*

Velten, Jeff. *Recombinant DNA Techniques*. Boca Raton, FL: CRC Press, 1994. 128 p. $34.95. ISBN 0849386845.

A combination textbook and manual covering DNA protocols.

Zyskind, Judith W. and Sanford I. Bernstein. *Recombinant DNA Laboratory Manual*. Revised ed. San Diego, CA: Academic Press, 1992. 224 p. $29.95 (comb bound). ISBN 0127844015.

Includes protocols for a number of procedures. For full annotation (p. 130), see Chapter 6, "Molecular and Cellular Biology."

Periodicals

Acta Biotechnologica: Journal of Biotechnology in Environmental Protection, Industry, Agriculture, and Health Care. v. 1– , 1981– . Berlin: Akademie Verlag. Bimonthly. $279.00. ISSN 0138-4988.

"Publishes original papers, short communications, reports and reviews of biotechnology in environmental protection, industry, agriculture, and health care. The journal exists to promote the establishment of biotechnology as a new and integrated scientific field."

American Journal of Human Genetics. v. 1– , 1948– . Chicago: University of Chicago Press. Monthly. $275.00. ISSN 0002-9297.

A "record of research and review relating to heredity in humans". The journal of the American Society of Human Genetics.

Animal Genetics. v. 1– , 1970– . Olney Mead, UK: Blackwell Scientific. Bimonthly. $250.00. ISSN 0268-9146.

"The official Journal of the International Society for Animal Genetics. The Journal covers the fields of immunogenetics, biochemical genetics and molecular genetics." Full-length articles, short communications, mini-reviews, and technical advances are included.

Annales de Génétique. v. 1– , 1958– . Expansion Scientifique Française. Quarterly. $283.00. ISSN 0003-3995.

Specializes in human and medical genetics.

Annals of Human Genetics. v. 1– , 1925– . Cambridge University Press. Quarterly. $179.00. ISSN 0003-4800.

"Publishes material directly concerned with human genetics or the application of scientific principles and techniques to any aspect of human inheritance."

Applied Biochemistry and Biotechnology. v. 1– , 1976– . Clifton, NJ: Humana Press. Semi-monthly. $575.00. ISSN 0273-2289.

"Areas emphasized in the journal are genetic engineering; enzyme technology; monoclonal and tissue culture technology; immobilized biochemicals, cells, cell organelles, and bacterial and their applications; fermentation technology; bioenergy; solid-phase synthesis assay; and characteristics of antibodies, receptors, RNA, DNA, proteins, and other biologically significant compounds; and engineering and scale-up studies." Includes reviews, lists of patents, and news items as well.

Applied Microbiology and Biotechnology. v. 1– , 1975– . Berlin:
    Springer-Verlag. Monthly. $1,597.00. ISSN 0175-7598.

Publishes short papers and mini-reviews in areas of applied microbiology and
biotechnology, including applied genetics, food biotechnology, and environmental biotechnology. Theoretical papers are excluded.

Behavior Genetics. v. 1– , 1970– . New York: Plenum. Bimonthly.
    $350.00. ISSN 0001-8244.

"A journal dealing with the inheritance and evolution of behavioral characters
in man and other species." Also publishes critical reviews.

Biochemical Genetics. v. 1– , 1967– . New York: Plenum. Bimonthly.
    $350.00. ISSN 0006-2928.

"A journal for the reporting of original research in biochemical genetics of any
organism, from virus to man. Papers will deal with the molecular aspects of
genetic variation and evolution, mutation, gene action and regulation, immunogenetics, somatic cell genetics, and nucleic acid function."

Biotechniques. v. 1– , 1983– . Nantick, MA: Eaton Publishing Co.
    Monthly. $95.00. ISSN 0736-6205.

Bio/Technology: The International Monthly for Industrial Biology. v. 1– ,
    1983– . New York: Nature Publishing. Monthly. $113.00. ISSN
    0733-222X.

Published by the same company as *Nature*, with a similar mix of reviews, news,
articles, and research papers dealing with biotechnology.

Biotechnology Advances. v. 1– , 1983– . New York: Pergamon. Quar-
    terly. $345.00. ISSN 0734-9750.

Covers patent abstracts, government reports on biotechnology, and reviews in
"all areas of biotechnology including relevant aspects of its disciplinary under-
pinnings in biology, chemistry and engineering." Updates Pergamon's *Comprehensive Biotechnology* [see Moo-Young (p. 48) in Dictionaries and Encyclopedias section].

Biotechnology and Applied Biochemistry. v. 1– , 1979. London:
    Portland Press. Bimonthly. $170.00. ISSN 0885-4513.

"An international journal devoted to the publication of papers presenting original results, review articles, and other features in the field of applied biochemistry. Results of fundamental studies directly related to the description of
new biotechnology, the improvement of existing biochemical processes, and the
preparation, utility, conversion, or application of biological materials will also
be published."

Biotechnology and Bioengineering. v. 1– , 1958– . New York: John Wiley and Sons. Semi-monthly. $1,190.00. ISSN 0006-3592.

Covers "original articles, reviews, and mini-reviews that deal with all aspects of applied biotechnology."

Biotechnology Letters. v. 1– , 1979– . Northwood, Middlesex, UK: Science and Technology Letters. Monthly. $300.00. ISSN 0141-5492.

"The international monthly for the rapid publication of new results in all aspects of process biotechnology."

Biotechnology Progress. v. 1– , 1985– . Washington, DC: American Chemical Society. Bimonthly. $325.00. ISSN 8756-7938.

"Top areas of interest include application of chemical and engineering principles in fields such as kinetics, transport phenomena, control theory, modeling, and material science to phenomena in areas such as molecular biology, genetics, biochemistry, cellular biology, physiology, applied microbiology, and food science." Also includes short reviews.

Biotechnology Techniques. v. 1– , 1987– . Northwood, Middlesex, UK: Science and Technology Letters. Bimonthly. $160.00. ISSN 0951-208X.

"The journal of rapid publication and permanent record for methods and techniques that are new and generally useful for biotechnology in all its aspects." Published in conjunction with *Biotechnology Letters.*

Chromosoma. v. 1– , 1939– . Heidelberg, Germany: Springer-Verlag. Monthly. $1,123.00. ISSN 0009-5931.

Covers eukaryotic chromosome structure and function, nuclear organization and function, molecular biology of eukaryotic genomes, and reviews.

Critical Reviews in Biotechnology. v. 1– , 1983– . Boca Raton, FL: CRC Press. Quarterly. $325.00. ISSN 0738-8551.

"Provides a forum of critical evaluation of recent and current publications and, periodically, for state-of-the-art reports from various geographic areas around the world." Covers biotechnological techniques for all areas, including academe and industry.

Critical Reviews in Eukaryotic Gene Expression. v. 1– , 1990– . Boca Raton, FL: CRC Press. Quarterly. $195.00. ISSN 1045-4403.

"Presents timely concepts and experimental approaches that are contributing to rapid advances in our understanding of gene regulation, organization, and structure. Provides detailed critical reviews of the current literature."

Current Genetics. v. 1– , 1980– . Berlin: Springer-Verlag. Monthly. $916.00. ISSN 0172-8083.

"Devoted to the rapid publication of original articles of immediate importance on genetics of eukaryotes, with emphasis on yeasts, other fungi, protists and cell organelles."

Cytogenetics and Cell Genetics. v. 1– , 1962– . Basel, Switzerland: S. Karger AG. Monthly. $960.00. ISSN 0301-0171.

3 volumes per year. "Original research reports in mammalian cytogenetics, molecular genetics including gene cloning and sequencing, gene mapping, cancer genetics, comparative genetics, gene linkage, and related areas."

Developmental Genetics. v. 1– , 1979– . New York: Wiley-Liss. Bi-monthly. $296.00. ISSN 0192-253X.

Covers "original research devoted to explorations of gene function and gene regulation during biological development. The journal welcomes contributions from researchers working on developmental genetics of microbes, plants, or animals, and utilizing genetical, molecular, physiological, or morphological approaches." Includes review articles.

DNA and Cell Biology. v. 1– , 1981– . New York: Mary Ann Liebert. Monthly. $310.00. ISSN 1044-5498.

Publishes papers and reviews in "any subject dealing with eukaryotic or pro-karyotic gene structure, organization, expression, or evolution." Formerly *DNA*.

Environmental and Molecular Mutagenesis. v. 1– , 1979– . New York: Wiley-Liss. Bimonthly. $260.00. ISSN 0893-6692.

"Publishes original research papers on mutation and mutation-related topics." Review articles also included. The journal of the Environmental Mutagen Society.

Enzyme and Microbial Technology: Biotechnology Research and Reviews. v. 1– , 1979– . Stoneham, MA: Butterworth-Heinemann. Monthly. $580.00. ISSN 0141-0229.

Includes papers "relevant to the directed use of enzymes and microorganisms. The term microorganism is considered to include bacteria, fungi, algae, protozoa, viruses, animal cells and plant cells." Also includes short communications and reviews.

Gene. v. 1– , 1977– . Amsterdam: Elsevier/North Holland. Biweekly. $3,328.00. ISSN 0378-1119.

"An international journal focusing on gene cloning and gene structure and function."

Genes and Development. v. 1– , 1987– . Cold Spring Harbor, NY: Cold Spring Harbor Laboratory Press. Monthly. $385.00. ISSN 0890-9369.

Publishes research articles in molecular biology, molecular genetics, and related areas.

Genetica. v. 1– , 1919– . Dordrecht: Kluwer. Bimonthly. $437.00. ISSN 0016-6707.

"Rapid publication of full-length papers and short communications describing the results of original research in genetics and related scientific disciplines." Occasionally publishes special issues based on a particular theme.

Genetic Analysis: Techniques and Applications. v. 1– , 1984– . Amsterdam: Elsevier/North Holland. Bimonthly. $165.00. ISSN 1052-3862.

Publishes "new methods, materials, and instruments for molecular biology, cell biology, biochemistry and genetics. Recent developments in gene cloning and nucleic acid analysis are emphasized."

Genetical Research. v. 1– , 1960– . Cambridge, UK: Cambridge University Press. Bimonthly. $259.00. ISSN 0016-6723.

Covers "all aspects of genetics, or in any field of research which has an important bearing on genetics." Also includes some reviews.

Genetic Engineering: Principles and Methods. v. 1– , 1979– . New York: Plenum Press. Irregular. Price varies. ISSN 0196-3716.

"The purpose of these volumes is to follow closely the explosion of new techniques and information that is occurring as a result of the newly-acquired ability to make particular kinds of precise cuts in DNA molecules."

Genetics. v. 1– , 1916– . Bethesda, MD: Genetics Society of America. Monthly. $240.00. ISSN 0016-6731.

"A periodical record of investigation into heredity and variation." The journal of the Genetics Society of America.

Genetics, Selection, Evolution. v. 1– , 1969– . Paris: Elsevier. Bimonthly. $227.00. ISSN 0999-193X.

"Open to original research papers in fields of animal and evolutionary genetics. Areas of interest include cytogenetics, biochemistry and factorial genetics, genetic analysis of natural or experimental populations, quantitative genetics and animal breeding."

Genetika. v. 1– , 1965– . Moscow: Nauka Science Publishers. Monthly. $311.00. ISSN 0016-6758.

Also, see *Soviet Genetics* (p. 168).

Genome. v. 1– , 1969– . Ottawa: National Research Council of
 Canada. Bimonthly. $219.00. ISSN 0831-2796.

"Publishes, in English or French, results from research in transmission and population genetics and research in mechanisms of inheritance and evolution at the molecular, chromosomal, and cellular level." The principle medium for the publication of scientific papers for the Genetics Society of Canada.

*Genomics: International Journal for Analyses of the Human and Other
 Genomes*. v. 1– , 1988– . San Diego, CA: Academic Press.
 Monthly. $600.00. ISSN 0888-7543.

"The journal emphasizes molecular cloning and sequencing of mammalian genes and their regulatory elements, mapping their location in the human genome, comparative mapping in key experimental organisms, construction of large-scale integrated genetic and physical maps, and understanding the role of individual genes in human disease." Includes short communications, full-length articles, and reviews.

Hereditas. v. 1– , 1920– . Lund, Sweden: Mendelian Society of Lund.
 Bimonthly. $164.00. ISSN 0018-0661.

"*Hereditas* is a journal for the publication of original research in genetics." Most contributors are European.

Heredity. v. 1– , 1947– . Oxford, UK: Blackwell Scientific. Monthly.
 $305.00. ISSN 0018-067X.

"Publishes articles in all areas of genetics, focusing on the genetics of eukaryotes. The traditional strengths of the journal lie in the fields of ecological and population genetics; biometrical and statistical genetics; animal and plant breeding; and cytogenetics." Published for the Genetical Society of Great Britain.

Human Genetics. v. 1– , 1964– . New York: Springer-Verlag.
 Monthly. $2,540.00. ISSN 0340-6717.

Includes review articles, research articles, short communications, and case reports dealing with human genetics. Topics covered include immunogenetics, cytogenetics, biochemical genetics, population genetics, and genetic diagnosis and counseling. A rare genetic variant register is also included.

Human Heredity: International Journal of Human and Medical Genetics.
 v. 1– , 1950– . Basel: S. Karger. Bimonthly. $276.00. ISSN
 0001-5652.

"Publishes papers reporting on original investigations in the field of human and medical genetics."

International Biodeterioration and Biodegradation. v. 1– , 1965– .
Barking, UK: Elsevier. Monthly. $472.00. ISSN 0964-8305.

"Original research papers and reviews on biological causes of deterioration or degradation." The official journal of the Biodeterioration Society and groups affiliated to the International Biodeterioration Association.

Japanese Journal of Genetics (Idengaku Zasshi). v. 1– , 1921– .
Tokyo: Japan Scientific Societies Press. Bimonthly. $99.00. ISSN 0021-504X.

Journal of Applied Bacteriology. v. 1– , 1938– . Oxford, UK: Blackwell Scientific Publications. Monthly. $562.00. ISSN 0021-8847.

Covers "all aspects of applied microbiology e.g. agriculture, biotechnology, environment, food, genetics, medical and veterinary, pharmaceutical, taxonomy, soil, systematic bacteriology and water," including review articles. The journal of the Society for Applied Bacteriology.

Journal of Biotechnology. v. 1– , 1984– . Amsterdam: Elsevier.
Monthly. $1,350.00. ISSN 0168-1656.

"Provides a medium for the rapid publication of both full-length articles and short communications on all aspects of biotechnology . . . . Papers presenting information of a multi-disciplinary nature, that would not be suitable for publication in a journal devoted to a single discipline, are particularly welcome."

Journal of Chemical Technology and Biotechnology. v. 1– , 1951– .
Barking, UK: Elsevier. Monthly. $590.00. ISSN 0268-2575.

"Concerned primarily with studies related to the conversion of scientific discoveries into products and processes in the areas of biotechnology and chemical technology," especially areas such as safety and environmental issues, fermentation, biodegradation, and industrial applications of recombinant DNA.

Journal of Genetics. v. 1– , 1910– . Bangalore: Indian Academy of Sciences. $75.00. ISSN 0022-1333.

Three issues per year.

Journal of Heredity. v. 1– , 1910– . New York: Oxford University Press. Bimonthly. $112.00. ISSN 0022-1503.

Research and review articles in "organismic genetics." The official journal of the American Genetic Association.

Journal of Medical Genetics. v. 1– , 1964– . London: BMJ Publishing
    Group. Monthly. $262.00. ISSN 0022-2593.

"Publishes original research on all areas of medical genetics, along with reviews, annotations, and editorials on important and topical subjects." The audience includes clinical geneticists and researchers in a variety of basic genetical disciplines.

Letters in Applied Microbiology. v. 1– , 1985– . Oxford, UK: Blackwell
    Scientific Publications. Monthly. Free. ISSN 0266-8254.

"A journal for the rapid publication of short papers (up to six typed pages) of high scientific standard in the broad field of applied microbiology . . . . Advances in rapid methodology will be a particular feature." A free companion publication to the *Journal of Applied Bacteriology*, available only with a subscription to that journal.

Molecular and General Genetics. v. 1– , 1908– . Berlin: Springer-
    Verlag. Monthly. $2,194.00. ISSN 0026-8925.

Continues *Zeitschrift fur Vererbungslehre.* "Provides publication in all areas of general and molecular genetics—developmental genetics, somatic cell genetics, and genetic engineering—irrespective of the organism." "The first journal on genetics."

Molecular Biotechnology. v. 1– , 1994– . Totowa, NJ: Humana Press.
    Bimonthly. $150.00. ISSN 1073-6085.

"Committed to the rapid publication of detailed laboratory protocols, original research papers on the use of these protocols in both basic and applied research, regular review articles, hints and tips, book reviews, software updates, new product reviews, and more."

Mutation Research. Amsterdam: Elsevier. $4011.00. ISSN 0921-8262.

In seven parts, each of which is available separately or as part of the whole. The sections are:

DNAging: Genetic Instability and Aging. v. 1– , 1989– . Bimonthly.
    $214.00. ISSN 0921-8734.

DNA Repair. v. 1– , 1983– . Bimonthly. $428.00. ISSN 0921-
8777.

*Environmental Mutagenesis and Related Subjects including Method
    ology.* v. 1– , 1988– . Bimonthly. $428.00. ISSN 0921–

Genetic Toxicology Testing. v. 1– , 1988– . Monthly. $641.50.
    ISSN 0165-1218.

Mutation Research Letters. v. 1– , 1988– . Monthly. $641.50. ISSN 0165-7992.

Reviews in Genetic Toxicology. v. 1– , 1988– . Bimonthly. $428.00. ISSN 0165-1110.

Nature Genetics. v. 1– , 1992– . Washington, DC: Macmillan Magazines, Ltd. Monthly. $495.00. ISSN 1061-4036.

Covers all areas of genetics, especially the human genome project and genetic diseases.

Oncogene: An International Journal. v. 1– , 1987– . Avenel, NJ: Macmillan Press. Monthly. $830.00. ISSN 0950-9232.

Publishes "full and detailed papers as well as short communications relevant to all aspects of oncogene research including the following topics: cellular oncogenes and their mechanisms of activation, structure and functional aspects of their encoded proteins, oncogenes in RNA and DNA tumour viruses, presence of oncogenes in human tumours, relevance and biology, cell cycle control, immortalisation, cellular senescence, regulatory genes and 'antioncogenes', growth factors and receptors."

Plant Cell, Tissue and Organ Culture: An International Journal on the Cell Biology of Higher Plants. v. 1– , 1981– . Dordrecht, the Netherlands: Kluwer Academic. Monthly. $180.00. ISSN 0167-6857.

"Publishes original results of fundamental studies on the behavior of plant cells, tissues and organs in vitro" including biotechnology and genetics.

Plasmid. v. 1– , 1977– . San Francisco: Academic Press. Bimonthly. $186.00. ISSN 0147-619X.

"Reports of original research into the biology of extrachromosomal gene systems and mobile genetic elements in both prokaryotic and eukaryotic organisms including their biological behavior, their molecular structure and genetic function, their gene products, and their use as genetic tools." Also includes short review articles.

Somatic Cell and Molecular Genetics. v. 1– , 1975– . New York: Plenum. Bimonthly. $425.00. ISSN 0740-7750.

Publishes in "cellular and molecular genetics of higher eukaryotic systems. The primary emphasis is on studies with animal or plant cells in the following areas: gene expression and regulation, gene transfer into cultured cells or embryos, gene isolation, gene mapping, gene therapy, molecular biology of inherited diseases, recombination, mutation, chromosome replication, and the genetics of subcellular organelles." Formerly *Somatic Cell Genetics*.

Soviet Genetics. v. 1– , 1966– . New York: Plenum. Monthly. $1.175.00. ISSN 0038-5409.

An English translation of the Russian journal *Genetika* (see p. 163). The translation appears about six months after the original Russian issue and has the same number and date as the Russian issue.

Teratogenesis, Carcinogenesis, and Mutagenesis. v. 1– , 1980– . New York: Wiley-Liss. Bimonthly. $324.00. ISSN 0270-3211.

"Publishes original research on the evaluation and characterization of teratogens, carcinogens, or mutagens." Also includes review articles.

Theoretical and Applied Genetics. v. 1– , 1929– . Berlin: Springer-Verlag. Monthly. $1,897.00. ISSN 0040-5752.

"Original articles in the following areas: general fundamentals of plant and animal breeding; physiological fundamentals of plant and animal breeding; applications of cell genetics to breeding."

Newsletters

Biotechnology News. v. 1– , 1981– . Maplewood, NJ: CTB International Publishers. 30/yr. $498. ISSN 0273-3226.

Covers the biotechnology industry, including company news, regulations, trends, etc.

DNA and Protein Engineering Techniques. v. 1– , 1988– . New York: Wiley-Liss. Bimonthly. $60.00. ISSN 0894-7937.

"Provides rapid exchange of information on new and improved experimental methods in molecular genetics and protein engineering . . . . The newsletter is a forum for the submission and circulation of short descriptions of new techniques, special tricks, and improvements of techniques."

Genetic Engineering News. v. 1– , 1981– . New York: Mary Ann Liebert. Biweekly. $190.00. ISSN 1270-6377.

"*Genetic Engineering News* serves its readers as a forum for the discussion of issues related to biotechnology including the publication of minority and conflicting points of view, rather than only presenting the majority view." Includes feature articles, company news, academic programs, and a calendar of events.

Industrial Bioprocessing: A Monthly Intelligence Service. v. 1– , 1979– . Englewood, NJ: Technical Insights. Monthly. $505.00. ISSN 1056-7194.

Provides industry news, new products, patents, events, and market forecasts.

Reviews of the Literature

Advances in Applied Microbiology. v. 1– , 1959– . New York:
    Academic Press. Annual. Price varies. ISSN 0065-2164.
See annotation (p. 206) in Chapter 8, "Microbiology and Immunology."

Advances in Biochemical Engineering/Biotechnology. v. 1– , 1972– .
    New York: Springer-Verlag. Irregular. Price varies. ISSN 0724-
    6145.
Each volume has review articles on a particular topic in biotechnology. Recent
topics include bioseparation, modern biochemical engineering, and applied
molecular genetics.

Advances in Biotechnological Processes. v. 1– , 1985– . New York:
    Wiley-Liss. Annual. Price varies. ISSN 0736-2293.
Review articles covering areas in biotechnology, such as upstream and
downstream processes, monoclonal antibodies, waste treatment, and bacterial
vaccines.

Advances in Genetics. v. 1– , 1947– . New York: Academic Press.
    Annual. Price varies. ISSN 0065-2660.
A series of review articles presenting critical summaries of outstanding genetic
problems, both theoretical and practical.

Advances in Human Genetics. v. 1– , 1970– . New York: Plenum.
    Irregular. Price varies. ISSN 0065-275X.
A series of critical reviews covering methodologies and results from various
disciplines which relate to human genetics.

Annual Review of Genetics. v. 1– , 1967– . Palo Alto, CA: Annual Re-
    views. Annual. $44.00. ISSN 0066-4197.
Qualified authors are invited to contribute "critical articles reviewing significant
developments" in genetics.

Biotechnology and Genetic Engineering Reviews. v. 1– , 1984– . New-
    castle-upon-Tyne, UK: Intercept. Annual. $147.25. ISSN
    02648725.
Contains original review articles covering industrial, agricultural, and medical
biotechnology, especially genetic engineering.

Current Opinion in Biotechnology. v. 1– , 1990– . Philadelphia, PA:
    Current Biology. Bimonthly. $575.00. ISSN 0958-1669.
"Reviews of all advances, evaluation of key references, comprehensive listing

of papers and patents." A combination of a review journal and abstracting service for biotechnology. Each issue covers a particular topic such as analytical biotechnology or protein engineering.

Current Opinion in Genetics and Development. v. 1– , 1989– .
   Philadelphia, PA: Current Biology. Bimonthly. $390.00. ISSN 0958-1669.

A companion to *Current Opinion in Biotechnology* (p. 169). Topics covered include genomes and evolution, viral genetics, and pattern formation.

Oxford Surveys on Eukaryotic Genes. v. 1– , 1984– . New York: Oxford University Press. Annual. $50.00. ISSN 0265-0738.

Provides "a forum for authoritative reviews of particular genes or gene families . . . whose structure or function is better understood as a result of recent experimental results."

Progress in Nucleic Acid Research and Molecular Biology. v. 1– , 1963– . San Diego: Academic Press. Irregular. Price varies. ISSN 0079-6603.

Contains review articles covering a variety of topics.

Trends in Biotechnology. v. 1– , 1983– . Cambridge, UK: Elsevier. Monthly. $490.00. ISSN 0167-7799.

"One of the most widely cited journals in biotechnology . . . focuses on innovative biotechnology R&D, identifying and reviewing key trends in a lively and readable style." Also includes upcoming meetings.

Trends in Genetics. v. 1– , 1985– . Cambridge, UK: Elsevier. Monthly. $490.00. ISBN 0168-9525.

Publishes review articles in areas of genetics of current interest; also includes upcoming meetings.

Societies

American Genetic Association
   PO Box 39, Buckeystown, MD 21717-0039.

Founded: 1903. 1,500 members in 1993. Emphasis on applied areas. Publications: *Journal of Heredity.*

American Society of Human Genetics
   9650 Rockville Pike, Bethesda, MD 20814-3998.

Founded: 1948. 4,200 members in 1993. Physicians, genetic councelors, researchers interested in human genetics. Publications: *American Journal of Human Genetics, Guide to Human Genetics Training Programs in North America,* biennial *Membership Directory.*

Genetics Society of America
	9650 Rockvill Pike, Bethesda, MD 20814-3998.

Founded: 1931. 3,700 members in 1993. All areas of genetics. Publications: *Genetics, Career Opportunities in Genetics,* biennial *Membership Directory.*

Texts/General Works

Assessing Ecological Risks of Biotechnology. Lev R. Ginzburg, ed.
	Boston: Butterworth-Heinemann, 1991. 379 p. $79.95. ISBN
	049901997.

Provides information on the concerns of scientists in assessing ecological risks. Covers the ecology of microbes, mathematical models in risk assessment, European biotechnology regulations, and more.

Ayala, Francisco J. and John A. Kiger, Jr. *Modern Genetics,* 2nd ed.
	Menlo Park, CA: Benjamin/Cummings Publishing Co., 1984. 923 p.
	$50.50. ISBN 0805303162.

Becker, Jeffrey M., Guy A. Caldwell, and Eve Ann Zachgo. *Biotechnology: A Laboratory Course.* San Diego, CA: Academic Press,
	1990. 232 p. (paper). ISBN 0120845601.

This is a manual for a laboratory course in biotechnology for upper-level undergraduate and graduate students, introducing techniques such as Southern transfer, gel electrophoresis, and various assays.

Benchmark Papers in Genetics. Stroudsburg, PA: Hutchinson Ross.

The series editors have selected "papers and portions of papers that demonstrate both the development of knowledge and the atmosphere in which that knowledge was developed." Each individual selection is critiqued by the volume editor.

	v. 1: Ballonoff, P. A., Compiler. *Genetics and Social Structure: Mathematical Structuralism in Population Genetics and Social Theory.* 1974.

	v. 2: Wagner, R. P., Editor. *Genes and Proteins.* 1975.

	v. 3: Weiss, K. M. and P. A. Ballonoff, Editors. *Demographic Genetics.* 1975.

v. 4: Drake, J.W. and R. E. Koch, Editors. *Mutagenesis*. 1976.

v. 5: Bajema, C. J., Editor. *Eugenics: Then and Now*. 1976.

v. 6: Phillips, R. L. and C. R. Burnham, Editors. *Cytogenetics*. 1977.

v. 7: Li, W. H., Editor. *Stochastic Models in Population Genetics*. 1977.

v. 8: Jameson, D. L., Editor. *Evolutionary Genetics*. 1977.

v. 9: Jameson, D. L., Editor. *Genetics of Speciation*. 1977.

v. 10: Schull, W. J. and R. Chakraborty, Editors. *Human Genetics*. 1979.

v. 11: Levin, D. A. *Hybridization: An Evolutionary Perspective*. 1979.

v. 12: Jackson, R. C. and D. P. Huber, Editors. *Polyploidy*. 1983.

v. 13: Milkman, R., Editor. *Experimental Population Genetics*. 1983.

v. 14: Davidson, R. L., Editor. *Somatic Cell Genetics*. 1984.

v. 15: Hill, W. G., Editor. *Quantitative Genetics. Part I: Explanation and Analysis of Continuous Variation. 1984.*

v. 16: Will, W. G., Editor. *Quantitative Genetics. Part II: Selection. 1984.*

v. 17: Holliday, R., Editor. *Genes, Proteins, and Cellular Aging*. 1986.

Berg, Paul and Maxine Singer. *Dealing with Genes: The Language of Heredity*. Mill Valley, CA: University Science Books, 1992. 288 p. $34.00. ISBN 0935702695.

Cook, L. M. *Genetic and Ecological Diversity: The Sport of Nature*. New York: Chapman and Hall, 1991. 192 p. $35.00 (paper). ISBN 0412356201 (paper).

This slim volume is an introduction to population genetics, and attempts to bring together the work of ecologists and population geneticists on within- and between-species diversity.

Crueger, Wulf and Anneliese Crueger. *Biotechnology: A Textbook of Industrial Microbiology*, 2nd ed. Sunderland, MA: Sinauer, 1990. 368 p. $44.95. ISBN 0878931317.

Drlica, K. *Understanding DNA and Gene Cloning*, 2nd ed. New York: Wiley, 1992. 262 p. $31.95 (paper). ISBN 0471622257 (paper).

Textbook for undergraduates and non-scientists.

Fincham, J. R. S. and J. R. Ravetz. *Genetically Engineered Organisms*.

Milton Keynes, UK: Open University Press, 1991. 158 p. ISBN
0335096190, 0335096182 (paper).

A more European viewpoint of the potential problems of releasing genetically
engineered organisms into the environment. Has overviews of DNA manipulation, the use of engineered organisms in industry and agriculture, and a
discussion of risk assessment.

Gardner, Eldon John, Michael J. Simmons, and D. Peter Snustad. *Principles of Genetics*, 8th ed. New York: Wiley, 1991. 649 p. $82.95,
$40.50 (paper). ISBN 0471504874; 0471504874 (paper).

Gillespie, John H. *The Causes of Molecular Evolution*. New York: Oxford
University Press, 1991. 336 p. $35.00. ISBN 0195068831.

Covers population genetics and the evolution of proteins and DNA.

Glick, Bernard R. and Jack J. Pasternak. *Molecular Biotechnology: Principles and Applications of Recombinant DNA*. Washington, DC: ASM
Press, 1994. 500 p. $39.95 (paper). ISBN 1555810713 (paper).

Advanced undergraduate and graduate text, covering both scientific principles
and applications of biotechnology.

Gonick, Larry and Mark Wheelis. *The Cartoon Guide to Genetics*. New
York: Harper Perennial, 1991. 214 p. $12.00 (paper). ISBN
0062730991.

Certainly one of the most unusual genetics texts; has understandable and
scientifically accurate information in cartoon format.

Goodenough, Ursula. *Genetics*, 3rd ed. Philadelphia: Saunders College
Publishing, 1984. 894 p. $39.00. ISBN 0030582121.

Griffiths, Anthony J. F. and Joan McPherson. *100+ Principles of
Genetics*. W. H. Freeman, 1989. 387 p. $18.95 (paper). ISBN
0716720167 (paper).

Designed as an easily understood overview of genetics. Each principle is
discussed in 1–2 pages, with cross-references to other principles.

Hartl, Daniel L. *Primer of Population Genetics*, 2nd ed. Sunderland,
MA Sinauer, 1987. 305 p. $23.95 (paper). ISBN 0878933018
(paper).

Designed as a text for students and workers in areas such as wildlife management or anthropology who need a basic understanding of the principles of
population genetics. Not intended as a text for use in population genetics
courses. See Hartl and Clark (p. 174) for a population genetics text.

Hartl, Daniel L. and Andrew G. Clark. *Principles of Population Genetics*, 2nd ed. Sunderland, MA: Sinauer, 1989. 682 p. $46.50. ISBN 0878933026.

Joset, F., J. Guespin-Michel, and L. Butler. *Prokaryotic Genetics: Genome Organization, Transfer and Plasticity*. Oxford, England: Blackwell Scientific, 1993. 464 p. ISBN 0632027282.

Kammermeyer, Karl and Virginia L. Clark. *Genetic Engineering Fundamentals: An Introduction to Principles and Applications*. New York: Marcel Dekker, Inc., 1989. 290 p. $125.00. ISBN 0824780698.

Kornberg, Arthur and Tania A. Baker. *DNA Replication*, 2nd ed. New York: W. H. Freeman and Co., 1991. 931 p. $64.95. ISBN 0716720035.

Levitan, Max. *Textbook of Human Genetics*, 3rd ed. New York: Oxford University Press, 1988. 475 p. $42.50. ISBN 0195049357.

Lewin, Benjamin. *Genes V*. New York: Oxford University Press, 1994. 1272 p. $65.00. ISBN 0198542879.

Mange, Arthur P. and Elaine Johansen Mange. *Genetics: Human Aspects*, 2nd ed. Sunderland, MA: Sinauer, 1990. 591 p. $47.95. ISBN 0878935010.
Covers basic genetics as well as human genetics.

Marx, Jean L., ed. *A Revolution in Biotechnology*. New York: Published for the International Council of Scientific Unions by Cambridge University Press, 1989. 227 p. $44.50. ISBN 0521327490.
Intended for the general public. The chapters are written by experts and cover a number of areas, including nitrogen fixation, improving crop plants, detecting genetic diseases, and regulation.

Milestones in Biotechnology: Classic Papers on Genetic Engineering. Julian Davies and William S. Reznikoff, eds. Boston: Butterworth-Heinemann, 1992. 570 p. $59.95 (paper). ISBN 0750692510 (paper).
Provides the full text of major papers in the history of genetic engineering and biotechnology.

Modern Microbial Genetics. Edited by U. N. Streips and R. E. Yasbin. New York: Wiley, 1991. 548 p. $88.95. ISBN 0471568457.
Graduate-level textbook.

Molecular Biology and Biotechnology. Edited by Chris A. Smith and E. J.

Wood. New York: Chapman and Hall, 1991. 247 p. $32.95
(paper). ISBN 0412407507.

Covers mostly role of DNA and RNA, including transcription, induction and repression, and mutation. Also includes recombinant DNA technology and cell culture; little direct discussion of biotechnology.

Mooney, Harold A. and Giorgio Bernardi, eds. *Introduction of Genetically Modified Organisms into the Environment.* New York: Wiley, 1990. (SCOPE Report 44). 201 p. $110.00. ISBN 0471926779.

These papers are the result of a meeting to consider issues relating to the release of genetically engineered organisms. Topics include a historical overview, possible consequences, and assessment and regulation.

Nicholl, Desmond S. T. *An Introduction to Genetic Engineering.* New York: Cambridge University Press, 1994. (Studies in biology). 168 p. $39.95. ISBN 0521430542, 0521436346 (paper).

Introductory text, suitable for undergraduates and high school students.

Old, R. W. *Principles of Gene Manipulation: An Introduction to Genetic Engineering.* 5th ed. Boston: Blackwell Scientific Publications, 1994. 474 p. $65.00. ISBN 0632037121.

For advanced undergraduates and graduate students.

Plant Population Genetics, Breeding, and Genetic Resources. A. H. D. Brown et al., eds. Sunderland, MA: Sinauer, 1990. 449 p. $65.00, $39.95 (paper), ISBN 0878931163, 0878931171 (paper).

"Evolved from the International Symposium on Population Genetics and Germplasm Resources in Crop Improvement." Three sections, the first discussing the kinds of genetic diversity found in plants, the second discussing genetic variation and microevolutionary processes, and the third discussing the application of plant population genetics to practical problems. There is an extensive bibliography.

Pollack, Robert. *Signs of Life: The Language and Meanings of DNA.* Boston: Houghton Mifflin, 1994. 212 p. $19.95. ISBN 0395644984.

An introduction to modern genetics written for non-scientists, based on the extended metaphor of the genome as book, with chapter headings such as "The Molecular Word Processor." Very well written and easily understood, covering both scientific and ethical aspects of genetics.

Rothwell, Norman. *Understanding Genetics: A Molecular Approach.* New York: Wiley-Liss, 1993. 672 p. $39.95. ISBN 0471594156.

The fourth edition of this text was published by Oxford University Press in 1988.

Spiess, E. B. *Genes in Populations*, 2nd ed. New York: John Wiley and Sons, 1990. 790 p. $95.00. ISBN 0471849731.

Strickberger, Monroe W. *Genetics*, 3rd ed. New York: Macmillan Publishing Co., 1985. 842 p. ISBN 002418120X.

An undergraduate text; includes problems in each chapter. An *Answer Manual for Genetics, Third Edition* is also available.

Suzuki, David T. and Peter Knudtson. *Genethics: The Ethics of Engineering Life*. Cambridge, MA: Harvard University Press, 1990. 372 p. $12.95 (paper). ISBN 0674345665.

Discusses the ethical aspects of genetic engineering, with emphasis on human genetic engineering. Includes case studies such as the link between XYY males and aggression (now discredited).

Wagner, Robert P., Marjorie P. Maguire, and Raymond L. Stallings. *Chromosomes: A Synthesis*. New York: Wiley-Liss, 1993. 458 p. $104.00. ISBN 047156124X.

For upper-level undergraduates and graduate students.

Watson, James D. et al. *Molecular Biology of the Gene*, 4th ed. Menlo Park, CA: Benjamin/Cummings Publishing Co., 1987. 2 v. $52.75 (v. 1), $41.95 (v. 2). ISBN 0805396128 (v. 1), 0805396136 (v. 2).

Volume 1 covers general principles (history, DNA structure and replication, regulation, etc); volume 2 covers specialized aspects (gene function in specialized systems, cancer, evolution of the gene). Designed as an undergraduate text and reference.

Watson, James D., Jan Witkowski, Michael Gilman, and Mark Zollner. *Recombinant DNA*, 2nd ed. New York: Scientific American Books, 1992. 626 p. $49.95. ISBN 0716719940, 0716722828 (paper).

Designed for undergraduate and graduate students and others interested in recombinant DNA; more of a resource than a text. Includes many color illustrations.

Wegner, Harold C. *Patent Law in Biotechnology, Chemicals and Pharmaceuticals*. New York: Stockton Press, 1992. 540 p. $170.00. ISBN 1561590487.

Discusses patent law as it relates to biotechnology. Intended for non-experts in patent law, including lawyers, scientists and businessmen. Among the topics covered are international patent law, the patenting of living organisms and tissue cultures, and the patenting of DNA sequences.

Witt, Steven C. *Biotechnology, Microbes and the Environment*. San Francisco: Center for Science Information, 1990. 219 p.

Written in very simple language for non-scientists; discusses risks and benefits of biotechnology. Clearly intended to lower anxiety levels.

Statistical Genetics

Falconer, D. S. *Introduction to Quantitative Genetics*, 3rd ed. New York: Halstead Press, 1989. 438 p. $39.95 (paper). ISBN 0582016428 (paper).

Fisher, R. A. *Statistical Methods for Research Workers*, 14th ed. Riverside, NJ: Hafner, 1973. 362 p.

Mather, Kenneth and John L. Jinks. *Biometrical Genetics: The Study of Continuous Variation*, 3rd ed. New York: Chapman and Hall, 1982. 396 p. $62.50. ISBN 0412228904.

Nei, Masatoshi. *Molecular Evolutionary Genetics*. New York: Columbia University Press, 1987. 512 p. $86.00, $31.50 (paper). ISBN 0231063202, 0231063210 (paper).

Covers statistical methods for analyzing many types of genetic data.

Weir, Bruce S. *Genetic Data Analysis: Methods for Discrete Population Genetic Data*. Sunderland, MA: Sinauer, 1990. 377 p. ISBN 0878938710, 0878938729 (paper).

Bibliography

Courteau, Jacqueline. 1991. Genome databases, *Science* 254(5029): 201–207.

Fuchs, Rainer and Graham N. Cameron. 1991. Molecular biological databases: The challenge of the genome era, *Progress in Biophysical and Molecular Biology* 58(3): 215–245.

McGraw-Hill Dictionary of Scientific and Technical Terms, 4th ed. Sybil P. Parker, Editor-in-Chief. New York: McGraw-Hill, 1989. 2088 p. ISBN 0070452709.

Nucleic Acids Research. 1994. Database issue, 221(17).

8

Microbiology and Immunology

This chapter includes reference sources useful for microbiology and immunology. For purposes of this discussion, microbiology encompasses bacteria and viruses; fungi are included with plant biology in Chapter 10, "Plant Biology." Because microbial systems are convenient and effective for studying a whole range of life processes, there is a significant overlap between this chapter and Chapters 5–7, covering biochemistry and biophysics, molecular and cellular biology, and genetics, respectively. Although medical microbiology and immunology are not comprehensively discussed in this chapter, some basic materials are included that pertain to the study of pathogenic microbiology and diagnostic immunology.

For an interesting discussion of the microbiological literature, see "The Microbiology Literature—Languages of Publication and Their Relative Citation Impact" in *FEMS Microbiology Letters*, Dec. 15, 1992, v. 100 (1–30): 33–37. This study by Eugene Garfield and A. Welljamsdorof examined trends in the number of papers published annually in various languages in 78 microbiology journals indexed in *Science Citation Index*, 1981–1991. Results showed that English is the language of microbiological research and that the impact of English-language papers was greater than that of other languages by factors ranging from 2.4 to 14.4.

Abstracts and Indexes

Abstracting and indexing serials are annotated in Chapter 4 and general sources useful for both microbiologists and immunologists are annotated in Chapter 3. Basically, the three great abstracting/indexing serials for microbiology and immunology are *Biological Abstracts*, *Chemical Abstracts*, and *Index Medicus*. Also of special value are the Cambridge Scientific Abstracts titles listed below. All are available as part of the **Life Sciences Collection**, which is annotated (p. 78) in Chapter 4, "Abstracts and Indexes."

Cambridge Microbiology Abstracts, Section B: Bacteriology. v. 1– , 1966– . Bethesda, MD: Cambridge Scientific Abstracts. Monthly. $985/yr. ISSN 0300-8398.

This abstracting journal covers all areas of bacteriology, including taxonomy, genetics, medical and veterinary bacteriology, antimicrobials, and others.

Cambridge Microbiology Abstracts, Section C: Algology, Mycology, and Protozoology. v. 1– , 1972– . Bethesda, MD: Cambridge Scientific Abstracts. Monthly. $845/yr. ISSN 0301-2328.

Covers all aspects of the study of algae, fungi, protozoa, and lichens.

Immunolgy Abstracts. v. 1– , 1967 . Bethesda, MD: Cambridge Scientific Abstracts. Monthly. $945/yr. ISSN 0307-112X.

Covers studies in humans and animals, including molecular immunology, tumor immunology, histocompatability, and disorders of the immune system. See *Virology and AIDS Abstracts* (below) for more extensive coverage of AIDS.

Virology and AIDS Abstracts. v. 1– , 1967– . Bethesda, MD: Cambridge Scientific Abstracts. Monthly. $845/yr. ISSN 0896-5919.

Covers all aspects of virology, including studies on plants, animals, and humans. Also has comprehensive coverage of AIDS, including drug tests, transmission, molecular aspects, and immunology.

Classification, Nomenclature, and Systematics

Look in the Methods section (p. 193) for techniques relevant to classification/ systematics.

Approved Lists of Bacterial Names. Edited by V. B. D. Skerman et al., on behalf of the Ad Hoc Committee of the Judicial Commission of the International Committee on Systematic Bacteriology. [Amended ed.] Washington, DC: American Society for Microbiology, 1989. 188 p.

Reprinted from *International Journal of Systematic Bacteriology*; v. 30, 1980, with corrections. Available only as a set with Moore and Moore, also listed in this section (p. 182).

Bergey's Manual of Determinative Bacteriology, 9th ed. Edited by John G. Holt. Baltimore, MD: Williams & Wilkins, 1993. 500 p. $49.95. ISBN 0683006037.

Based on data in *Bergey's Manual of Systematic Bacteriology*, this manual also includes new genera and species, new combinations, and new taxa published through the Jan. 1992 issue of *International Journal of Systematic Bacteriology*. Contains information on shape and size, gram reaction, morphological features, motility and flagella, relation to oxygen, basic type of metabolism, carbon and energy sources, habitat and ecology.

Bergey's Manual of Systematic Bacteriology. Editor-in-Chief, John G. Holt. Baltimore, MD: Williams & Wilkins, 1989. 4 v. $370/set. ISBN 0683041126.

Definitive reference for classification, nomenclature, and identification of bacteria. Each generic listing includes antigenic structure, pathogenicity, ecology, enrichment and isolation procedures, maintenance procedures, methods for testing special characters, etc. Complemented by *Methods for General and Molecular Bacteriology* published by the American Society for Microbiology; see the Methods section (p. 197).

BIOSIS Register of Bacterial Nomenclature (BRBN). Philadelphia: BIOSIS. Available through a license agreement with BIOSIS for a basic lease fee, 1995: $5,160 revised continuously; tape updates are sent semiannually.

This is a computerized database containing details on 16,000 bacterial names, preferred synonyms and taxonomic levels; citations to original descriptions of bacterial names.

Classification and Nomenclature of Viruses: Fifth Report of the
   International Committee on Taxonomy of Viruses for Virology
   Division of the International Union of Microbiological Socie-
   ties/Division of the International Union of Microbiological
   Societies. Edited by R. I. B. Francki et al. New York: Springer-
   Verlag, 1991. (Archives of Viro-logy. Supplementum 2) 450 p.
   $69.00. ISBN 038782260.

Summarizes the decisions reached by the ICTV held at the International Congresses of Virology, 1984–1990.

Computer-Assisted Bacterial Systematics. Edited by Michael Goodfellow
   et al. Orlando: Academic Press, 1985. 443 p. (Special publications of the Society for General Microbiology, 15). $117.00.
   ISBN 0122896653.

Papers presented at a symposium entitled "Twenty-five years of numerical taxonomy" held at the University of Warwick in April 1983.

Fraenkel-Conrat, Heinz. *The Viruses: Catalogue, Characterization, and
   Classification*. New York: Plenum Press, 1985. 266 p. ISBN
   0306417669. Bibliographies.

List of well-established and studied viruses, in alphabetical order. Identified in taxonomic terms, illustrated by electron micrographs. Divided into sections for animal and plant viruses, and phages of prokaryotes.

Handbook of New Bacterial Systematics. Edited by Michael Goodfellow
   and A. G. O'Donnell. San Diego: Academic Press, 1993. 560 p.
   $155.00. ISBN 0122896726.

Comprehensive up-to-date review of concepts and identification methods for the "new" bacterial systematics. Extensively referenced, this handbook follows Sam Cowan's concept of taxonomic "trinity." It is divided into sections dealing with classification, nomenclature, and identification, presenting principles and applications of chemical, numerical and molecular methods.

*Illustrated Glossary of Protoctista. Vocabulary of the Algae, Api-
   complexa, Ciliates, Foraminifera, Microspora, Water Molds, Slime
   Molds, and Other Protoctists*. Edited by Lynn Margulis, Heather I.
   McKhann, and Lorraine Olendzenski. Boston: Jones and Bartlett,
   1993. 288 p. $50.00. ISBN 0867200812.

An abbreviated version of the *Handbook of Protoctista* [annotated (p. 188) in the Handbooks section]. The main body of this book lists a general glossary of terms, an alphabetical glossary of taxa at family and higher levels not contained in the original *Handbook*, and classifications from phyla to families.

International Code of Nomenclature of Bacteria; and, *Statutes of the International Committee on Systematic Bacteriology*; and, *Statutes of the Bacteriology and Applied Microbiology Section f the International Union of Microbiological Societies: Bacteriological Code*. 1990 revision. Edited by S. P. Lappage et al. Washington, DC: Published for the International Union of Microbiological Societies by the American Society for Microbiology, 1992. 189 p. $47.00. ISBN 155581039X.

"Approved by the Judicial Commission of the International Committee on Systematic Bacteriology, the International Committee on Systematic Bacteriology, the International Union of Microbiological Societies, and the Plenary Session of the International Congress of Bacteriology and Mycology, Osaka, Japan, September 1990."

International Journal of Systematic Bacteriology. v. 1– , 1951– . Birmingham, AL: American Society for Microbiology. Quarterly. $164/yr. ISSN 0020-7713. Also available in CD-ROM format, $200/yr.

The official journal of the International Committee on Systematic Bacteriology of the International Union of Microbiological Societies. This journal includes articles on the systematics of bacteria, yeasts, and yeastlike organisms, including taxonomy, nomenclature, identification, characterization, and culture preservation.

Moore, W. E. C. and Lillian V. H. Moore. *Index of the Bacterial and Yeast Nomenclatural Changes: Published in the International Journal of Systematic Bacteriology since the 1980 Approved Lists of Bacterial Names: (1 January 1980 to 1 January 1989)*. Washington, DC: American Society for Microbiology, 1989. 72 p. $55.00. ISBN 1555810144. Sold as a set with *Approved Lists . . .* (p. 180).

Stedman's Bergey's Bacteria Words. John G. Holt, Editor-in-Chief, et al. Baltimore: Williams & Wilkins, 1992. 354 p. $28.00. ISBN 068307945X.

Based in part on *Bergey's Manual of Systemic Bacteriology*, 1984–1989. Contains over 11,000 entries of validly published bacteria names, synonyms, and taxonomy.

Stedman's ICTV Virus Words. Edited by Charles H. Calisher and Claude M. Fauquet. Baltimore, MD: Williams & Wilkins, 1992. 271 p. $28.00. ISBN 0683079565.

5,500 entries of validly published virus names, taxonomic status, and accepted synonyms.

Dictionaries and Encyclopedias

Cruse, Julius M. and Robert E. Lewis. *Illustrated Dictionary of Immunology*. Boca Raton, FL: CRC, 1994. 352 p. $59.95. ISBN 84934557X.

Written for students, clinicians, and scientists, this dictionary provides a thorough treatment of contemporary immunological definitions, including 650 illustrations. An appendix supplies the latest cluster of differentiation designations on leukocyte surface markers.

Dictionary of Immunology, 3rd ed. Edited by W. J. Herbert et al. Oxford: Blackwell Scientific, 1985. 240 p. ISBN 0632009845.

Aim is to include terms of current immunological usage to satisfy the needs of any biologist, clinician or biochemist.

Encyclopedia of Immunology, Editor-in-Chief, Ivan M. Roitt. San Diego: Academic Press, 1992. 3 v. $475.00 (set). ISBN 0122267605.

Largest comprehensive reference source of current immunological knowledge available. Written by leaders in the field, the encyclopedia is arranged in a similar fashion to *Encyclopedia of Microbiology*.

Encyclopedia of Microbiology, Joshua Lederberg, Editor-in-Chief. San Diego: Academic Press, 1992. 4 v. $675.00 (set). ISBN 0122268903. Index.

Written by some of the world's leading scientists, this comprehensive encyclopedia covers traditional fields as well as the latest microbiological research. Each article provides a table of contents, a glossary of terms, an in-depth presentation of the topic, and a bibliography.

Encyclopedia of Virology. Edited by Robert G. Webster and Allan Granoff. San Diego: Academic Press, 1993. 3 v. $475.00 (set). ISBN 0122269608.

The format is similar to that of the *Encyclopedia of Microbiology* by the same publisher, and is the largest single reference source of current virological knowledge. It includes 270 articles by leading researchers appropriate for both general and specialist readers.

Hull, Roger et al. *Virology: Directory & Dictionary of Animal, Bacterial and Plant Viruses.* London: Macmillan Reference, 1989. 326 p. $80.00. ISBN 0333390636.

Names of animal, plant, and bacterial viruses, plus terms commonly used in the virological literature. Lists cell lines, describes chemical and techniques, includes equations and formulae.

Littrell, Helen E. *Immunologic and AIDS Word Book.* Springhouse, PA: Springhouse Corp., 1992. 166 p. ISBN 0874344751.

Terminology for immunology and AIDS.

Rosen, Fred S. et al. *Dictionary of Immunology*, reprinted with corrections. New York: Stockton Press, 1989. 223 p. $50.00. ISBN 0935859586.

Authoritative, comprehensive dictionary.

Singleton, Paul and Diana Sainsbury. *Dictionary of Microbiology and Molecular Biology*, 2nd ed. New York: Wiley, 1993. 1,019 p. $50.00. ISBN 0471940526.

Comprehensive and easy to use.

Directories

For other association directories, consult the subject sections of *Encyclopedia of Associations* and the Societies section (p. 211). Also, there are various Internet news groups and electronic bulletin boards for microbiologists and immunologists that provide a forum for discussion that include information about members, announcements of meetings, funding sources, and job opportunities.

American Type Culture Collection (ATTC).

The American Type Culture Collection is a nonprofit, private organization established to acquire, preserve, and distribute biological cultures for the international research community. Over the years the organization has initiated various catalogs, manuals, guides, and workshops relevant to the life sciences community.

Especially useful is the ATCC *Microbes and Cells at Work*, 2nd ed., 1991. This is an alphabetic index to special applications for the microorganisms, cell cultures, recombinant DNA materials, and viruses in ATCC's collec-tion of over 50,000 strains of biological cultures. Available in print ($55.00) and electronic version ($60.00).

Another useful compilation is the ATCC's *Catalogue of Bacteria and Bacteriophage*, 18th ed., free in the United States; the IBM-compatible 3.5- or 5.25-

inch diskette version is $25.00, postpaid to U.S. addresses. For more information about the American Type Culture Collection, see the entry (p. 47) in Chapter 3, "General Sources."

For additional information about culture collections, see the MICROBIAL STRAIN DATA NETWORK (below).

ASM Membership Directory, 1995. Washington: American Society for Microbiology, 1995. About 548 p. $25.95 (paper).

Lists names, addresses, degrees, and division affiliation for over 41,000 ASM members in the United States and many other countries. Provides descriptions of ASM programs, divisions, awards, and branch organizations, as well as headquarters staff information.

Linscott's Directory of Immunological and Biological Reagents, 8th ed. Santa Rosa, CA: Linscott's Directory, 1994/95. 280 p. $75.00. ISSN 0740-7394. Also available on IBM-compatible floppy disks.

Catalog for more than 47,000 biological products, reagents, and immune serums. Kept up to date by supplements.

MICROBIAL STRAIN DATA NETWORK (MSDN); a Global Communications Network for Life Sciences Information. PO Box 2636, Kensington, MD 20891-2636, e-mail: msdn0004@bdt.ftpt.br

This non-profit organization provides specialized information and communications services for life scientists worldwide. It is run from a Secretariat in the UK with part-time U.S. support and an international Management Committee. MSDN provides a unique package of database and bulletin boards covering microbiology, biotechnology, and biodiversity. There is information on microbial strains, culture collections, biotechnology publications and meetings, and environmental matters. The scope includes cell lines, hybridomas, molecular probes and recombinant materials as well as microorganisms. Access to MSDN is through public data networks and the Internet.

Guides to Internet Resources

There is a bionet.microbiology FAQ available on the Internet that is compiled from suggestions and input from participants in the microbiology newsgroup. The FAQ is posted monthly to bionet.microbiology. The FAQ provides information on accessing bionet.microbiology, on retrieving old articles, how and what to post, related newsgroups and bulletin boards, other Internet sources of interest to microbiologists, answers to frequently asked questions, etc. Immunologists, also, have a newsgroup, named bionet.immunology. Other Internet resources are accessible using Gopher and World Wide Web servers; for example, Cold Spring Harbor Laboratory has announced access to its WWW server with information including its book catalogs, indexes of current journal issues, announcements about courses and programs, home pages, and the like.

Handbooks, Manuals, and Databases

All kinds and formats of handbooks and manuals are included in this section, from style manuals to handbooks on laboratory safety. Many of the handbooks or databanks that microbiologists use are in electronic form that can be accessed online, on CD-ROM, or via the Internet. For example, molecular sequence databases are available on the Internet via Gopher: **GenBank** is one of largest and most heavily used, and it is completely described (p. 146) in Chapter 7, "Genetics."

Two publishers that should be especially acknowledged are the American Society for Microbiology in Washington, DC, and the Cold Spring Harbor Laboratory Press in Plainview, NY. Both of these publishers issue authoritative laboratory methods manuals and handbooks of importance to microbiologists, immunologists, and molecular biologists.

Consult Chapters 5–7 (on biochemistry, molecular biology, and genetics) for other databases relevant to microbiology and immunology.

ASM Style Manual for Journals and Books. Washington, DC: American
    Society for Microbiology, 1991. 199 p. $28.00.

"The new manual supersedes the 1985 edition and all ASM handbooks published previously." Authorized by the American Society of Microbiology, this style manual covers report writing, dissertation and manuscript preparation, technical writing and printing for microbiologists.

Atlas, Ronald M. *Handbook of Microbiological Media*. Edited by
    Lawrence C. Parks. Boca Raton, FL: CRC Press, 1993. 400 p.
    $90.00. ISBN 0849329442.

A comprehensive reference to the formulations and applications of more that 1,500 microbiological media used for isolation, cultivation, identification, and maintenance. Includes instruction for preparation and use; accepted name of media, and synonyms, from commercial manufacturers and the literature.

Atlas of Invertebrate Viruses. Edited by Jean R. Adams and Jean R.
    Bonami. Boca Raton, FL: CRC Press, 1991. 684 p. $299.95.
    ISBN 0849368065.

Useful reference for invertebrate pathologists, virologists, and electron microscopists. Illustrated with black and white photographs; each chapter has a lengthy bibliography. Appendices include information on techniques for light and electron microscopy, staining techniques, morphological guide, fixation, and embedding protocols.

Bergey's Manuals.

See the entry (p. 180) in the Classification, Nomenclature, and Systematics section.

Biosafety in Microbiological and Biomedical Laboratories: Working with Infectious Agents in Laboratory Settings, 3rd ed. Edited by Jonathan Y. Richmond and Robert W. McKinney. Upland, PA: Diane Publishing, 1994. 177 p. $45.00. ISBN 0788105485.

Standards and safety measures for microbiological and medical laboratories, originally published under the auspices of U. S. Department of Health and Human Services, Public Health Service, Centers for Disease Control and National Institutes of Health.

Biotechnology Handbooks, 1987– . New York: Plenum.

These handbooks contain data useful for the practice of microorganisms in the service of biotechnology. The price varies.

Vol. 1: *Penicillium and Acremonium*. 1987. 297 p. ISBN 030642356.

Vol. 2: *Bacillus*. 1989. 414 p. ISBN 0306431378.

Vol. 3: *Clostridia*. 1989. 304 p. ISBN 0306432617.

Vol. 4: *Saccharomyces*. 1991. 327 p. ISBN 0306436345.

Vol. 5: *Methane and Methanol Utilizers*. 1992. 304 p. ISBN 030643878X.

Vol. 6: *Photosynthetic Prokaryotes*. 1992. 275 p. $69.50. ISBN 0306438798.

Vol. 7: *Aspergillus*. 1994. 273 p. $69.60. ISBN 030644545X.

CRC Handbook of Microbiology, 2nd ed. Edited by Allen I. Laskin and Hubert A. Lechevalier. Boca Raton, FL: CRC Press, 1977–1987. 9 v. $1,700 (set).

Data on properties of microorganisms, their composition, products and activities. Contents of Vol. 1: *Bacteria*; Vol. 2: *Fungi, Algae, Protozoa, and Viruses*; Vol. 3 and 4: *Microbial Composition*; Vol. 5: *Microbial Products*; Vol.

6: *Growth and Metabolism*; Vol. 7: *Microbial Transformation*; Vol. 8: *Toxins and Enzymes*; Vol. 9: *Antibiotics* and *Antimicrobial Inhibitors*.

Cellular Immunology LABFAX. Edited by P. J. Delves. San Diego, CA: Academic Press, 1994. 250 p. $59.95. ISBN 0122088859.

Another volume in the *LABFAX* series that lists information on cells of the immune system, including data on the development, structure, function, preparation, and assay of immune system tissues, cells, and molecules.

Clark, William A. et al. *Identification of Unusual Pathogenic Gram-Negative Aerobic and Facultatively Anaerobic Bacteria*. Atlanta, GA: Centers for Disease Control, 1984. $20.00. Available from the Superintendent of Documents, Washington, DC 20402 (stock no. 017-023-00149-3).

Based on a collection of more than 60,000 bacteria, this very highly regarded manual provides information on media, reagents, and methods used by the Special Bacteriology Laboratory at the Centers for Disease Control.

Color Atlas and Textbook of Diagnostic Microbiology, 4th ed. Elmer W. Koneman et al. Philadelphia: Lippincott, 1992. 1,154 p. ISBN 0397512015. Bibliography. Index.

Laboratory manual for current rapid techniques and emerging technologies in the laboratory diagnosis of infectious diseases.

DIFCO Manual: Dehydrated Culture Media and Reagents for Microbiology, 10th ed. Detroit, MI: DIFCO Laboratories, 1984 reprint. 1,115 p. $25.00. ISBN 0961316993.

Provides comprehensive information about products used in microbiology from a respected pioneer in bacteriological culture media.

GENBANK. 1982– . Bethesda, MD: National Center for Biotechnology Information.

Database of nucleotide and protein sequences. Available online, on CD-ROM, and through the Internet. For other relevant genetic databanks, consult Chapters 6 and 7, on molecular biology and gentics.

Handbook of Protoctista: The Structure, Cultivation, Habitats, and Life Histories of the Eukaryotic Microorganisms and Their Descendants Exclusive of Animals, Plants, and Fungi: A Guide to the Algae, Ciliates, Foraminifera, Sporozoa, Water Molds, Slime Molds, and the Other Protoctists. Edited by Lynn Margulis et al. Boston: Jones and Bartlett, 1990. 914 p. $225.00. ISBN 0867200529.

Indispensable for anyone dealing with protoctists. This authoritative handbook provides information on protoctist classification, nomenclature, distribution, evolutionary history, life cycles, maintenance, cultivation, identification, and references to the literature.

IMI Descriptions of Fungi and Bacteria. 1964– . Kew, Surrey, England: CAB International Mycological Institute. 4 sets/yr. ISSN 0009-9716.

Title varies, *Commonwealth Mycological Institute Descriptions of Fungi and Bacteria*; continuation of *CMI Descriptions of Pathogenic Fungi and Bacteria*. Provides standardized, illustrated descriptions of organisms including disease caused, geographical distribution, physiological specialization, transmission, references to key literature. Since 1986 the descriptions are first published in *Mycopathologia* and sets are reprinted for separate sale.

Immunochemistry LABFAX. Edited by M. A. Kerr and R. Thorpe. Boston: Blackwell Scientific, 1993. 256 p. $50.00. ISBN 1872748058.

Detailed compendium of essential information on plasma proteins, immunoglobulin properties and purification, antibody products, labeling and derivatization, and data on techniques such as ELISA, blotting, and immunolocalization.

Laboratory Safety: Principles and Practices, 2nd ed. Edited by Diane O. Fleming et al. Washington, DC: American Society for Microbiology, 1994. 400 p. $49.00. ISBN 1555810470.

Authoritative compilation produced under the auspices of the largest microbiological society in the United States. Sections cover Hazard assessment in the laboratory, hazard control in the laboratory, safety program management, and appendices including the CDC/NIH *Biosafety in Microbiological and Biomedical Laboratories*, the latest OSHA guidelines on blood-borne pathogens and chemical handling in labs, and a brief guide to first aid.

Manual of Industrial Microbiology and Biotechnology. Edited by Arnold L. Demain and Nadine A. Solomon. Washington, DC: American Society for Microbiology, 1986.

Annotated (p. 156) in Chapter 7, "Genetics."

Miller, Jeffrey H. *A Short Course in Bacterial Genetics: A Laboratory Manual and Handbook for* Escherichia coli and Related Bacteria. Plainview, NY: Cold Spring Harbor Laboratory Press, 1992. 2 v. ISBN 0879693495. Manual and handbook: $110.00. Kit: $350.00, ISBN 0879694009.

Contents include physical and genetic maps of various bacteria, clone banks and libraries, databases, sequences, genetic codes and codon usage in selected organisms, properties of amino acids, atomic weights, formulas, procedures, commercial suppliers, etc. A kit of 44 bacterial strains and phage lysates is available for the set of experiments described in *A Short Course* . . . . Also, see a related manual listed under Stanley R. Maloy (p. 196) in the Methods section.

Quality Control: Principles and Practice in the Microbiology Laboratory.
   Edited by J. J. S. Snell et al. London: Public Health Laboratory
   Service, 1992. 172 p. (paper). $24.95. ISBN 0901144312.

The U.K. version of the U.S. Centers for Disease Control manual, listed under the title *Biosafety in Microbiological and Biomedical Laboratories* . . . (p. 187).

The Prokaryotes, 2nd ed. Edited by Albert Balows et al. New York:
   Springer-Verlag, 1992. $1,980.00 (4 vol. set). ISBN 0387972587.

"A Handbook on the Biology of Bacteria: Ecophysiology, Isolation, Identification, Applications." Provides a comprehensive survey of all established and proposed bacterial genera for which adequate data are available.

Virology LABFAX. Edited by D. Harper. Boston: Blackwell Scientific
   (distributed in the United States by Academic Press), 1993. 345 p.
   $59.95 (spiral casebound). ISBN 1872748511.

Key data reference book in the LABFAX series for virologists. Includes information on virus taxonomy, electron microscopy, viral diseases, cell culture, immunology, monoclonal antibodies, sequence data, safety, vaccines, journals, etc. For other relevant LABFAX titles consult Chapters 5 and 6, on biochemistry and molecular and cell biology.

Histories

Brock, Thomas D., ed. *Milestones in Microbiology*. Englewood Cliffs,
   NJ: Prentice-Hall, 1961. 275 p. $25.00.

Historically important papers selected to demonstrate the development of microbiology. The editor comments on each article putting the experiments and the scientists into historical and scientific perspective.

Bulloch, William. *The History of Bacteriology*. New York: Oxford University Press, 1938. 422 p.

This authoritative, classic history of medical bacteriology includes "bibliographical notices of some of the early workers in bacteriology."

Burnet, F. M. and D. O. White. *Natural History of Infectious Disease*, 4th ed. New York: Cambridge University Press, 1972.

Out of print, but still a classic.

Clark, Paul Franklin. *Pioneer Microbiologists of America*. Madison: University of Wisconsin Press. 1961. 369 p.

History of early American microbiology.

Collard, Patrick. *The Development of Microbiology*. New York: Cambridge University Press, 1976. Reprinted by Books on Demand. 201 p. $52.50. ISBN 0317077406.

This book presents "the development of certain ideas in microbiology, relating the views held at different times to the contemporaneous state of knowledge in other fields and showing how successive models grew out of the internal contradictions of their predecessors."

Dawes, E. A. The Federation of European Microbiological Societies—An Historical Review. *FEMS Microbiological Letters* 100: 15-23, 1992.

De Kruif, Paul. *Microbe Hunters*. New York: Pocket Books, 1940. 342 p.

Fascinating history that has seen many reprints. There is also a series of audiocassettes that contains a complete and unabridged reading of de Kruif's account of the early microbe hunters (New London, CT: Sound Writings, 1992, 8 audiocassettes, running time 12 hours, $39.00).

Grafe, Alfred. *A History of Experimental Virology*; translated by Elvira Reckendorf. New York: Springer-Verlag, 1991. 343 p. $69.00. ISBN 387519254. Glossary. Bibliography.

Comprehensive, compact survey of virology and the knowledge achieved during the past century.

Grainger, Thomas H. *A Guide to the History of Bacteriology*. (Chronica Botanica, no. 18). New York: Ronald Press, 1958. 210 p.

Citations and annotations are provided to the reference tools for the literature of bacteriology, to the history of bacteriology, and to biographical references and biographies of bacteriologists.

Hughes, Sally Smith. *The Virus: A History of the Concept*. New York: Science History Publications, 1977. 140 p. $12.00. ISBN 0882021680. Bibliography. Index.

Written in nontechnical language for people with a general interest in the history of science.

Immunology 1930–1980: Essays on the History of Immunology. Edited by Pauline M. H. Mazumdar. Toronto: Wall & Thompson, 1989. 307 p. $49.50. ISBN 092133219X.

"Most of the chapters in this volume are based on papers that were originally presented as part of the Sixth International Congress of Immunology, held at the University of Toronto in July 1986."

Lechevalier, Hubert A. *Three Centuries of Microbiology.* New York: McGraw-Hill, 1974. 536 p.

Although out of print, this history is useful for quotations from classical papers stressing the main lines of historical development.

Milestones in Immunology. Edited by Debra Jan Bibel. New York: Springer-Verlag, 1988. 330 p. $35.00. ISBN 0910239150.

Consists of reprints of articles from various sources.

Phage and the Origins of Molecular Biology, expanded edition. Edited by J. Cairns, G. S. Stent, and J. D. Watson. Cold Spring Harbor, NY: Cold Spring Harbor, 1992. 366 p. $40.00. ISBN 0879694076.

A new, expanded edition of the landmark collection of autobiographical essays honoring Max Delbruck's 60th birthday, originally published in 1966.

Portraits of Viruses: A History of Virology. Edited by F. Fenner and A. Gibbs. New York: Karger, 1988. 344 p. $170.00. ISBN 3805548192.

Articles originally published in *Intervirology* between 1979 and 1986.

Silverstein, Arthur M. *A History of Immunology.* San Diego: Academic Press, 1989. 422 p. $60.00. ISBN 012643770X.

About half of the chapters were published in abbreviated form in the journal *Cellular Immunology.*

Waterson, A. P. and Lise Wilkinson. *An Introduction to the History of Virology.* New York: Cambridge University Press, 1978. 237 p. ISBN 0521219175.

The underlying theme of the book is the evolution of the present concept of a virus. A very useful 30-page section gives brief biographies of scientists including their dates, memberships, important discoveries, and references to additional information.

Woese, Carl R. "Prokaryote Systematics: The Evolution of a Science." In *The Prokaryotes,* 2nd ed. p. 3–18. New York: Springer-Verlag, 1992.

A history of bacterial systematics.

Methods

Techniques useful for microbiology may be found also in Chapters 5–7, on biochemistry, molecular biology, and genetics.

Bacteriological Analytical Manual, 7th ed. Division of Microbiology, Center for Food Safety and Applied Nutrition, U. S. Food and Drug Administration. Arlington, VA: Association of Official Analytical Chemists, 1992. 529 p. (loose-leaf). ISBN 0935584498.

Methods used by the Food and Drug field laboratories effective for microorganisms in foods. The manual is loose-leaf so that it can be updated easily between editions.

Bryant, Neville J. *Laboratory Immunology & Serology*, 3rd ed. Philadelphia: Saunders, 1992. 387 p. ISBN 0721642128.

Example of a useful bench manual that successfully introduces the student to techniques for diagnostic immunology.

Chemical Methods in Prokaryotic Systematics. Edited by Michael Goodfellow and Anthony G. O'Donnell. New York: Wiley, 1994. (Modern Microbiological Methods). 576 p. $129.95. ISBN 0471941913.

Written by leading specialists, this laboratory manual offers comprehensive and up-to-date, reliable techniques for the classification and identification of prokaryotes: archaea and bacteria. Protocols, difficulties, limitations, and interpretations are discussed.

Collins and Lyne's Microbiological Methods, 6th ed. Edited by C. H. Collins et al. Boston: Butterworths, 1989. 450 p. ISBN 0407008853.

Revised edition of *Microbiological Methods*, 5th ed., 1984. Standard guide to bacteriological and microbiological methods and techniques.

Compendium of Methods for the Microbiological Examination of Foods, 3rd ed. Edited by Carl Vanderzant and Don F. Splittstoesser; compiled by he APHA Technical Committee on Microbiological Methods for Foods. Washington, DC: American Public Health Association, 1992. 1,219 p. $90.00. ISBN 0875531733. Index.

Syllabus of microbiological methods, general and specialized laboratory procedures, indicator microorganisms and pathogens that are involved in food processing, spoilage, and food safety. Provides information on culture media, reagents, stains, equipment, and references to the literature.

Computers in Microbiology: A Practical Approach. Edited by T. N. Bryant et al. New York: IRL Press, 1989. (Practical Approach Series). 214 p. $40.00 (paper). ISBN 0019630151.

Deals with applications of computers to a number important areas in microbiology: data and image analysis; fermentation measurement and control; modelling and simulation; taxonomy and systematics; and teaching.

Cowan and Steel's Manual for the Identification of Medical Bacteria, 3rd ed. Edited by G. I. Barrow and R. K. A. Feltham. New York: Cambridge University Press, 1993. 331 p. $69.95. ISBN 0521326117.

Essential for the clinical microbiology laboratory for the rarely encountered or unusual organism.

Current Methods for Classification and Identification of Microorganisms. Edited by R. R. Colwell and R. Grigorova. Orlando, FL: Academic Press, 1987. (Methods in Microbiology, v. 19). 518 p. ISBN 0125215193.

Provides methodology for techniques for identification and classification of microorganisms, including relatively simple procedures, classical techniques, and procedures for DNA and RNA sequencing.

Current Protocols in Immunology. Edited by John E. Coligan. New York: Wiley Interscience, 1991– . 2 v. ISBN 0471522767. Available in looseleaf ($295 with one year of quarterly updates) and CD-ROM ($450 with one year quarterly updates). Can be renewed annually.

Indispensable to immunology research labs; kept up to date by semiannual supplements.

Diagnostic Molecular Microbiology; Principles and Applications. Edited by David H. Persing et al. Herndon, VA: ASM Press, 1993. 660 p. $59.00. ISBN 155581056X.

Covers both principles and applications of molecular diagnostic methods pertaining to infectious diseases. Provides theoretical and practical framework for understanding the powerful uses of nucleic acid amplification technologies for the rapid detection and characterization of bacterial, viral, fungal, and parasitic pathogens in the clinical laboratory. Part I summarizes the basic scientific theory underlying molecular diagnostics; Part II provides 66 protocols, or molecular recipes, from leading labs around the world.

Enzyme and Microbial Techniques. v. 1– , 1979– . Woburn, MA: Butterworth-Heinemann. Monthly. $735/yr. ISSN 0141-0229.

An international biotechnology journal relevant to basic and applied aspects of the use of enzymes and microorganisms. Contributions may be in the form of research papers, rapid communications, or reviews. Patent reports are a regular section for each issue.

Handbook of Methods in Aquatic Microbial Ecology. Edited by Paul F. Kemp et al. Boca Raton, FL: Lewis, 1993. 777 p. $129.95. ISBN 0873715640.

Eighty-six chapters and 95 authors provide a compilation of techniques, with detailed recipes, for many new methods and molecular techniques useful for aquatic microbiology.

Identification Methods in Applied and Environmental Microbiology. Oxford, England: Blackwell, 1992. (Society for Applied Bacteriology Technical Series, v. 29). 334 p. ISBN 0632033797. Index.

This review supplies up-to-date information on methods used to characterize and identify important bacterial genera.

Immunochemical Protocols. Edited by Margaret M. Manson. Totowa, NJ: Humana Press, 1992. (Methods in Molecular Biology, v. 10). 472 p. $69.50. ISBN 0896032043.

Fifty readily reproducible protocols and 62 illustrations provide a convenient reference manual for a range of antibody techniques.

Immunocytochemistry. Edited by J. E. Beesley. Oxford, England: IRL Press, 1993. (Practical Approach Series). 248 p. $45.00. ISBN 0199632693. Index.

A comprehensive guide to practical immunocytochemistry. General principles are covered as well as detailed techniques, including immunolabeling for light and electron microscopy.

ImmunoMethods, v. 1– , 1992– . San Diego: Academic Press. Bimonthly. $148/yr. ISSN 1058-6687.

Provides focused, detailed, and authoritative reports on immunological methods and their applications.

Johnson, Ted R. and Christine L. Case. *Laboratory Experiments in Microbiology*, 3rd ed., brief ed. Redwood City, CA: Benjamin/ Cummings, 1992. 372 p. $23.75. ISBN 080538488X.

Laboratory manual for beginning students.

Journal of Immunological Methods. v. 1– , 1971/72– . Amsterdam: Elsevier. Monthly. $2,330/yr. ISSN 0022-1759.

Covers techniques, and articles on novel methods for all aspects of immunology.

Journal of Microbiological Methods. v. 1– , 1983– . Amsterdam: Elsevier Science. Monthly. $498/yr. ISSN 0167-7012.

Original articles, short communications, and review articles on novel methods, or significant improvements to an existing method, in all aspects of microbiology excluding virology.

Journal of Virological Methods. v. 1 – , 1980– . Amsterdam: Elsevier. Monthly. $1,173/yr. ISSN 0166-0934.

Publishes original papers and invited reviews covering techniques on all aspects of virology.

Laboratory Methods in Immunology. Edited by Heddy Zola. Boca Raton, FLA: CRC Press, 1990. 2 v. $190.00. v. 1, ISBN 0849344816.

". . . cookbook-style recipes for immunological techniques, interspersed with reviews on strategies required . . . ."

Laboratory Techniques in Biochemistry and Molecular Biology.

Well-established series useful for microbiologists; annotated (p. 99) in Chapter 5, "Biochemistry."

Levin, Morris A. et al. *Microbial Ecology: Principles, Methods, and Applications.* New York: McGraw-Hill, 1992. (McGraw-Hill Environmental Biotechnology Series). 945 p. $79.95. ISBN 0070375062.

Compilation of 50 articles by leading scientists describing current concepts and measurement techniques in microbial ecology.

Lloyd, David. *Flow Cytometry in Microbiology.* New York: Springer-Verlag, 1993. 188 p. $119.00. ISBN 0387197966.

Techniques of interest to micro-biologists; includes applications.

Maintenance of Microorganisms and Cultured Cells: A Manual of Laboratory Methods. Edited by B. E. Kirsop and A. Doyle, 2nd ed. London: Academic Press, 1991. 308 p. $47.50 (spiral bound). ISBN 0124103510.

Cultures, culture media, and laboratory preservation techniques for bacteria, yeasts, protozoa, and algae. There are chapters on animal and cell culture, and a list of suppliers is included in an appendix.

Maloy, Stanley R. *Experimental Techniques in Bacterial Genetics.* Boston: Jones and Bartlett, 1990. (Jones and Bartlett Series in Biology). 180 p. $37.50. ISBN 0867201185.

Designed for a beginning university course in bacterial genetics, this manual provides practical information for the novice.

Manual of Clinical Microbiology, 5th ed. Editor-in-chief, Albert Balows et al. Washington, DC: American Society for Microbiology, 1991. 1,364 p. $84.00. ISBN 1555810292.

Documents currently accepted practices in clinical microbiology for detection, isolation, and identification of important agents of infectious diseases.

Manual of Laboratory Immunology, 2nd ed. Philadelphia: Lea & Febiger, 1990. ISBN 08121113195.

Revised edition of *Manual of Laboratory Immunology* by Julia E. Peacock and Russell H. Tomar, published in 1980.

Methods for General and Molecular Bacteriology, Editor-in-Chief, Philipp Gerhardt. Washington, DC: American Society for Microbiology, 1993. 825 p. $69.00. ISBN 1555810489.

Revision of *Manual of Methods for General Bacteriology* (1981) which complements the systematics treatise *Bergey's Manual of Systematic Bacteriology*. This important manual is designed "to meet the need for a compact, moderately priced handbook of reliable, basic methods for practicing general bacteriology in the laboratory." It covers all kinds of bacteria, archaeobacteria as well as eubacteria, complementing general textbooks and systematics treatises. New chapters cover molecular biology as well as antigen–antibody reactions, photo-graphy, and records and reports. A classic from the American Society for Microbiology.

Methods in Enzymology. v. 1– , 1955– . New York: Academic Press.

Widely used series fully annotated in the general sources chapter. There are many volumes of interest to microbiologists or immunologists, e.g., *Bacterial Pathogenesis*, Pt. A and B. Vol. 235 ($105.00) and vol. 236 ($99.00) of the series published in 1994.

Methods in Microbiology. v. 1– , 1969– . San Diego: Academic Press. Irregular. ISSN 0580-9517.

Reviews devoted to a single topic in each volume. Vol. 21: *Plasmid Technology*, 2nd ed., 1988; Vol. 22: *Techniques in Microbial Ecology*, 1990; Vol. 23–24: *Techniques for the Study of Mycorrhiza*, 1991–92.

Methods of Immunological Analysis. v. 1– , 1993– . New York: VCH. $265 (v. 1); $2,580 (set). ISBN (v. 1) 352729067.

The projected set of twelve volumes, scheduled to be complete in 1996, is aimed at both specialists and non-specialists who carry out qualitative and quantitative immunoanalysis. The first three volumes provide an introduction, and the remaining eight are laboratory protocols of immunoanalytical techniques.

Methods and Techniques in Virology. Edited by Pierre Payment and
     Michel Trudel. New York: Dekker, 1993. 309 p. $125.00. ISBN
     0824791010.

This reference provides a comprehensive set of methods and detailed protocols ranging from basic tissue culture, virus production and titration, to advanced molecular methods for analysis of viral proteins and nucleic acids. Besides a general discussion of up-to-date techniques, there are also applications of specific methods and developments in viral techniques.

Microbes as Tools for Cell Biology. Edited by David G. Russell. San
     Diego: Academic Press, 1994. (*Methods in Cell Biology*, v. 45).
     360 p. $45.00 (comb bound). ISBN 0126040400.

A laboratory guide that provides a microbial tool kit for biologists who wish to use microbes as probes for basic cellular functions. Three sections cover culture and genetic manipulation of microbes, assays for pathogen-host recognition, and analysis of intracellular parasitism.

Miller, J. Michael. *Handbook of Specimen Collection and Handling in
     Microbiology*, 2nd ed. Atlanta, GA: Centers for Disease Control,
     Revised 1985. 53 p. Supt. of Docs. no.: HE 20.7008/2:M
     58/2/985.

Centers for Disease Control laboratory manual outlines handling techniques, and collection and preservation of laboratory specimens.

Modern Techniques for Rapid Microbiological Analysis. Edited by Wilfred
     H. Nelson. New York: VCH, 1991. 263 p. $69.50. ISBN
     1560810017.

Rapid identification methods for bacteria and phytoplankton.

Molecular Virology. Edited by Andrew J. Davison and Richard M. Elliott.
     Oxford, England: IRL Press, 1993. (Practical Approach no. 127).
     256 p. $45.00. ISBN 0199633584.

Application of modern techniques of molecular analysis to important virus groups. As is usual with most of the Practical Approach volumes, there is an appendix listing suppliers for specialist items.

Molecular Virology Techniques, Pt. A. San Diego, CA: Academic Press,
     1994. 401 p. $85.00.

Vol. 4 of the series *Methods in Molecular Genetics* listed (p. 158) in Chapter 7, "Genetics."

Nucleic Acid Techniques in Bacterial Systematics. Edited by Erko
   Stackebrandt and Michael Goodfellow. (Modern Microbiological
   Methods). New York: Wiley, 1991. 329 p. $135.00. ISBN
   0471929069.

Comprehensive laboratory manual of revolutionary nucleic acid–based techniques for characterizing, classifying, and identifying bacteria.

Penn, C. W. *Handling Laboratory Microorganisms*. Philadelphia: Open
   University Press, 1991. 160 p. $69.00. ISBN 0335092047.

Microbiological technique for beginners.

Plasmids; A Practical Approach, 2nd ed. Edited Kimber G. Hardy.
   Oxford, England: IRL Press, 1993. (Practical Approach no. 138).
   272 p. $45.00. ISBN 0199634459.

Provides protocols for studying bacterial plasmids, and for using both plasmids and phagemids as vectors. Techniques are included, as well as methods for using plasmid vectors in important groups of bacteria.

Rapid Methods and Automation in Microbiology and Immunology. Edited
   by R. C. Spencer, E. P. Wright, and S. W. B. Newsom. Andover,
   UK: Intercept, 1994. 502 p. $89.00. ISBN 0946707782.

"Tomorrow's Techniques." This book is the result of the 7th International Congress on Rapid Methods and Automation in Microbiology and Immunology. It summarizes recent developments in microbiological methods in the fields of medicine, food production, and the environment.

Society for Applied Bacteriology Technical Series. v. 1– , 1966– .
   Boston: Blackwell Scientific.

Useful series. Some recent titles:

   Immunological Techniques in Microbiology, 1987.

   ATP Luminescence: Rapid Methods in Microbiology, 1989.

   *Rapid Microbiological Methods for Food, Beverages & Pharmaceu-
      ticals*, 1990.

   Identification Methods in Applied & Environmental Microbiology,
      1992.

Towner, K. J. and A. Cockayne. *Molecular Methods for Microbial Identification and Typing*. New York: Chapman & Hall, 1993. 202 p.
   $45.00. ISBN 041249390X.

A source book and introduction to techniques for microbiological identification and typing.

Periodicals

Antonie Van Leeuwenhoek International Journal of General and Molecular Microbiology. v. 1– , 1935– . Dordrecht, Netherlands: Kluwer Academic. Monthly. $195/yr. ISSN 003-6072.

Publishes papers on fundamental and applied aspects of microbiology. Topics of particular interest include: taxonomy, structure and development, biochemistry and molecular biology, physiology and metabolic studies, genetics, ecological studies, medical mycology and molecular biological aspects of microbial pathogenesis and parasitology. The importance of microorganisms to biotechnology receives special attention.

Applied and Environmental Microbiology. v. 1– , 1953– . Birmingham, AL: American Society for Microbiology. Monthly. $274/yr. Also available in CD-ROM format, $302/yr. ISSN 0099-2240.

Significant current research in industrial microbiology and biotechnology, food microbiology, and microbial ecology.

Applied Microbiology and Biotechnology. v. 1– , 1975– . Berlin: Springer-Verlag. Monthly. $1,597/yr. ISSN 0175-7598.

Publishes original papers, short contributions, and mini-reviews on biotechnology, biochemical engineering, applied genetics and regulation, applied microbial and cell physiology, food biotechnology, and environmental biotechnology.

Archives of Microbiology. v. 1– , 1939– . Berlin: Springer-Verlag. Monthly. ISSN 0302-8933. $1,924/yr.

Founded in 1930 as *Archiv fuer Mikrobiologie.* Covers basic results on molecular aspects of structure, function, cellular organization and ecophysiological behavior of prokaryotic and eukaryotic microorganisms. Papers should be submitted in English.

Archives of Virology. v. 1– , 1939– . New York: Springer-Verlag. Semi-monthly. ISSN 0304-8608. $1,364/yr.

Official journal of the Virology Division of the International Union of Microbiological Societies. Publishes papers from all branches of research on viruses, virus-like agents, and virus infections of humans, animals, plants, insects, and bacteria.

ASM News. v. 1– , 1935– . Birmingham, AL: American Society for Microbiology. Monthly. $25/yr. ISSN 0044-7897.

Timely information on scientific and policy issues of concern to microbiologists. Current topics, in-depth features on life sciences, updates on regulatory affairs, opinion pieces, book reviews, extensive calendar of pertinent meetings, employment listings and reports on ASM activities.

Canadian Journal of Microbiology, Revue canadienne de microbiologie. v. 1– , 1954– . Ottawa, Canada: National Research Council Canada. Monthly. $241/yr. ISSN 0008-4166.

Papers in English or French in any area of microbiology.

Cellular Immunology. v. 1– , 1970– . San Diego: Academic Press. Monthly. $1,120/yr. ISSN 0008-8749.

Original investigations on immunological activities of cells in experimental or clinical situations for *in vivo* and/or *in vitro* studies.

European Journal of Immunology. v. 1– ,1971– . Weinheim, Germany: VCH. Monthly. $750/yr. ISSN 0014-2980.

Associated with the European Federation of Immunological Societies. Publishes papers on various aspects of immunological research from the fields of experimental and human immunology, molecular immunology, immunobiology, immunopathology, immunogenetics, and clinical immunology.

FEMS Immunology and Medical Microbiology. v. 1– , 1988– . Amsterdam: Elsevier. Monthly. $673/yr. ISSN 0928-8244.

Published on behalf of the Federation of European Microbiological Societies (FEMS). Publishes papers dealing with aspects of the interaction of microorganisms with the host, including microbial factors responsible for pathogenesis and disease, and also for triggering the host immune responses; basic mecha-nisms of the immune repertoire; and molecular approaches to the development of antimicrobials and their application.

FEMS Microbiology Ecology. v. 1– , 1985– . Amsterdam: Elsevier. Semi-quarterly. $449/yr. ISSN 0168-6496.

Published on behalf of the Federation of European Microbiological Societies (FEMS). Original articles on fundamental aspects of the ecology of microorganisms in natural soil, air or aquatic environments, or in artificial or managed environments.

FEMS Microbiology Letters. v. 1– , 1977– . Amsterdam: Elsevier. Semi-monthly. $2,216. ISSN 0378-1097.

Published on behalf of the Federation of European Microbiological Societies (FEMS). All aspects of microbiology are covered in the form of research papers or mini-reviews.

Immunobiology. v. 156– , 1979– . New York: Fischer. Semi-quarterly. $795/yr. ISSN 0171-2985.

Continues *Zeitschrift fuer Immunitatsforschung* founded in 1909 by Paul Ehrlich. Publishes papers in clinical immunology, immunochemistry, tumor and immunopathology, leucocyte physiology, viral and bacterial immunology, cell-mediated immunity, and immunogenetics and transplantation.

Immunogenetics. v. 1– , 1974– . Berlin: Springer-Verlag. Monthly. $571/yr. ISSN 0093-7711.

Publishes articles, brief communications, and reviews in the following areas: immunogenetics of cell interaction, immunogenetics of tissue differentiation and development, phylogeny of alloantigens and of immune response, genetic control of immune response and disease susceptibility, and genetics and biochemistry of alloantigens.

Immunology. v. 1– , 1958– . Oxford, England: Blackwell Scientific. Monthly. $499/yr. ISSN 0019-2805.

Official journal of the British Society for Immunology. Original work in all areas of immunology including cellular immunology, immunochemistry, immunogenetics, allergy, transplantation immunology, cancer immunology and clinical immunology. Review articles are published occasionally.

Infection and Immunity. v. 1– , 1970– . Birmingham, AL: American Society for Microbiology. Monthly. $368/yr. ISSN 0019-9567. Also available in CD-ROM format.

Articles of interest to microbiologists, immunologists, epidemiologists, pathologists, and clinicians.

International Journal of Food Microbiology. v. 1– , 1984– . Amsterdam: Elsevier. Monthly. $704/yr. ISSN 0168-1605.

Covers all aspects of microbiological safety, quality and acceptability of foods.

Journal of Applied Bacteriology. v. 1– , 1938– . Oxford, England: Blackwell Scientific. Monthly. $500/yr. ISSN 0021-8847.

Published for the Society for Applied Bacteriology on all aspects of applied microbiology, e.g., agriculture, biotechnology, environment, food, genetics, medical and veterinary, pharmaceutical, taxonomy, soil, systematic bacteriology and water. Associate publication to the *Journal* is *Letters in Applied Microbiology*.

Journal of Bacteriology. v. 1– , 1916– . Birmingham, AL: American
Society for Microbiology. Semi-monthly. $391/yr. Also available in
CD-ROM format, $400/yr. ISSN 0021-9193.

Articles include new information on genetics and molecular biology, structure
and function, plant microbiology, plasmids and transposons, eukaryotic cells, cell
surfaces, physiology nd metabolism, enzymes and proteins, and bacterio-phages.
Each issue also contains a mini-review on a selected topic.

Journal of General and Applied Microbiology. v. 1– , 1955– . Tokyo:
Microbiology Research Foundation. Bimonthly. $85/yr. ISSN
0022-1260.

Devoted to the publication of original papers pertaining to general and applied
microbiology.

Journal of General Virology. v. 1– , 1967– . Reading, UK: Society for
General Microbiology. Monthly. $810/yr. ISSN 0022-1317.

Publishes papers describing original, fundamental research in virology. Full-
length papers, short communications and review articles are included.

Journal of Immunology. v. 1– , 1916. Baltimore, MD: American Association of Immunologists. Bimonthly. $300/yr. ISSN 0022-1767.

Official journal of the American Association of Immunologists. Publishes
original articles on immunochemistry, transplantation and tumor immunology,
molecular biology, molecular genetics, cellular immunology, clinical
immunology, immunopathology, and microbial and viral immunology.

Journal of Medical Microbiology. v. 1– , 1968– . Edinburgh: Churchill
Livingstone. Monthly. $429/yr. ISSN 0022-2615.

A journal of the Pathological Society of Great Britain and Ireland. Contains
papers on all aspects of microbiology relevant to human or veterinary medicine.

Journal of Virology. v. 1– , 1967– . Birmingham, AL: American
Society for Microbiology. Monthly. $380/yr. ISSN 0022-538X.
Also available in CD-ROM format.

This journal deals with broad-based concepts concerning viruses of plants,
animals, bacteria, protozoa, fungi, and yeasts.

Letters in Applied Microbiology. v. 1– , 1985– . Oxford, England:
Blackwell Scientific, for the Society for Applied Bacteriology.
Monthly. Free to the subscribers of the *Journal of Applied Bacteriol-
ogy*. ISSN 0266-8254.

A journal for the rapid publication of short papers in the broad field of applied microbiology, including environmental, food, agricultural, pharmaceutical and veterinary microbiology, taxonomy, soil, systematic bacteriology, water and bio-deteriorations. Advances in rapid methodology is a particular feature; *Letters* reflects developments in biotechnology.

Microbial Ecology; An International Journal. v. 1– , 1974– . New York: Springer-Verlag. Bimonthly. $250/yr. ISSN 0095-3628.

Features articles of original research and mini-review articles on those areas of ecology involving microorganisms, including prokaryotes, eukaryotes, and viruses.

Microbiology. v. 1– , 1947– . Reading, UK: Society for General Microbiology. (Formerly *Journal of General Microbiology* until 1994, and continues *JGM* volume numbering) Monthly. $810/yr. ISSN 1350-0872.

Includes quality research papers across the whole spectrum of microbiology. The journal also features short reviews on rapidly expanding or especially significant areas.

Microbiology and Immunology. v. 21– , 1977– . Tokyo: Japanese Society for Bacteriology. Monthly. $255/yr. ISSN 0385-5600.

"In conformance with the aims of the Japanese Society for Bacteriology, the Society of Japanese Virologists, and the Japanese Society for Immunology, this journal is devoted to the world-wide dissemination of advanced knowledge in bacteriology, virology, immunology, and related fields."

Molecular and Cellular Biology. v. 1– , 1981– . Birmingham, AL: American Society for Microbiology. Monthly. $379/yr. ISSN 0270-7306. Also available in CD-ROM format.

Premier journal covering all aspects of the molecular biology of eukaryotic cells, including regulation of gene expression, transcription, cell growth and development, oncogenesis, cell and organelle structure and assembly, DNA replication and recombination, chromosome structure, and somatic cell genetics.

Molecular Immunology. v. 16– , 1979– . Oxford, England: Pergamon Press. Biweekly. $1,255/yr. ISSN 0161-5890.

Continues *Immunochemistry.* A leading journal publishing immunological knowledge delineated at the molecular level. It publishes research reports, short communications, structural data reports, review articles, as well as summaries of meetings, announcements, and letters to the editor. Supplements accompany some volumes beginning with v. 30, 1993.

Molecular Microbiology. v. 1– , 1987– . Oxford, England: Blackwell
 Scientific. Biweekly. $1,090/yr. ISSN 0950-382X. Absorbed
 Microbiological Sciences.

The most highly cited primary research journal in microbiology. It publishes
original research articles addressing any microbiological question at a molecular
level. *MicroReviews* and *MicroCorrespondence* are regular sections of the journal.

Research in Immunology. v. 1– , 1887– . Amsterdam: Elsevier. 9
 issues annually. $809/yr. ISSN 0923-2494.

Established in 1897 as the *Annales de l'Institut Pasteur.* Contains full-length
articles on all aspects of immunology including immunochemistry, cellular immunology, immunogenetics, and transplantation immunopathology. Overriding
criteria for publication are originality, high scientific quality, and up-to-date
relevance.

Research in Microbiology. v. 1– 1887– . Amsterdam: Elsevier.
 Monthly. $350/yr. ISSN 0923-2508.

Established in 1887 as the *Annales de l'Institut Pasteur.* Covers all aspects of
microbiology, including general and molecular microbiology, physiology and
microbial genetics, environmental and applied microbiology, industrial
microbiology, mycology and medical microbiology.

Research in Virology. v. 1– 1887– . Amsterdam: Elsevier. Bimonthly.
 $250/yr. ISSN 0923-2516.

Publishes original reports covering all aspects of virology, including molecular
virology, virus–cell interactions, viral oncogenesis, medical virology and epidemiology. Studies on the entire range of human, animal, and plant viruses and
bacteriophages are welcome.

Scandinavian Journal of Immunology. v. 1– , 1972– . Oxford, England:
 Blackwell Scientific. Monthly. $400/yr. ISSN 0300-9475.

An international journal reporting results of original work on cellular and
molecular immunology, including investigative techniques which are often published as supplements.

Systematic and Applied Microbiology. v. 1– , 1980– . Stuttgart:
 Gustav Fischer. Quarterly. $412/yr. ISSN 0723-2020.

Continues *Zentralblatt fuer Bakteriologie, Mikrobiologie und Hygiene. 1. Abt.
Originale C, Allgemeine, angewandte und okologische Mikrobiologie.* Covers
comparative physiology and biochemistry, systematics, applied and ecological
microbiology.

Virology. v. 1– , 1955– . San Diego: Academic Press. Semimonthly.
    $1,200/yr. ISSN 0042-6822.

Publishes basic research in all branches of virology, including viruses of vertebrates and invertebrates, plants, bacteria, and yeasts/fungi. In particular, articles
on the nature of viruses, molecular biology of virus multiplication, molecular
pathogenesis, and molecular aspects of the control and prevention of viral infections are invited.

Virus Research. v. 1– , 1984– . Amsterdam: Elsevier. Monthly.
    $925.00. ISSN 0168-1702.

Rapid publication for original papers on fundamental research concerning virus
structure, replication, and pathogenesis. Occasional review articles, book
reviews, and meeting reports are also included.

Yeast; A Forum for Yeast Researchers. v. 1– , 1985– . Chichester,
    England: Wiley. Monthly. $745/yr. ISSN 0749-503X.

Contains original research and review articles on *Saccharomyces* and other yeast
genera. Also has section titles "Current Awareness on Yeast," which indexes articles, books, and proceedings on yeast published since the previous issue of *Yeast.*

Reviews of the Literature

Advances in Applied Microbiology. v. 1– , 1959– . San Diego:
    Academic Press. Irregular. $80/vol. ISSN 0065-2164.

This series was designed "to publish critical and definitive reviews in those areas
of microbiology which are of interest to the practical microbiologist."

Advances in Immunology. v. 1– , 1961– . San Diego: Academic Press.
    Irregular. Vol. 57, $80.00. ISSN 0065-2776.

A scholarly review of research and valuable reference work covering current
work in immunology.

Advances in Microbial Ecology. v. 1– , 1977– . New York: Plenum.
    Irregular. $95/vol. ISSN 0147-4863.

Up-to-date research on the roles of microorganisms in natural and artificial
ecosystems, emphasizing microbial processes and interactions, the effects of
environmental factors on microbial populations, and the economic impact of
these organisms.

Advances in Microbial Physiology. v. 1– , 1967– . San Diego: Aca-
    demic Press. Irregular. $99/vol. ISSN 0065-2911. Author and
    subject indexes.

This series aims to include articles from the wide range of "specialized interests that constitute microbial physiology."

Advances in Virus Research. v. 1– , 1953– . San Diego: Academic Press. Irregular. Vol. 44: $99.00. ISSN 0065-3527.

Critical review articles are selected to cover all types of viruses from many different aspects, focusing on the virus not the disease.

Annual Review of Immunology. v. 1– , 1983– . Palo Alto, CA: Annual Reviews, Inc. Annual. $48/yr. ISSN 0732-0582.

Analytical articles reviewing significant developments within the discipline.

Annual Review of Microbiology. v. 1– , 1947– . Palo Alto, CA: Annual Reviews, Inc. Annual. $48/yr. ISSN 0066-4227.

Similar to other Annual Review publications reviewing topics of current and enduring interest in microbiology.

Bacterial Cell Wall. Edited by J.-M. Ghuysen. (New Comprehensive Biochemistry, v. 27). Amsterdam: Elsevier, 1994. 606 p. $200.00. ISBN 04444880941.

Integrated collection of contributions providing a fundamental reference for anyone interested in bacterial cell wall research.

Biochemistry of Archaea (Archaebacteria). Edited by M. Kates, D. J. Kushner, and A. T. Matheson. Amsterdam: Elsevier, 1993. 582 p. (New Comprehensive Biochemistry, v. 26). $191.50. ISBN 0444817131. Index and bibliographical references.

This volume brings together recent knowledge concerning general metabolism, bio-energetics, molecular biology and genetics, membrane lipid and cell-wall structural chemistry and evolutionary relations, of the three major groups of archaea.

Bulletin de l'Institut Pasteur. v. 1 – , 1903– . Paris: Elsevier. Quarterly. $171.00. ISSN 0020-2452.

In English or French. State-of-the-art reviews in microbiology, immunology, and infectious diseases. In addition, each article includes a comprehensive bibliography.

Concepts in Virology: From Ivanovsky to the Present. Edited by Brian W. J. Mahy and Dmitri K. Lvov. Langhorne, PA: Harwood Academic Publishers, 1993. 438 p. $90.00. ISBN 3718605686.

Proceedings of the 1992 centenary symposium celebrating the work of Dmitri Ivanovsky. Thirty-nine papers reporting state-of-the-art human and animal viral research, with an historical perspective, on significant developments in U.S. and Soviet virology.

Critical Reviews in Biochemistry and Molecular Biology. v. 1– 1972– .
    Boca Raton, FL: CRC Press.
Annotated (p. 107) in Chapter 5, "Biochemistry and Biophysics."

Critical Reviews in Immunology. v. 1– , 1979– . Boca Raton, FL: CRC
    Press. Quarterly. $245/yr. ISSN 1040-8401.
The journal publishes timely and critical review articles in various aspects of contemporary immunology, opinions/hypotheses, letters to the editor, news and comments, book reviews, and a calendar of events.

Critical Reviews in Microbiology. v. 1– , 1971– . Boca Raton, FL: CRC
    Press. Quarterly. $245/yr. ISSN 1040-841X.
Reviews in all areas of microbiology, including bacteriology, virology, phycology, mycology, and protozoology.

Current Opinion in Immunology. v. 1– , 1988– . Philadelphia: Current
    Biology. Monthly. $390/yr. ISSN 0952-7915.
Includes primary papers, review articles with accompanying annotated references, and a comprehensive listing of the previous year's papers. This journal will be available online in 1995 as part of the OCLC Electronic Journals Online services.

Current Topics in Microbiology and Immunology. v. 1– , 1914– .
    Berlin: Springer-Verlag. Irregular. Price varies for each volume.
    ISSN 0070-217X.
Continues *Ergebnisse der Mikrobiologie, Immunitatsforschung, experimentellen Therapie, Bacteriologie, und Hygiene.* Each volume presents current knowledge on a particular topic; recent volumes include vol. 183: *Neutralization of Animal Viruses* (1993); and vol. 184: *Adhesion in Leukocyte Homing and Differentiation* (1993).

Developments in Biological Standardization. v. 23– , 1974– Basel,
    Switzerland: Karger. Supersedes *Progress in Immunobiological
    Standardization.* Irregular. ISSN 0301-5149.
Formed by the union of *Progress in Immunobiological Standardization* and the Symposia series from the International Association of Biological Standardi-

zation, presenting reports and reviews from the Association meetings on microbiology and vaccines.

Environmental Microbiology. Edited by Ralph Mitchell. New York: Wiley/Liss, 1992. (Wiley Series in Ecological and Applied Microbiology). 411 p. $96.00. ISBN 0471506478.

Fifteen chapters reviewing a wide variety of topics dealing with environmental microbiology.

FEMS Microbiology Reviews. v. 1– , 1985– . Amsterdam: Elsevier. Quarterly. $432/yr. ISSN 0168-6445.

Published on behalf of the Federation of European Microbiological Societies (FEMS). Includes comprehensive reviews and current interest mini-reviews of the entire field of microbiology.

Frontiers in Microbiology. Edited by Graham C. Walker and Dale Kaiser. Herndon, VA: American Society for Microbiology, 1993. 155 p. $11.95 (paper). ISBN 1555810578.

A collection of minireviews published in the *Journal of Bacteriology* in 1992.

Immunological Reviews. v. 1– , 1977– . Copenhagen: Munksgaard. Bimonthly. $306/yr. ISSN 0105-2896.

Comprehensive and analytical reviews within the fields of clinical and experimental immunology.

Immunology Today. v. 1– , 1980– . Amsterdam: Elsevier. Monthly. $490/yr. ISSN 0167-5699.

Monitors advances in various fields of immunology with succinct review articles. In January 1995 this journal will go online on OCLC's Electronic Journals Online (EJO) system.

Microbiological Reviews. v. 42– , 1978– . Birmingham, AL: American Society for Microbiology. Quarterly. $130/yr. Also available in CD-ROM format, $155/yr. ISSN 0146-0749. 1973–77: *Bacteriological Reviews*.

Review journal covering all aspects of microbiology, including bacteriology, virology, mycology, and parasitology.

Molecular and Cellular Biology of the Yeast Saccharomyces. Edited by James R. Broach, John R. Pringle, and Elizabeth W. Jones.

Plainview, NY: Cold Spring Harbor Laboratory, 1991. 3 v. $97.00. ISBN 087969355X (v. 1).

Valuable reference for researchers using yeast as an experimental organism.

Progress in Immunology VIII, Proceedings of the 8th International Congress of Immunology, Budapest, 1992. Berlin: Springer-Verlag, 1993. 925 p. $147.00. ISBN 9637922784.

This volume presents the most important papers and up-to-date reports from the triennial Congress of Immunology on basic and applied immunology.

Progress in Industrial Microbiology. v. 1– , 1958– . Amsterdam: Elsevier. Irregular. Price varies.

Each volume reviews a particular topic: vol. 29: *Aspergillus: 50 Years On.* 1994. 851 p. $328.75. ISBN 0444416688. This volume celebrates the anniversary of *Aspergillus nidulans* as a tool for genetics.

Seminars in Immunology. v. 1– , 1989– . San Diego: Academic Press. Bimonthly. $177/yr. ISSN 1044-5323.

Each volume devoted to a highly topical and important subject in the field.

Seminars in Virology. v. 1– , 1990– . San Diego: Academic Press. Bimonthly. $177/yr. ISSN 1044-5773.

Another in the *Seminars* review series, this publication presents authoritative topical reviews edited by internationally acknowledged experts.

Trends in Microbiology; Virulence, Infection and Pathogenesis. v. 1– , 1993– . Cambridge, UK: Elsevier Trends Journals. Monthly. $490/yr. ISSN 0966-842X.

The newest *Trends* journal. Identifies key advances and reviews all areas of infection. Also includes news and comment, opinion pieces, book reviews, and a calendar of events.

The Viruses. Series editors: Heinz Fraenkel-Conrat and Robert R. Wagner. New York: Plenum, 1982– . Priced separately.

"This series is designed to provide a comprehensive review of the significant current areas of research in virology. Individual volumes or groups of volumes deal with a single virus family or group and cover all aspects of these viruses ranging from physicochemistry to pathogenicity and ecology" (from a volume jacket). A 1993 volume covers *The Arenaviridae.* Edited by Maria S. Salvato.

Societies

Consult the multivolume *Encyclopedia of Associations*, including the *International Organizations* volume, for additional information.

American Academy of Microbiology (AAM)
  1325 Massachusetts Ave., NW, Washington, DC 20005.

Founded: 1955. 1000 members. Professional arm of the American Society for Microbiology concerned with microscopic and submicroscopic organisms. Encourages exchange of information among members. Conducts recognition program and other professional programs, bestows awards and fellowships. Publications: triennial *Directory of Fellows of the American Academy of Microbiology*. Annual convention, usually in May.

American Association of Immunologists (AAI)
  9650 Rockville Pike, Bethesda, MD 20014.

Founded 1913. 5,500 members. Scientists engaged in immunological research, including aspects of virology, bacteriology, biochemistry, genetics, and related disciplines. Promotes interaction between laboratory investigators and clinicians; conducts training courses, symposia, workshops, and lectures; maintains library and placement services; bestows awards; compiles statistics. Computerized mailing lists; online databases. Committees: Awards, Clinical Immunology, Education, Historian, IUIS Advisory, Minority Affairs, Nominations Program, Public Affairs, Status of Women, Veterinary Immunology. Publications: quarterly *AAI Newsletter*; periodic *Directory*; *Journal of Immunology*. Annual conference in conjunction with the Federation of American Societies for Experimental Biology.

American Society for Microbiology (ASM)
  1325 Massachusetts Ave., NW, Washington, DC 20005.

Founded: 1899. 39,000 members with 36 local groups. Formerly the Society of American Bacteriologists. Scientific society of microbiologists promoting advancement of scientific knowledge in order to improve education in microbiology. Encourages the highest professional and ethical standards and the adoption of sound legislative and regulatory policies affecting the discipline of microbiology at all levels. Maintains numerous committees and 22 divisions, placement services, archives, compiles statistics. Sponsors competitions, bestows awards. Computerized database of mailing information. Affiliated with the International Union of Microbiological Societies. Publications: over twenty scientific journals available in alternate formats; *ASM News*; other professional pamphlets. Annual convention.

Association of Microbiological Diagnostic Manufacturers (AMDM)
    555 13th St. NW, Ste. 7W-403, Washington, DC 20004.

Founded: 1976. This association includes medical device manufacturers, distributors, and users. Represents members at legislative hearings, and conducts seminars. Publications: biennial *Newsletter*.

Committee on the Status of Women in Microbiology (CSWM)
    c/o Dr. Anne Morris Hooke, Miami University, Dept. of Microbiology, Oxford, OH 45056.

Founded: 1972. A committee of the American Society for Microbiology who investigate the status of women in microbiology in relation to their male counterparts in the workplace and within their professional society. Reports findings and conducts seminars at the annual meeting of ASM. Affiliated with the Federation of Organizations for Professional Women. Publications: *The Communicator*, a quarterly newsletter.

Foundation for Microbiology (FFM)
    c/o Byron H. Waksman, 300 E. 54th St., Ste. 5K, New York, NY 10022.

Founded: 1951. Encourages research in microbiology by establishing and maintaining professorships in microbiology; promotes meetings for discussion, exchange, and information disseminations; funds courses in microbiology; aids in publishing of scientific works. Publications: annual and quinquennial *Report*.

International Association of Biological Standardization (IABS)
    Case Postale 456, CH-1211 Geneva 4, Switzerland.

Founded: 1956. 530 members. Multinational. Representatives of medical, veterinary, and scientific fields; pharmaceutical companies; state controllers; private and university research workers. Works to regulate the standardization of medical and veterinary biological products. Conducts symposia. Publications: quarterly *Journal of Biological Standardization*; semi-annual *Newsletter*.

International Committee on Economic and Applied Microbiology (ICEAM)
    c/o Arnold Demain, Massachusetts Institute of Technology, Dept. of Biology, Cambridge, MA 02139.

Founded: 1970. Multinational. National microbiology societies. Promotes research and education in microbiology, biotechnology, and industrial production by means of microorganisms; sponsors conferences, seminars, and symposia. Publications: periodic *PAG Guidelines*. Biennial meeting with symposium.

International Committee on Food Microbiology and Hygiene (ICFMH)
    c/o Institute of Hygiene and Toxicology, Federal Research Centre for Nutrition, Engesserstrasse 20, W-7500 Karlsruhe, Germany.

Founded: 1953. Multinational. Microbiological societies in 30 countries. Furthers the academic practice of food micrology, especially with regard to safety and quality. Organizes professional training activities, arranges annual workshop. Affiliated with the American Society for Microbiology. Publications: bimonthly *International Journal of Food Microbiology*. Triennial symposium.

International Committee on Microbial Ecology (ICOME)
    Ocean Research Institute, University of Tokyo, Minamidai 1-15-1, Nakano-ku, Tokyo, Japan.

Founded: 1970. Multinational. Microbiology societies dealing with microbial ecology. Affiliated with the International Union of Biological Sciences and the International Union of Microbiological Societies. Publications: annual *Advances in Microbial Ecology*. Triennial meeting with symposium.

International Committee on Systematic Bacteriology (ICSB)
    c/o Prof. M. Goodfellow, Dept. of Microbiology, Medical School, University of Newcastle, Newcastle-upon-Tyne, Tyne and Wear NE2 4HH, England.

Founded: 1930. Multinational. A committee of the International Union of Microbiological Societies. Sponsors international collaboration and research in systematic bacteriology. Publications: quarterly *International Journal of Systematic Bacteriology*; *International Code of Nomenclature of Bacteria*. Quadrennial congress.

International Committee on Taxonomy of Viruses (ICTV)
    c/o Prof. K. W. Buck, Dept. of Biology, Imperial College of Science and Medical Technology, Prince Consort Road, London SW7 2BB, England.

Founded: 1966. Multinational. A committee of the International Union of Microbiological Societies. Seeks to develop a standard, internationally accepted system of virus classification and nomenclature. Publications: triennial *ICTV Reports*.

International Union of Immunological Societies (IUIS)
    Weizmann Institute of Sciences, 76100 Rehovot, Israel

Founded: 1969. National professional societies of basic and applied immunologists. Encourages the orderly development and utilization of the science of immunology; promotes the application of new developments to clinical and veterinary problems and standardizes reagents and nomenclature; conducts educational symposia and scientific meetings. Publications: triennial *International Union of Immunological Societies*. Triennial congress.

International Union of Microbiological Societies (IUMS)
    c/o Dr. Marc H. van Regenmortel, IBMC, 15, rue Descartes,
    F-67084 Strasbourg, France.

Founded: 1930. Multinational. National microbiological societies in 62 countries representing 100,000 microbiologists. Numerous commissions and committees, divisions, federations. Publications: five scientific journals; *IUMS Directory*. Annual congress.

Society for Applied Bacteriology (SAB)
    c/o Alastair Campbell, Polytechnic Southwest, Dept. of Food Seale-Hayne Faculty, Newton Abbot, Devon, England.

Founded: 1931. 1,600 members. Multinational. Individuals involved in the study of microbiology whose purpose is to promote and advance the study of microbiology, particularly bacteriology, in its application to agriculture, industry, and the environment. Publications: two scholarly journals; annual symposium series; a technical series. Holds three meetings each year.

Society for General Microbiology (SGM)
    Harvest House, 62 London Road, Reading, Berks., RG1 5AS,
    England.

Founded: 1945. 4,825 members; 2 regional groups. Multinational. Works to advance the study of general microbiology. Bestows awards and grants. Affiliated with the International Union of Microbiological Societies. Publications: three scholarly journals and an annual symposium series. Three meetings each year.

Society for Industrial Microbiology (SIM)
    PO Box 12534, Arlington, VA 22209-8534

Founded: 1948. 2,122 members and 2 local groups. Mycologists, bacteriologists, biologists, chemists, engineers, zoologists, and others interested in biological processes as applied to industrial materials and processes of microorganisms. Maintains placement service; conducts surveys and scientific workshops in industrial microbiology; presents awards. Computerized mailing list. Affiliated with the American Institute of Biological Sciences. Publications: scientific journals; *Membership Directory*; *SIM News*. Annual conference with symposium; also holds annual international conference.

UNEP/UNESCO/ICRO Panel on Microbiology
    c/o Dr. E. DaSilva, UNESCO, Div. of Science, Research, and Higher Education, Paris, France.

Founded: 1977. Multinational. Purpose is to preserve microbial gene pools and to make them available to developing nations. Publications: periodic *MIRCEN Journal of Applied Microbiology and Biotechnology*; *MIRCEN News*; surveys, reports, digests.

World Federation for Culture Collections (WFCC)
   c/o Barbara Kirsop, Microbial Strain Data Network, 307 Huntingdon
   Road, Cambridge CB3 OJX, England.

Founded: 1970. Multinational. Microbiologists in 55 countries working in research, education, and industry. Encourages the study of procedures for the isolation, culture, characterization, conservation, and distribution of micro-organisms. Works to establish a network of individuals and institutions possessing collections of microorganism cultures and cell lines and to facilitate communication between collection owners and users. Sponsors, with other organizations, the Microbial Strain Data Network Services including databases, electronic mail, bulletin boards, links to other networks, specialized software, training and consultancy.

Texts/General Works

Abbas, Abul K., Andrew H. Lichtman, and Jordan S. Pober. *Cellular and Molecular Immunology*, 2nd ed. Philadelphia: Saunders, 1994. 457 p. ISBN 072165505X.

An up-to-date text for understanding modern immunology.

Austyn, Jonathan M. and Kathryn J. Wood. *Principles of Cellular and Molecular Immunology*. New York: Oxford University Press, 1993. 735 p. $95.00. ISBN 0198542976.

Comprehensive introduction to cellular and molecular immunology.

Bacillus subtilis *and Other Gram-Positive Bacteria: Biochemistry, Physiology, and Molecular Genetics*. Edited by Abraham L. Sonenshein. Washington, DC: American Society for Microbiology, 1993. 987 p. $125.00. ISBN 1555810535.

Fundamental information about gram-positive bacteria including their physiology, genetics, biochemistry and regulation. This is a companion volume to *Escherichia coli and Salmonella typhimurium: Cellular and Molecular Biology*. Edited by Frederick C. Neidhardt. Washington, DC: American Society for Microbiology, 1987. This latter two-volume set does for gram-negative bacteria what *Bacillus subtilis* does for gram-positive bacteria.

The Bacteria; A Treatise on Structure and Function. Edited by I. C. Gunsalus et al. New York: Academic Press, 1960– . Multivolume, each volume priced separately.

Now in its twelfth volume, this time-honored treatise covers a range of topics in bacteriology. Volume 12, 1990: *Bacterial Energetics*. Edited by Terry A. Krulwich. 569 p. $143.00. ISBN 0123072123.

Birge, E. A. *Bacterial and Bacteriophage Genetics*, 3rd ed. New York: Springer-Verlag, 1994. 440 p. $49.50. ISBN 038794270X.

For students of microbiology, bacteriology, and genetics.

Brock, Thomas D. et al. *Biology of Microorganisms*, 7th ed. Englewood Cliffs, NJ: Prentice Hall, 1994. 909 p. $82.00. ISBN 0130421693.

Over 25 years the goal of this text is "to cover the field of microbiology as it is currently understood while at the same time emphasizing basic microbiological principles."

Cann, Alan J. *Principles of Molecular Virology*. San Diego, CA: Academic Press, 1993. $29.95. 234 p. ISBN 012158531X.

A general text suitable for undergraduates.

Clark, William R. *The Experimental Foundations of Modern Immunology*, 4th ed. New York: Wiley, 1991. 506 p. ISBN 0471517070.

Text designed to teach the language of immunology and to describe the most important experiments during early development in the field. Recent progress is described from an experimental point of view.

Comprehensive Virology. Edited by Heinz Fraenkel-Conrat and Robert R. Wagner. New York: Plenum Press, 1974–1984. 19 v. Index in volume 19.

Individual volumes have distinct titles.

Fundamental Immunology, 3rd ed. Edited by William E. Paul. New York: Raven Press, 1993. 1,490 p. $95.00. ISBN 0781700221.

Standard text and reference in immunology.

Genetic Engineering of Microorganisms. Weinheim, Germany: VCH, 1993. 200 p. $30.00. ISBN 3527300392.

An introduction to the industrial relevant fundamentals of genetic engineering.

Golub, Edward S. and Douglas R. Green. *Immunology: A Synthesis*, 2nd ed. Sunderland, MA: Sinauer, 1991. 744 p. $45.00. ISBN 878932631.

Up-to-date synthesis of modern immunology spanning molecular, cellular, and clinical aspects.

Hudson, Leslie and Frank C. Hay. *Practical Immunology*, 3rd ed. Oxford: Blackwell Scientific, 1989. 507 p. ISBN 0632014911. Index.

Primary source book for immunologists, serving as an introduction to the techniques and ideas of this science.

Levine, Arnold J. *Viruses*. New York: Scientific American Library, 1991. 239 p. ISBN 0716750317.

Levy, Jay A., Heinz Fraenkel-Conrat, and Robert A. Owens. *Virology*, 3rd ed. Englewood Cliffs, NJ: Prentice Hall, 1994. 447 p. ISBN 0139537538.

Comprehensive coverage of the field of virology, including pathology and history.

Logan, Miall A. *Bacterial Systematics*. Boston: Blackwell Scientific, 1994. 263 p. $32.95. ISBN 063203775X.

This introductory text describes and explains the theory and practice of bacterial classification and identification in a comprehensive review, appropriate at the undergraduate level. There are organism and subject indexes.

Maloy, Stanley R., John E. Cronan, Jr., and David Freifelder. *Microbial Genetics*, 2nd ed. Boston: Jones & Bartlett, 1994. 484 p. $50.00. ISBN 0867202483.

An authoritative text by well-known authors and scientists.

Microbiology, 4th ed. By Bernard D. Davis et al. Philadelphia: Lippincott, 1990. 1215 p. ISBN 0397506899.

Revised edition of *Microbiology,* including *Immunology and Molecular Genetics*, 3rd ed., 1980.

Mims, Cedric A. et al. *Medical Microbiology*. St. Louis: Mosby, 1993. 1 v. (various paging). $35.95 (paper). ISBN 0397446314.

"A slide atlas of medical microbiology, based on the contents of this book, is available."

Microbiology—Concepts and Applications. By Michael J. Pelczar, Jr. et al. New York: McGraw-Hill, 1993. ISBN 00704492581.

Neidhardt, Frederick C. et al. *Physiology of the Bacterial Cell: A Molecular Approach*. Sunderland, MA: Sinauer, 1990. 507 p. $48.50. ISBN 0878936084.

Analysis of the molecular devices in bacteria, explaining the composition and structure of the bacterial cell, how metabolism leads to the synthesis of a new cell, and how metabolism is regulated and coordinated.

Postgate, John. *Microbes and Man*, 3rd ed. New York: Cambridge
    University Press, 1992. 297 p. $49.95. ISBN 0521412595.

An excellent book discussing the impact of microorganisms, especially bacteria, on humans and the environment.

Priest, F. and B. Austin. *Modern Bacterial Taxonomy*, 2nd ed. London:
    Chapman & Hall, 1994. 240 p. $29.95. ISBN 041246120X.

This text covers molecular systematics, the construction of phylogenetic trees, typing of bacteria, DNA probes, and the use of the polymerase chain reaction in bacterial systematics.

Raven, Peter H. et al. *Biology of Plants*, 5th ed. New York: Worth,
    1992. 791 p. $24.95. ISBN 0879015322.

This excellent introduction to plant biology also discusses viruses, bacteria, and protists.

Rheinheimer, Gerhard. *Aquatic Microbiology*, 4th ed. Translated by
    Norman Walker. New York: Wiley, 1993. $79.50. ISBN
    0471926957.

The book serves well as a text and as a reference work.

Roitt, Ivan M. *Essential Immunology*, 8th ed. Boston: Blackwell Scientific,
    1994. 448 p. $32.95. ISBN 0632033134.

A fine immunology primer.

Salyers, Abigail A. and Dixie D. Whitt. *Bacterial Pathogenesis: A
    Molecular Approach*. Washington, DC: American Society for
    Microbiology, 1994. 418 p. $44.95 (paper). ISBN 1555810705.

After a comprehensive introduction, this text discusses the application of molecular techniques to the study of bacteria–host interaction, and the molecular basis of infectious diseases. Each chapter concludes with a glossary, a set of questions, and selected readings.

Schlegel, Hans G. *General Microbiology*, 7th ed. Translated by M. Kogut.
    New York: Cambridge University Press, 1993. 655 p. $100.00.
    ISBN 052143372X.

Established text with strength in providing a comprehensive description of the physiology and biochemistry of microorganisms.

Topley and Wilson's Principles of Bacteriology, Virology and Immunology.
    Edited by M. T. Parker and L. H. Collier. 8th ed. London: Edward
    Arnold, 1990. 5 v. ISBN 0713145943.

A standard in the fields of immunology and medical bacteriology.

Towner, K. J. and A. Cockayne. *Molecular Methods for Microbial Identification and Typing*. London: Chapman & Hall, 1993. 208 p. $52.95. ISBN 041249390X.

Introductory text considers applications and compares their advantages and disadvantages.

Voyles, Bruce A. *The Biology of Viruses*. St. Louis: Mosby, 1993. 386 p. $46.95. ISBN 0801663911.

Introduction to the viruses that illustrates the common features of their life-styles.

Weir, D. M. and John Stewart. *Immunology*, 7th ed. New York: Churchill Livingstone, 1993. 372 p. ISBN 0443046603.

Also, see *Handbook of Experimental Immunology*, 4th ed. Edited by D. M. Weir et al. Boston: Blackwell Scientific, 1986. 4 v. ISBN 0632014997 (set).

9

Ecology, Evolution, and Animal Behavior

This chapter covers materials for the allied fields of ecology, evolution, and animal behavior. As defined in the *McGraw-Hill Dictionary of Scientific and Technical Terms* (1989), ecology is "A study of the interrelationships which exist between organisms and their environment." Conservation biology and environmentalism are closely related but not extensively covered in this chapter. Evolution is "The processes of biological and organic change in organisms by which descendants come to differ from their ancestors." See also Chapter 7, "Genetics," for related materials.

Animal behavior here encompasses all biological subdisciplines including ethology ("The study of animal behavior in a natural context"), sociobiology ("A discipline that applies evolutionary biology to the study of animal social behavior, including human behavior; considered a synthesis of ethology, ecology, and evolution, in which social behavior is viewed as the result of natural selection and other biological processes"), and behavioral ecology ("The branch of ecology that focuses on the evolutionary causes of variation in behavior among populations and species"). Human behavior and comparative psychology are largely excluded from consideration.

Abstracts and Indexes

Animal Behavior Abstracts. v. 1– , 1972– . Bethesda, MD: Cambridge
    Scientific Abstracts. Quarterly. $515.00. ISSN 0301-8695.

Covers all aspects of animal behavior, including psychological as well as biological studies. Also available as part of the Life Sciences Collection online from DIALOG and STN, and on CD-ROM from SilverPlatter.

Current Advances in Ecological and Environmental Sciences. v. 1– ,
    1975– . Tarrytown, NY: Pergamon Press. Monthly. $1,040.00.
    ISSN 0955-6648.

Formerly *Current Advances in Ecological Sciences*. Covers over 2,000 periodicals of interest to ecologists, arranged by subject. Also has cross-references and author index. Available online as part of CABS from DIALOG.

Ecological Abstracts. v. 1– , 1974– . Norwich, UK: Elsevier. Monthly.
    $925.00. ISSN 0305-196X.

Covers all aspects of ecology, including aquatic, terrestrial, and applied. About 1,400 abstracts from 700 journals are added per year, as well as material from books, proceedings, and other sources. Has regional and organism indices. Available online as part of Geobase (1980 to the present) from DIALOG.

Ecology Abstracts. v. 1– , 1975– . Bethesda, MD: Cambridge
    Scientific Abstracts. Monthly. $985.00. ISSN 0143-3296.

Formerly *Applied Ecology Abstracts.* Covers all aspects of ecology. Available online from DIALOG and STN (1978 to the present) and on CD-ROM from SilverPlatter as part of the Life Sciences Collection and PolTox I (1981 to the present).

Environment Abstracts. v. 1– , 1970– . Bethesda, MD: Congressional Information Service, Inc. Monthly. $1,070. ISSN 0093-3287.

Covers over 800 journals plus proceedings, select government and other reports, monographs, newsletters, and other sources in areas including management, technology, biology, and law relating to the environment. Available online as ENVIROLINE from DIALOG (1971 to the present) and on CD-ROM as Environment Abstracts from CIS (1975 to the present). The majority of the documents indexed in *Environment Abstracts* are available in the full-text *Envirofiche* collection ($8,985 for 1994, backfiles to 1975 also available).

Environmental Periodicals Bibliography. v. 1– , 1972– . Santa Barbara, CA: Environmental Studies Institute. ISSN 0145-3815.

Indexes over 300 periodicals in the fields of human ecology, water resources, and others. Available online from DIALOG (1973 to the present) as Environmental Bibliography.

Pollution Abstracts. v. 1– , 1970– . Bethesda, MD: Cambridge Scientific Abstracts. Monthly. $885.00 (with annual index, $745.00 without). ISSN 0032-3624.

Covers both scientific research and government policies. Also available online from DIALOG and STN (1970 to the present) and on CD-ROM from Silver-Platter as part of PolTox I (1981 to the present).

See also *Biological Abstracts, Chemical Abstracts, Current Contents/Agriculture, Biology, and Environmental Sciences,* and *Zoological Record* in Chapter 4, "Abstracts and Indexes."

Dictionaries and Encyclopedias

Allaby, Michael. *Dictionary of the Environment,* 3rd ed. New York: New York University Press, 1991. 423 p. $75.00. ISBN 081470591X.

There are brief definitions of terms dealing with ecology, the environment, and pollution. One nice feature is a list of 22 environmental disasters which occurred since the mid-1950's, such as Chernobyl, the *Torrey Canyon* oil spill, and Love Canal.

Art, Henry W., general ed. *Dictionary of Ecology and Environmental Science*. New York: Holt, 1993. 632 p. $60.00. ISBN 0805020799.

Designed for a wide range of specialized and non-specialized users. There are over 8,000 entries with brief definitions. A number of appendices are also included, though not listed in a table of contents. They include subjects such as the periodic table of elements, a USDA hardiness zone map, leaf description terms used to identify plants, and the Beaufort wind scale.

Ashworth, William. *Encyclopedia of Environmental Studies*. New York: Facts on File, 1991. 480 p. $60.00. ISBN 0816015317.

Intended for environmental activists and other non-scientists, this encyclopedia has entries on environmental groups and policy-making entities as well as definitions of terms from geology, ecology, meteorology, environmental engineering, and other scientific disciplines.

Axelrod, Alan and Charles Phillips. *The Environmentalists: A Biographical Dictionary from the 17th Century to the Present*. New York: Facts on File, 1993. 258 p. $45.00. ISBN 0816027153.

Covers both individuals and organizations, with about 600 entries ranging in length from a paragraph to a couple of pages. Some surprising individuals appear, such as Saddam Hussein (for his role in the destruction of Kuwait's environment). Most of the entries are for friends of the environmental movement, however, and include the Animal Liberation Front, Alexander von Humboldt, and John Muir.

Beacham's International Threatened, Endangered and Extinct Species. Washington, DC: Beacham Publishing, 1994. CD-ROM. $495.00, annual updates $95.00.

A multimedia CD-ROM presentation of information on over 1,200 species of plants an animals worldwide. Includes maps, photos, some sounds, glossary, and articles on each species in addition to a 10,000-item bibliography, CITES list, and complete text of the U.S. Fish and Wildlife Service's *Technical Bulletins* from 1976 to the present.

Cambridge Encyclopedia of Human Evolution. Steve Jones, Robert Martin, and David Pilbeam, eds. New York: Cambridge University Press, 1992. 506 p. $95.00. ISBN 0521323703.

The entries in this work are not alphabetical, but are arranged by subject in several broad categories. The emphasis here is as much on non-human primate evolution, behavior, language, and ecology as on humans, though there are discussions of early human behavior and ecology and human populations past and present, in addition to the human fossil record and other evolutionary topics.

Appendices include a who's who in human evolution, a geological timescale, and a world map of important fossil sites.

Concise Oxford Dictionary of Ecology. Michael Allaby, ed. New York: Oxford University Press, 1994. 415 p. $12.95 (paper). ISBN 0192116894, 0192861603 (paper).

Another of Oxford University Press's excellent concise dictionaries. Covers over 5,000 terms in ecology and conservation, as well as relevant terms from fields such as animal behavior, physiology, climatology, and glaciology.

Dictionary of Environmental Science and Technology. Edited by Andrew Porteous. Bristol, PA: Distributed by Taylor & Francis for the Open University Press, UK, 1991. 403 p. ISBN 0335092314.

Written for the general reader to introduce a working knowledge of the scientific and technical language associated with current environmental issues and areas of study. Although emphasis is on the UK, it includes information on major international and U.S. organizations as well.

Dictionary of Ethology and Animal Learning. Rom Harre and Roger Lamb, eds. 1st MIT Press ed. Cambridge, MA: MIT Press, 1986. 171 p. ISBN 0262580764 (paper).

This dictionary consists of extracts from the *Encyclopedic Dictionary of Psychology*. Most definitions are of paragraph length, though some are longer. Both biological and psychological terms are included.

Dictionary of Substances and Their Effects. Mervyn L. Richardson and Sharat Gangolli, eds. England: Royal Society of Chemistry, 1992– . $350.00/vol. ISBN 0851863310 (v. 1), 0851863418 (v. 2).

To be published in seven volumes, expected completion in 1995. This set provides a guide to the effects of over 5,000 substances, including identifiers (registry numbers, molecular formula, etc.), physical properties, occupational exposure, ecotoxicity, environmental fate, mammalian and avian toxicity, legislation, and comments. There are also references where applicable. The final volume of the set will include an index of chemical names, CAS registry numbers, and molecular formulas.

Encyclopedia of Animal Behavior. Peter J. B. Slater, ed. New York: Facts on File Publications, 1987. 144 p. $29.95. ISBN 0816018162.

This attractive book, like the other Facts on File encyclopedias listed on p. 224, is suitable for undergraduate students or the general public. It consists of discussions of animal behavior, including ethology, the behavior of individual animals, and social behaviors.

Encyclopedia of Animal Ecology. Peter D. Moore, ed. New York: Facts
   on File Publications, 1987. 144 p. $29.95. ISBN 0816018189.

This attractive book covers world zoogeographical areas, and discusses man and
nature.

Encyclopedia of Animal Evolution. R. J. Berry and A. Hallam, eds. New
   York: Facts on File Publications, 1987. 144 p. $29.95. ISBN
   0816018197.

Also suitable for undergraduates, this encyclopedia covers the prehistoric world,
the consequences and mechanisms of evolution, and a brief discussion of human
evolution.

Environmental Dictionary, 2nd ed. Compiled by James J. King.
   Executive Enterprises Pub. Co., Inc. 977 p. ISBN
   0781601711. $89.95.

This dictionary covers environmental/regulatory terminology and may also be
used as a locator for regulations that apply to specific terms.

Environmental Encyclopedia. William P. Cunningham, et al., eds. Detroit:
   Gale, 1994. 981 p. $195.00. ISBN 0810388561.

Has numerous illustrations and short bibliographies for further reading in a
number of the over 1,300 articles and definitions. Written in non-technical
language.

Glossary of Environmental Terms and Acronym List. Environmental
   Protection Agency, Office of Communications and Public Affairs.
   Washington, DC: US Environmental Protection Agency, 1989. 29 p.

This glossary is designed to help the general public define the most common
terms found in EPA documents. There is also an extensive list of acronyms.

Grzimek, Bernhard. *Grzimek's Encyclopedia of Ecology.* New York: Van
   Nostrand Reinhold, 1976. 705 p.

While slightly dated, this encyclopedia is still valuable for its historical coverage
and numerous illustrations. It is a companion volume to the *Grzimek's Animal
Life Encyclopedia* (see entry on p. 396 in Chapter 13, "Zoology").

Grizmek, Bernhard. *Grzimek's Encyclopedia of Evolution.* New York:
   Van Nostrand Reinhold, 1976. 560 p.

Similar in scope and utility to the *Grzimek's Encyclopedia of Ecology*, above.

Heymer, Armin. *Ethologisches Worterbuch. Ethological Dictionary.
   Vocabulaire Ethologique.* Hamburg, Germany: Verlag Paul Parey,
   1977. 237 p. ISBN 3489663365.

This trilingual German/English/French dictionary has short definitions of ethological terms in order of the German term, followed by the English and French terms and their definitions. There are also English and French indices.

Immelman, Klaus and Colin Beer. *A Dictionary of Ethology*. Cambridge, MA: Harvard University Press, 1989. 336 p. $35.00. ISBN 0674205065.

This is a translation of *Worterbuch der Verhaltensforschung*. Has mostly paragraph long definitions of terms used in ethology and animal behavior.

Lincoln, Roger J., Geoffrey Allan Boxshall, and Peter F. Clark. *A Dictionary of Ecology, Evolution, and Systematics*. New York: Cambridge University Press, 1983. 298 p. $24.95 (paper). ISBN 0521269024 (paper).

As well as brief definitions of terms used in the general area of natural history, this dictionary includes 21 appendices covering a range of topics including the geological time scale, zoogeographic areas, and transliterations for the Greek and Russian alphabets.

McGraw-Hill Encyclopedia of Environmental Science & Engineering. Sybil P. Parker and Robert A. Corbitt, eds., 3rd ed. New York: McGraw-Hill, 1993. 749 p. $85.50. ISBN 0070513961.

Topics such as meteorology, dam building, and ecology are covered in this encyclopedia. Many of the entries are taken from the *McGraw-Hill Encyclopedia of Science and Technology*.

Milner, Richard. *The Encyclopedia of Evolution: Humanity's Search for its Origins*. New York: Facts on File, 1990. 481 p. $45.00. ISBN 0816014728.

Most of the entries in this encyclopedia relate to either modern evolutionary concepts or the history of evolutionary theory. There are also many entries relating to what might be called Darwinian trivia, such as articles about Jemmy Buttons (a Fuegan Indian taken to England on the *Beagle*) and Darwin's Sandwalk path at his home in Sussex.

Official World Wildlife Fund Guide to Endangered Species of North America. Edited by David W. Lowe, et al. Washington, DC: Beecham Publishing, 1990. 2 vols. $195.00. ISBN 0933833172.

This publication has received excellent reviews. It provides descriptions for plants and animals that are federally listed as either endangered or threatened. Entries include a locator map, a black and white photograph, and information on habitat, food, reproduction, threats to the species, behavior, historic range, current distribution, and steps being taken towards recovery. A glossary, indexes of occurrence by state, common and scientific names. This comprehensive account is a valuable resource and highly recommended; updated periodically by additional volumes. Vol. 3: 1992, 1647 p. Species listed Aug. 1989–Dec. 1991. $85.00. ISBN 0933833296.

Stevenson, L. Harold and Bruce Wyman. *The Facts on File Dictionary of Environmental Science*. New York: Facts on File, 1991. 304 p. $24.95. ISBN 0816023174.

Covers over 3,000 entries from the environmental sciences, as well as legal and governmental terms. Written for both researchers and non-specialists.

Directories

Also see entries in the Societies section (p. 252).

Conservation Directory. 1956– . Washington, DC: National Wildlife Federation. Annual. ISSN 0069-911X.

Lists environmental departments, agencies, and offices for about 2,000 U.S. government agencies, as well as universities with environmental programs and, regional, national, and international conservation organizations. The directory includes indices of publications, persons mentioned, and subjects.

Directory and Register of Certified Ecologists, annual supplement to *Bulletin of the Ecological Society of America*.

Environmental Profiles; A Global Guide to Projects and People. Edited by Linda Sobel Katz et al. Hamden, CT: Garland Publishing, 1993. 1,400 p. $125.00. ISBN 0815300638.

This international guide identifies and describes important environmental undertakings as well as the people involved with them. It includes scientific solutions, potential allies in the political arena, speakers, educators, equipment, funding sources, and facts. Data is provided from 140 countries, 3000 people, and 7000 projects, and is extensively indexed.

Gale Environmental Sourcebook; A Guide to Organizations, Agencies, and Publications. Edited by Karen Hill and Annette Piccirelli. Detroit: Gale Research Inc., 1992. 688 p. $75.00. ISBN 0810384035.

Descriptive information for 8,634 environmental organizations, information services, programs, and publications for all aspects of the environment including advocacy and education, policy and enforcement, research and development, and consumer issues/products. Appendices list worldwide endangered or threatened species. Alphabetical and subject indexes.

Naturalists' Directory and Almanac (International). v. 43– , 1980– .
    Kinderhook, NY: World Natural History Publications. Irregular.
    $24.95. ISSN 0277-609X.

This directory is based on the PIFON (Permanent International File of Naturalists) database and provides the address and subject interests of naturalists who wished to be included in the directory. There are geographical and subject indices.

World Directory of Environmental Organizations, 4th ed. Sacra-
    mento, CA: California Institute of Public Affairs, 1992. 184 p.
    $35.00. ISBN 0912102977.

This useful directory is a comprehensive guide to 2,100 organizations in over 200 countries that are concerned with problems of the environment and natural resources. An index, a glossary, a list of landmark world events in environmental protection, and a bibliography of related directories and databases are included. The *World Directory* is a cooperative project of the California Institute of Public Affairs, the Sierra Club, and the International Union for Conservation of Nature and Natural Resources.

World Environmental Directory. v. 1– , 1974– . Silver Spring, MD:
    Business Publishers. ISSN 0094-4742.

Lists individuals, companies, agencies, and organizations involved in the environment in the United States, Canada, and worldwide.

General Works

Barbour, Michael G. and William Dwight Billings, eds. *North American
    Terrestrial Vegetation*. New York: Cambridge University Press,
    1990. 448 p. $37.95 (paper). ISBN 0521386780.

Describes the major plant formations of North America, such as grasslands, chaparral, and Alpine areas.

Behavior of Marine Animals: Current Perspectives in Research. Edited by
    Howard E. Winn and Bori L. Olla. New York: Plenum Press,

1972– . Irregular. v. 6: Shorebirds, migration and foraging behavior. 1984. $85.00. ISBN 0306415917.

Volumes to date include v. 1: Invertebrates. v. 2: Vertebrates. v. 3: Cetaceans. v. 4: Marine birds. v. 5: Shorebirds, breeding behavior and populations, and v. 6: Shorebirds, migration and foraging behavior.

Bekoff, Marc and Dale Jamieson, eds. *Interpretation and Explanation in the Study of Animal Behavior*. Volume 1: *Interpretation, Intentionality, and Communication*. Volume 2: *Explanation, Evolution, and Adaptation*. Boulder, CO: Westview Press, 1990. 2 vols. v. 1, $45.00 (paper). v. 2, $45.00 (paper). ISBN 0813377048 (v. 1, paper), 08113379792 (v. 2, paper).

These two volumes contain 37 essays covering a wide range of topics dealing with animal behavior. The emphasis is on issues, rather than on specific behaviors, and topics include moral issues, the minds of animals, communication, analysis of behaviors, and the relationship between animal behavior and artificial intelligence.

Benchmark Papers in Animal Behavior. 1974– . Stroudsburg, PA: Dowden, Hutchinson, and Ross.

> v. 1: Carter, C. S., compiler. *Hormones and Sexual Behavior*. 1974.

> v. 2: Stokes, A. W., compiler. *Territory*. 1974.

> v. 3: Schein, M. W., compiler. *Social Hierarchy and Dominance*. 1975.

> v. 4: Collias, N. E. and E. C. Collias, eds. *External Constructions in Animals*. 1976.

> v. 5: Hess, E. H. and S. B. Petrovich, eds. *Imprinting*. 1977.

> v. 6: Porges, S. W. and M. G. H. Coles, eds. *Psychophysiology*. 1976.

> v. 7: Tavolga, W. N., ed. *Sound Reception in Fishes*. 1976.

> v. 8: Banks, E. M., ed. *Vertebrate Social Organization*. 1977.

> v. 9: Tavolga, W. N., ed. *Sound Production in Fishes*. 1976.

> v. 10: Muller-Schwarze, D., ed. *Evolution of Play Behavior*. 1978.

> v. 11: Silver, R., ed. *Parental Behavior in Birds*. 1977.

> v. 12: Scott, J. P. *Critical Periods*. 1978.

Benchmark Papers in Behavior. 1974– . Stroudsburg, PA: Dowden, Hutchinson, and Ross.

v. 13: Satinoff, E., ed. *Thermoregulation*. 1980.

v. 14: Hendersen, R. W. *Learning in Animals*. 1982.

v. 15: Dewsbury, D. A. *Mammalian Sexual Behavior: Foundations for Contemporary Research*. 1981.

v. 16: Hirsch, J. and T. R. McGuire, eds. *Behavior-Genetic Analysis*. 1983.

v. 17: Dewsbury, D. A. *Foundations of Comparative Psychology*. 1984.

v. 18: Brozek, J. *Malnutrition and Human Behavior: Experimental, Clinical, and Community Studies. 1985.*

v. 19: Burghardt, G. M. *Foundations of Comparative Ethology*. 1985.

Benchmark Papers in Ecology. Stroudsburg, PA: Dowden, Hutchinson, and Ross.

v. 1: Pomeroy, L. R., compiler. *Cycles of Essential Elements*. 1974.

v. 2: Davis, D. E., compiler. *Behavior as an Ecological Factor*. 1974.

v. 3: Whittaker, R. H., compiler. *Niche: Theory and Application*. 1975.

v. 4: Wiegart, R. G., ed. *Ecological Energetics*. 1976.

v. 5: Golley, F. B., ed. *Ecological Succession*. 1977.

v. 6: McIntosh, R. P. *Phytosociology*. 1978.

v. 7: Tamarin, R. H. *Population Regulation*. 1978.

v. 8: Lieth, H. H. *Patterns of Primary Production in the Biosphere*. 1978.

v. 9: Shugart, H. H., ed. *Systems Ecology*. 1979.

v. 10: Jordan, C. F., ed. *Tropical Ecology*. 1981.

v. 11: Lidicker, W. Z. and R. L. Caldwell, eds. *Dispersal and Migration*. 1982.

v. 12: Young, G. L., ed. *Origins of Human Ecology*. 1983.

v. 13: Patrick, R., ed. *Diversity*. 1983.

The Book of Life. General editor, Stephen Jay Gould. New York: W. W. Norton, 1993. 256 p. $40.00. ISBN 0393035573.

A facinating, beautifully illustrated, and well-written book covering the history of life and of evolutionary thought. A coffee table book sure to please dinosaur lovers and other amateurs, but with lots of meat for the student.

Brooks, Daniel R. and Deborah A. McLennan. *Phylogeny, Ecology, and Behavior: A Research Program in Comparative Biology.* Chicago: University of Chicago Press, 1991. 434 p. $45.00, $21.00 (paper). ISBN 0226075710, 0226075729 (paper).

The authors wrote this volume to encourage the re-integration of the evolutionary perspective with the study of ecology and animal behavior.

Computational Neuroscience. Eric L. Schwartz, ed. Cambridge, MA: MIT Press, 1990. 456 p. $47.50, $25.00 (paper). ISBN 0262192918, 0262691647 (paper).

The purpose of this volume is to provide definions of the field of computational neuroscience. The volume is divided into several major sections, including overviews, the synaptic level, the network level, neural maps, and systems.

Conceptual Issues in Evolutionary Biology. Edited by Elliott Sober, 2nd ed. Cambridge, MA: MIT Press, 1994. 506 p. $55.00, $27.50 (paper). ISBN 0262193361, 0262691620 (paper).

An anthology of essays written by scientists and philosophers on issues in evolutionary biology, such as teleology, fitness, units of selection, and ethics. A very useful supplement for courses in evolution.

Dawkins, Richard. *The Selfish Gene.* New ed. New York: Oxford University Press, 1989. 352 p. $27.95, $10.95 (paper). ISBN 0192177737, 0192860925 (paper).

The author argues that the basic unit of selection is the gene, not the individual, as most other authorities believe. This is a classic, and very important (though controversial) book.

Ecosystems of the World. David W. Goodall, series ed. New York: Elsevier, 1977– . v. ISBN 0444417028 (series).

This multi-volume set covers the major ecosystems of the world, both terrestrial and aquatic, natural and managed. The set is planned with 29 titles, some titles with 2 volumes. Some examples of the ecosystems covered are *Mires: Swamp, Bog, Fen, and Moor* (2 volumes), *Managed Grasslands*, and *Tropical Savannahs*. Each title contains contributions from a number of individuals and includes discussions both of general topics and descriptions of particular biogeographical regions. The volumes are published out of sequence, and each costs about $200.

Ereshefsky, Marc, ed. *The Units of Evolution: Essays on the Nature of Species*. Cambridge, MA: MIT Press, 1992. 405 p. $27.50. ISBN 0262550202.

An anthology of previously published papers discussing the biological concept of species, a particularly thorny problem in evolutionary biology. The authors are all recognized experts, and the topics discussed range from biological concepts to philosophical problems.

Grant, Verne. *The Evolutionary Process: A Critical Study of Evolutionary Theory*, 2nd ed. New York: Columbia University Press, 1991. 487 p. $52.00. ISBN 0231073240.

The main objective of the author, a well-known evolutionary biologist, was to "provide a comprehensive and critical review of modern evolutionary theory." He covers topics in genetics and mutation, natural selection, acquired characters, speciation, macroevolution, and human evolution.

Griffin, Donald R. *Animal Minds*. Chicago: University of Chicago Press, 1992. 310 p. $24.95. ISBN 0226308634.

An accessible volume which takes the stance that animals (maybe even insects) do have consciousness; an interesting contrast is with Kennedy's *The New Anthropomorphism* (p. 233).

Hoelzel, A. Rus and Gabriel A. Dover. *Molecular Genetic Ecology*. New York: IRL Press, 1991. (In Focus series). 75 p. $14.95. ISBN 0199632650.

The authors present information on the applications of molecular biology to the study of genetic variation. The series is designed to help undergraduate and graduate students to keep up with fast-moving fields outside their own area of concentration.

Hutchinson, G. Evelyn. *A Treatise on Limnology*. New York: Wiley, 1957–1993. 4 v. $125.00 (v. 4). ISBN 0471542946 (v. 4).

The classical treatise on limnology. The author died before completing the fourth volume, leaving a proposed fifth volume unbegun. The existing volumes are Geography, physics and chemistry; Introduction to lake biology and the limnoplankton; Limnological botany; and The zoobenthos. The final volume would have covered productivity and various ecological topics.

Keller, Evelyn Fox and Elisabeth A. Lloyd, eds. *Keywords in Evolutionary Biology*. Cambridge, MA: Harvard University Press, 1992. 414 p. $45.00. ISBN 0674503120.

These short essays cover a range of topics in evolutionary biology, including topics such as adaptation, eugenics, fitness, natural selection, the niche, and units of selection. The essays are written by recognized experts in the field and are designed to provide general readers with an introduction to the concepts discussed.

Kennedy, J. S. *The New Anthropomorphism*. New York: Cambridge University Press, 1992. 194 p. $54.95, $17.95 (paper). ISBN 0521410649, 0521422671 (paper).

The author discusses the dangers of the modern version of anthropomorphism (the attribution of human emotions and thinking processes to animals).

Margulis, Lynn. *Symbiosis in Cell Evolution: Microbial Communities in the Archean and Proterzoic Eons*, 2nd ed. New York: Freedman, 1993. 452 p. ISBN 0716770288, 0716770296 (paper).

The author presents the thesis that eukaryotic organisms (having cells with nuclei) evolved from the symbiotic relationship between some prokaryotic cells (those without nuclei). While still controversial, the theory is gaining support.

Margulis, Lynn and Lorraine Olendzenski, eds. *Environmental Evolution: Effects of Life on Planet Earth*. Cambridge, MA: MIT Press, 1992. 405 p. $29.95. ISBN 0262132737.

This volume was developed out of a class given at Boston University and Amherst. The text discusses the evolutionary interactions between organisms and the environment, from the very earliest life to modern organisms. The authors also include discussion of the influence of living organisms on the environment, including the Gaia hypothesis.

Michod, Richard E. and Bruce R. Levin. *The Evolution of Sex: An Examination of Current Ideas*. 1988. $31.50. ISBN 0878934596.

The 17 essays in this book discuss various questions dealing with the evolution of sex as a form of genetic recombination. There is an extensive bibliography. It can be used as a text or a general resource.

Pickett, Steward T. A., Jurek Kolasa, and Clive G. Jones. *Ecological Understanding*. San Diego: Academic Press, 1994. 206 p. $59.95. ISBN 012554720X.

A philosophical discussion of theory in ecology and evolution and the need for integrating ideas from other fields and sub-disciplines. Intended for the use of practicing scientists.

Real, Leslie A. and James H. Brown. *Foundations of Ecology: Classic Papers with Commentaries*. Chicago: Published in association with

The Ecological Association of America by The University of Chicago
Press, 1991. 905 p. $70.00, $27.50 (paper). ISBN 0226705935,
0226705943 (paper).

An anthology of 40 important ecology papers published before 1977, each with
an explanatory essay.

Rittner, Don. *EcoLinking: Everyone's Guide to Online Environmental
Information.*

See the entry (p. 235) in the Guides to Internet Resources section.

Sigmund, Karl. *Games of Life: Models in Evolutionary Biology.* New
York: Oxford University Press, 1993. 240 p. $49.95, $17.95
(paper). ISBN 0198546653, 0198547838 (paper).

The author discusses mathematical models used in genetics, evolution, population biology, and animal behavior in a very clear and understandable, yet light-
hearted manner. Very readable.

Wilson, Edward O. *The Diversity of Life.* Cambridge, MA: Harvard
University Press, 1992. 424 p. $29.95. ISBN 0674212983.

Wilson discusses a variety of topics dealing with biodiversity on a level suitable
for the general public as well as for more advanced readers. One especially
useful chapter (Chapter 8, "The Unexplored Biosphere") discusses the estimation
of the total number of known species, which Wilson places at 1.4 million, give
or take a hundred thousand or so. Other chapters include essays on mass ex-
tinctions, the species concept, evolution, and several chapters on the human
impact on extinction, and the environmental ethic.

Wilson, Edward O. *Sociobiology: The New Synthesis.* Cambridge, MA:
Harvard University Press, 1975. 697 p. $45.00. ISBN
0674816218.

This massive compilation helped to launch the field of sociobiology, and is
probably one of the best-known works in the field.

Guides to Internet Resources

Ecology, evolution, and animal behavior are well represented in discussion
groups. Some noteworthy USENET examples include: bionet.biology.tropical,
sci.bio.conservation, sci.bio.ecology, sci.bio.ethology, and sci.bio.evolution.
Selected listservs include: biosph-l (listserv@ubvm.cc.buffalo.edu), discussion
of the biosphere and ecology; ecolog-l (listserv@umdd.umd.edu), the list for the
Ecological Society of America; ethology (listserv@searn.sunet.se), and humevo
(listserv@gwuvm.gwu.edu), group for human evolution.

There are a number of excellent biology Gophers which cover ecology and evolution, though animal behavior resources seem scarcer. Try the Smithsonian Natural History Gopher, the Harvard and Cornell Biodiversity and Biological Collections Gophers, or Stanford's Biological Information Servers. All provide access to other ecological Gophers. At the time of writing, there were fewer WWW sites than Gopher sites, although this is rapidly changing.

Ecolinking: Everyone's Guide to Online Environmental Information.
    Compiled by Don Ritter. Berkeley, CA: Peachpit Press, 1992.
    352 p. $18.95 (paper). ISBN 0938151355.

This guide provides information on computer access to global networks and commercial online services for academic, bibliographic, and scientific environmental issues and research.

Guides to the Literature

Beacham's Guide to Environmental Issues and Sources. Walton
    Beacham, ed. 5 v. $240.00. ISBN 0933833318.

Has more than 40,000 citations, many with annotations, to books, articles, reports, videos, and a number of other sources dealing with environmental issues. The set is divided into chapters, then subdivided into narrower topics. There are extensive references to topics such as ecology, biodiversity, wildlife conservation, and various ecosystems such as rainforests and deserts in addition to environmental topics such as recycling. A very useful and useable set.

Bibliographic Guide to the Environment. Boston, MA: G. K. Hall,
    1992. Annual. $165.00. ISSN 1063-6153.

Updated annually, this guide includes books and related materials covering law, urban planning, public health, the sciences, economics, and industry as they relate to environmental issues. Subject, author, and title access are provided for conservation, pollution, atmospheric trends, alternative and renewable energy, waste management, public policy issues, endangered species, and environmental laws and legislation.

Encyclopedia of Environmental Information Sources. Detroit: Gale Re-
    search, 1993. 1,813 p. $125.00. ISBN 0810385686.

This guide covers a wide variety of information sources, including abstracting and indexing services, almanacs, yearbooks, dirctories, encyclopedias, periodicals, resarch centers, and societies in the areas of the environment and ecology. Broken down into narrow topics such as bogs, nicotine, or New Mexico environmental agencies. Has 3,400 citations and over 1,100 topics. Also available on magnetic tape or diskette. Covers a wider range of types of information sources than *Beacham's*, above, but has fewer citations.

Handbooks

Calow, Peter, ed. *Handbook of Ecotoxicology*. Cambridge, MA:
Blackwell Scientific Publishers, 1993– . 2 v. $145.00 (v. 1).
ISBN 0632035730 (v. 1).

Volume 1 of this handbook provides information on tests for ecotoxicological effects, both field and laboratory, for all types of ecological systems. Volume 2 focuses on the toxicity of synthetic chemicals themselves. The handbook is intended for practitioners and provides extensive references.

Calow, Peter and Geoffrey E. Petts, eds. *The Rivers Handbook*. Cambridge, MA: Blackwell Scientific Publishers, 1992– . v. 1,
Hydrological and Ecological Principles. $185.00 (v. 1). ISBN
0632028327. v. 2, *Problems, Diagnosis, and Management* (due
spring 1994).

This two-volume set provides information on river management. The first volume discusses the basic scientific principles, including hydrology and geology, the flora and fauna of rivers, and nutrient and energy cycles. There are also 5 case studies, ranging from the highly managed Rhône River to the largely untouched Orinoco. According to the Preface of Volume 1, Volume 2 will develop "the principles and philosophy presented in Volume 1 into the management sphere, organizing the approach around *problems, diagnosis,* and *treatment.*"

Jorgensen, Sven Erik. *Handbook of Ecological Parameters and Ecotoxicology*. New York: Elsevier, 1991. 1,263 p. $328.00. ISBN
0444886044.

Contains data and parameters for the ecological and toxicological fields, including data on the composition and ecological parameters of organisms, the ecosphere and chemical compounds, equations for environmental processes, biological effects, equilibria and rate constants for environmentally important processes, and the effects of pesticides.

McFarland, David. *The Oxford Companion to Animal Behaviour*. New
York: Oxford University Press, 1987. 685 p. $49.95, $19.95
(paper). ISBN 0198661207, 0192819909 (paper).

This handbook has short essays covering a variety of topics in animal behavior, and is intended as a reference work for non-specialists. The authors also provide indexes of scientific names and common names of the species mentioned in the handbook, with cross-references to the essay in which the species is mentioned.

Histories

Bowlby, John. *Charles Darwin: A New Life*. New York: W. W. Norton, 1991. 511 p. $24.95. ISBN 0393029409.

One of many recent biographies of Darwin; see also Darwin (below) and Desmond and Moore (p. 238). Bowlby's main interest is in Darwin's life-long illness, which he attributes to hyperventilation syndrome, a psychosomatic illness.

Bowler, Peter J. *Evolution: The History of an Idea*. Revised ed. Berkeley: University of California Press, 1989. 432 p. $42.50, $13.95 (paper). ISBN 0520063856, 0520063864 (paper).

Covers the history of evolutionary theories, including pre-Darwinian, Darwinian, and post-Darwinian thought. Includes controversies, both within and outside of the main scientific line.

Bowler, Peter J. *The Norton History of the Environmental Sciences*. 1st American ed. New York: W. W. Norton and Co., 1993. 634 p. $35.00, $15.95 (paper). ISBN 0393038352, 0393310426 (paper).

This is a history of geography, geology, oceanography, meteorology, natural history, paleontology, evolution, and ecology, written by a well-known scholar of the history of evolutionary theory. The coverage ranges from the ancient world to the modern, with perhaps most emphasis on the 18th and 19th centuries, and is arranged by concepts such as the tree of life or plate tectonics.

Bramwell, Anna. *Ecology in the 20th Century: A History*. New Haven, CT: Yale University Press, 1989. 292 p. ISBN 0300043430.

The "ecology" of the title refers to the political or social environmental movement, not the scientific discipline. The author concentrates on England, Germany, and the United States and covers the time from Ernst Haeckel's coining of the word "ecology" in the 1860's to the modern Green, eco-socialist, neo-pagan, and back-to-the-land movements.

Darwin, Charles. *Autobiography and Selected Letters*. Edited by Francis Darwin. New York: Dover Publications, 1958. 365 p. $6.95. ISBN 0486204790.

Darwin's own version of his life.

The Darwin CD-ROM. Created by Pete Goldie and Michael T. Ghiselin. San Francisco: Lightbinders, Inc., 1992. $99.95.

This multimedia CD-ROM includes the full text, plus illustrations, from Darwin's major works, including *The Origin of Species*, *The Descent of Man*, and *The Voyage of the Beagle*. Darwin and Alfred Russel Wallace's jointly written article, "On the Tendency of Species to Form Varieties," is also included, along with the third edition of Ghiselin's *Triumph of the Darwinian Method* (a guide to the study of Darwin), a Darwin bibliography, and a Darwin timeline. Audio is also included for a number of the organisms (birds and mammals) pictured in *The Voyage of the Beagle*.

Desmond, Adrian and James Moore. *Darwin*. New York: Warner Books, 1991. 808 p. $35.00. ISBN 0446515892.

The alternate title of this biography about sums up the authors' premise: *Darwin: The Life of a Tormented Evolutionist*. The focus is on Darwin's personal, rather than his scientific, life. Less emphasis on Darwin's health than Bowlby (p. 237).

Dewsbury, Donald A., ed. *Leaders in the Study of Animal Behavior: Autobiographical Perspectives*. Lewisburg: Bucknell University Press, 1985. 512 p. $65.00. ISBN 0838750524.

This is a collection of autobiographical essays written by the major figures in the study of animal behavior, including luminaries such as Konrad Lorenz, John Maynard Smith, Edward O. Wilson, Niko Tinbergen, and Irenaus Eibl-Eibesfeldt.

Golley, Frank B. *History of the Ecosystem Concept in Ecology: More than the Sum of the Parts*. New Haven, CT: Yale University Press, 1993. 254 p. $30.00. ISBN 0300055463.

Eminent ecologist explains the ecosystem concept tracing its evolution and contributions from Americans and Europeans. Discusses the explosive growth of ecosystem studies.

Hagen, Joel B. *An Entangled Bank: The Origins of Ecosystem Ecology*. New Brunswick, NJ: Rutgers University Press, 1992. 245 p. $38.00, $16.00 (paper). ISBN 0813518237, 0813518245 (paper).

Covers the history of ecosystem ecology (as opposed to evolutionary ecology and other specialties) from the early days of ecology to the modern "spaceship Earth" and Gaia viewpoints. The emphasis is on the American schools of thought.

Hull, David L. *Darwin and His Critics: The Reception of Darwin's Theory of Evolution by the Scientific Community*. Chicago: University of Chicago Press, 1983. 473 p. $17.00 (paper). ISBN 0226360466 (paper). Originally published in 1973 by Harvard University Press.

In this book, Hull has collected the major reviews of the *Origin of Species* which were written shortly after its publication, offering a good source of information on what contemporary critics thought of the theory of evolution by natural selection.

Lorenz, Konrad Z. *The Foundations of Ethology*. New York: Springer-Verlag, 1981. 380 p. ISBN 0387816232.

This volume presents Lorenz's personal view of the history and development of ethology.

Mayr, Ernst. *One Long Argument: Charles Darwin and the Genesis of Modern Evolutionary Thought*. Cambridge, MA: Harvard University Press, 1991. 195 p. $19.95. ISBN 0674639057.

This well-written and favorably reviewed history, written by a well-known evolutionary biologist, covers the acceptance of Darwin's theory, from the first publication of the *Origin of Species*, through the Evolutionary Synthesis of the 1930s and 1940s. The book consists of both new and previously published essays and is intended for students and the general public, although evolutionary biologists will certainly also find material of interest.

Reid, Robert G. B. *Evolutionary Theory: The Unfinished Synthesis*. Ithaca, NY: Cornell University Press, 1985. 405 p. $38.95. ISBN 0801418313.

The author discusses the history of neo-Darwinism, the branch of evolutionary theory which holds that natural selection by itself is an insufficient force to explain evolution.

Worster, Donald. *Nature's Economy*, New ed. New York: Cambridge University Press, 1986. 404 p. $49.95, $13.95 (paper). ISBN 0521267927, 052131870X (paper).

Covers the changes in the general and scientific understanding of the natural world, from the Arcadian romance of Gilbert White's *Natural History of Selbourne*, through Thoreau, Darwin, American scientific ecology and its relationship to policy making, and the modern environmentalism. Emphasis is on the American scene, and includes discussion of the development of the science as well as the world view.

Methods

Bibby, C. J., Neil D. Burgess and David A. Hill. *Bird Census Techniques*. San Diego, CA: Academic Press, 1992. 257 p. $42.00. ISBN 0120958309.

A guide to the various techniques for making bird censuses, with detailed descriptions and examples.

Brown, Luther and Jerry F. Downhower. *Analyses in Behavioral Ecology:*

A Manual for Lab and Field. Sunderland, MA: Sinauer Associates, 1988. 194 p. $18.95. ISBN 0878931228 (paper).

This manual provides examples of experiments to elucidate animal behavior in four broad categories: sensory capabilities, feeding patterns, spacing patterns, and reproduction. They also provide discussions of the most commonly used statistical tests.

Buckland, S. T., D. R. Anderson, K. P. Burnham, and J. L. Laake. *Distance Sampling: Estimating Abundance of Biological Populations*. New York: Chapman and Hall, 1993. 446 p. $84.95, $34.95 (paper). ISBN 0412426609, 0412426706 (paper).

Covers the use of distance sampling for estimating population density. Distance sampling is usually used for vertebrate species or inanimate objects such as burrows and consists of surveying lines or points in the field, recording the distance between objects of interest. This is the only book available which concentrates on this important method.

Ecological Time Series. Edited by Thomas M. Powell and John H. Steele. New York: Chapman and Hall, 1994. 496 p. $85.00, $35.00 (paper). ISBN 0412051915, 0412052016 (paper).

Covers time series for population processes, community structure, and other subjects.

Haccou, Patsy and Evert Meelis. *Statistical Analysis of Behavioural Data: An Approach based on Time-Structured Models*. New York: Oxford University Press, 1992. 396 p. $59.95. ISBN 0198546637.

The analysis of continuous time records of behavior is discussed in this manual. The authors provide a number of different statistical methods, ranging from the relatively simple (different methods of graphing results, for instance) to the statistically sophisticated (such as continuous time Markov chain modelling). Tables for the most important statistical tests used in the book are also included.

Hairston, Nelson G. *Ecological Experiments: Purpose, Design, and Execution*. New York: Cambridge University Press, 1989. 370 p. $29.95 (paper). (Cambridge studies in ecology). ISBN 0521346924 (paper), 0521345960.

Emphasizes proper experimental design in preparing and conducting field studies. Includes chapters on conducting experiments in various environments.

Jorgensen, Sven Erik. *Fundamentals of Ecological Modelling*, 2nd ed. New York: Elsevier, 1994. 628 p. (Developments in Environmental Modelling, 19). $235.00. ISBN 0444815724, 0444815783 (paper).

The author offers a discussion of the use of modelling in ecology for both eco-logists and engineers. Topics covered include basic concepts, general models, conceptual and static models, and models for modelling population dynamics, biogeochemical processes, and ecosystems, as well as the application of models in environmental management. The second edition also includes a computer disk.

Krebs, Charles J. *Ecological Methodology*. New York: HarperCollins, 1989. 654 p. $56.00. ISBN 0060437847.

Provides statistical methods for ecological studies, such as estimating abundance and determining the optimal sample size.

Martin, Paul and Patrick Bateson. *Measuring Behaviour: An Introductory Guide*, 2nd ed. New York: Cambridge University Press, 1993. 238 p. $16.95. ISBN 0521446147.

This edition adds material on new technologies, as well as information on quan-titative studies of behavior.

Methods in Ecology Series. Cambridge, MA: Blackwell Scientific Publi-cations. Irregular. Price varies.

"The aim of this series is to provide ecologists with concise and authoritative books that will guide them in choosing and applying an appropriate methodology to their problem. New technologies are a feature of the series." Volumes in the series include *Geographical Population Analysis: Tools for the Analysis of Bio-diversity* by B. A. Maurer (1994, 144 p., $32.95, ISBN 0632037415) and *Mole-cular Methods in Ecology* by D. T. Parkin 1995, $34.95, ISBN 0632034378).

Scheiner, Samuel M. and Jessica Gurevitch, eds. *Design and Analysis of Ecological Experiments*. New York: Chapman and Hall, 1993. 445 p. $79.00, $35.00 (paper). ISBN 0412035510, 0412035618 (paper).

The editors have selected methods for use in designing and analyzing experi-ments which may not be well known to ecologists and students. The authors offer many examples, and computer code where applicable.

Skalski, J. R. and D. S. Robson. *Techniques for Wildlife Investigations: Design and Analysis of Capture Data*. San Diego: Academic Press, 1992. 237 p. $59.95. ISBN 0126476756.

Includes criteria for designing effective experiments, statistical methods for analyzing mark-recapture data, and many examples.

Spellerberg, Ian F. *Monitoring Ecological Change*. New York: Cambridge University Press, 1991. 334 p. $84.95, $29.95 (paper). ISBN 0521366623 (paper), 0521424070.

Covers techniques for studying changes in ecosystems and the status of species, caused by both man-made and natural effects. After opening chapters covering the scientific basis and the present status of long-term monitoring programs, the author goes on to discuss monitoring in practical terms, with examples for birds, freshwater organisms and ecosystems, and others. Useful for both students and practicing ecologists.

Video Techniques in Animal Ecology and Behaviour, 1st ed. Edited by
    Stephen D. Wratten. New York: Chapman and Hall, 1994. 211 p.
    $84.95. ISBN 0412466406.

Lists techniques for studying the behavior of various animals using video cameras. Examples given include flying insects, parasites, wild birds, farm animals and companion animals, and microscopic organisms.

Periodicals

Acta Oecologica: International Journal of Ecology. v. 1– , 1980– .
    Montrouge, France: Gauthier-Villars. Bimonthly. $245.00. ISSN
    1146-609X.

"Devoted to fundamental ecology and its applications."

Aggressive Behavior. v. 1– , 1975– . New York: Wiley and Sons.
    Bimonthly. $488.00. ISSN 0096-140X.

"A multidisciplinary journal devoted to the experimental and observational analysis of conflict in humans and animals."

American Midland Naturalist. v. 1– , 1909– . Notre Dame, IN:
    University of Notre Dame. Quarterly. $75.00. ISSN 0003-0031.

Publishes "articles reporting original research in any field of biological science and review articles of a critical nature on topics of current interest in biology."

American Naturalist. v. 1– , 1867– . Chicago: University of Chicago
    Press. Monthly. $198.00. ISSN 0003-0147.

The official journal of the American Society of Naturalists. *American Naturalist* publishes articles which "advance and diffuse the knowledge of organic evolution and other broad biological principles so as to enhance the conceptual unification of the biological sciences."

Animal Behaviour. v. 1– , 1952– . London, UK: Academic Press.
    Monthly. $595.00. ISSN 0003-3472.

Published for the Association for the Study of Animal Behaviour (UK) and the Animal Behavior Society (US and Canada). Publishes both original research and review articles. The original articles "should bear a fundamental relationship to the natural lives of animals."

Australian Journal of Ecology. v. 1– , 1976– . Melbourne, Australia: Blackwell Scientific Publications. Quarterly. $362.00. ISSN 0307-692X.

"Research papers, critical reviews, key-note articles and abstracts of Australian theses dealing with any aspect of pure of applied ecology are considered for publication."

Behavioral Ecology. v. 1– , 1990– . New York: Oxford University Press. Quarterly. $147.00. ISSN 1045-2249.

The official journal of the International Society for Behavioral Ecology. Publishes "original articles, reviews, and correspondence on all aspects of the field of behavioral ecology, encompassing both empirical and theoretical work and covering the range from invertebrates to humans."

Behavioral Ecology and Sociobiology. v. 1– , 1976– . Heidelberg, Germany: Springer-Verlag. Monthly. $1,125.00. ISSN 0340-5443.

"The journal publishes original contributions dealing with quantitative empirical and theoretical studies in the field of the analysis of animal behavior on the level of the individual, population and community."

Behaviour: An International Journal of Behavioural Biology. v. 1– , 1947– . Leiden, Netherlands: E. J. Brill. Monthly. $268.00. ISSN 0005-7959.

"*Behaviour* aims to publish substantial contributions to the biological analysis of the causation, ontogeny, function, and evolution of behaviour of all animal species, including humans."

Behavioural Processes. v. 1– , 1976– . Amsterdam: Elsevier. Monthly. $454.00. ISSN 0376-6357.

"The journal publishes experimental, theoretical and review papers dealing with fundamental behavioural processes through the methods of natural science. Experimental papers may deal with any species, from unicellular organisms to human beings. Sample topics are cognition in man and animals, the phylogeny, ontogeny and mechanisms of learning, animal suffering and the neuroscientific bases of behaviour."

Biochemical Systematics and Ecology. v. 1– , 1973– . Oxford, UK: Oxford University Press. Semiquarterly. $608.00. ISSN 0305-1978.

"Devoted to the publication of original papers and reviews, both submitted and invited, in two subject areas: (i) the application of biochemistry to problems relating to systematic biology of organisms (biochemical systematics); (ii) the role of biochemistry in interactions between organisms or between an organism and its environment (biochemical ecology)."

Biological Conservation. v. 1– , 1969– . Barking, UK: Elsevier Applied Science. Quarterly. $830.00. ISSN 0006-3207.

Publishes "original papers dealing with the preservation of wildlife and the conservation or wise use of biological and allied natural resources."

Brain, Behaviour and Evolution. v. 1– , 19– . Basel, Switzerland: Karger. Monthly. $968.00. ISSN 0006-8977.

Official organ of the J. B. Johnston Club. "Designed to focus on the structure, function and evolution of nervous systems."

Bulletin of the Ecological Society of America. v. 1– , 1917– . Tempe, AZ: Ecological Society of America. Quarterly. $25.00. ISSN 0012-9623.

"Contains announcements of meetings of the Society and related organizations, programs, awards, articles, and items of current interest to members." Also includes an annual directory (see the entry on p. 253 in the Societies section).

Canadian Field-Naturalist. v. 1– , 1897– . Ottawa, Canada: Ottawa Field-Naturalists' Club. Quarterly. $35.00. ISSN 0008-3550.

"A medium for the publication of scientific papers by amateur and professional naturalists or field-biologists reporting observations and results of investigations in any field of natural history provided that they are original, significant, and relevant to Canada."

Conservation Biology. v. 1– , 1987– . Cambridge, MA: Blackwell Scientific. Quarterly. $175.00. ISSN 0888-8892.

"Provides a forum for the discussion and dissemination of the critical ideas in conservation theory and management."

Ecography: Ecology in the Holarctic Region. v. 1– , 1978– . Copenhagen: Munksgaard International Publishers. Quarterly. $150.00. ISSN 0906-7590.

Formerly *Holarctic Ecology.* Publishes "papers in the areas of descriptive ecology as well as on ecological patterns."

Ecological Applications. v. 1– , 1991– . Tempe, AZ: Arizona State University Press. Quarterly. $75.00. ISSN 1051-0761.

An official publication of the Ecological Society of America. "Open to research and discussion papers that integrate ecological science and concepts with their applications and implications. Of special interest are papers that develop the basic scientific principles on which environmental decision-making should rest, and those that describe the applications of ecological concerns to environmental problem-solving, policies, and management."

Ecological Modelling. v. 1– , 1975– . Amsterdam: Elsevier. Monthly. $1,057.00. ISSN 0304-3800.

"This journal is concerned with the use of mathematical models and systems analysis for the description of ecosystems and for the control of environmental pollution and resource development."

Ecological Monographs. v. 1– , 1931– . Tempe, AZ: Arizona State University Press. Quarterly. $45.00. ISSN 0012-9615.

An official publication of the Ecological Society of America. Publishes "research and discussion papers that develop or test ecological theory with data from field and laboratory experiments, observations, or simulations." Covers longer papers than are included in *Ecology*, below.

Ecological Research. v. 1– , 19– . Melbourne: Blackwell Scientific Publications. 3/yr. $140.00. ISSN 0912-3814.

Published for the Ecological Society of Japan. "Original papers, critical reviews, keynote articles and short communications dealing with any aspects of pure or applied ecology are considered for publication in *Ecological Research*."

Ecologist. v. 1– , 1969– . Camelford, UK: Ecosystems Ltd. Bimonthly. $75.00. ISSN 0261-3131.

"*The Ecologist* has been at the forefront in pushing debates on issues such as the environmental effects of large dams, the policies of the World Bank, and deforestation."

Ecology. v. 1– , 1920– . Tempe, AZ: Arizona State University Press. Bimonthly. $210.00. ISSN 0012-9658.

An official publication of the Ecological Society of America. Publishes "research and discussion papers that develop or test ecological theory with data from field and laboratory experiments, observations, or simulations." Longer papers are covered in *Ecological Monographs*, above.

Ecotoxicology and Environmental Safety. v. 1– , 1977– . Orlando, FL: Academic Press. Bimonthly. $351.00. ISSN 0147-6513.

"Publishes manuscripts dealing with studies of the biologic and toxic effects caused by natural or synthetic chemical pollutants to ecosystems, whether animal, plant, or microbial."

Ethology. v. 1– , 1937– . Berlin: Paul Parey. Monthly. $696.00.
      ISSN 0179-1613.

Formerly *Zeitschrift für Tierpsychologie*. "*Ethology* publishes original work in the field of behavioural research (ethology), that should not have been published or submitted elsewhere."

Ethology and Sociobiology. v. 1– , 19– . New York: Elsevier. Bimonth-
      ly. $324.00. ISSN 0162-3095.

Publishes articles "primarily concerned with the publication of ethological and sociobiological data and theories . . . the primary focus of the journal is the human species."

Ethology, Ecology, and Evolution. v. 1– , 19– . Florence, Italy:
      Universita di Firenze. Quarterly. $150.00. ISSN 0269-7653.

Formerly *Monitore Zoologico Italiano*. Provides "rapid publication of research and review articles on all aspects of animal behaviour. Articles should emphasize the significance of the research for understanding the function, ecology, or evolution of behaviour."

Evolution. v. 1– , 1947– . Lawrence, KS: Allen Press. Bimonthly.
      $160.00. ISSN 0014-3820.

Published for the Society for the Study of Evolution. Publishes "significant new results of empirical or theoretical investigations concerning facts, processes, mechanics, or concepts of evolutionary phenomena and events."

Evolutionary Ecology. v. 1– , 1987– . London: Chapman and Hall.
      Quarterly. $399.00. ISSN 0269-7653.

"*Evolutionary Ecology* is a conceptually oriented journal of basic biology which seeks to publish papers of bold, original research. It defines evolutionary ecology broadly, including much behavioral and community ecology, ecological genetics, applications of optimality theory to ecology and many other topics."

Evolutionary Theory and Review: An International Journal of Fact and
      Interpretation. v. 1– , 1973– . Chicago: Biology Department,
      University of Chicago. Irregular. $30.00. ISSN 0093-4755.

"Subject: The evolutionary half of biology: that part of biology where the center of interest is on organisms and populations, i.e., on the phenotype and its various interrelations rather than on molecules and cells for their own sake." A low-budget journal dedicated to "speculations and critical discussion of books, papers, or ideas . . . . Papers that disagree with the editors' views have a higher probability of acceptance than those that agree."

Evolutionary Trends in Plants. v. 1– , 19– . Leamington Spa, UK: ETP. $300.00. 2/yr. ISSN 1011-3258.

"Contains features, research reports and reviews charting major advances across the spectrum of plant evolution, genetics, and ecology."

Functional Ecology. v. 1– , 1987– . Oxford, UK: Blackwell Scientific. Bimonthly. $335.00. ISSN 0269-8463.

"Publishes short, original papers in a wide range of ecological topics, but particularly emphasizing the fields of physiological, biophysical and evolutionary ecology."

Journal of Animal Ecology. v. 1– , 1932– . Oxford, UK: Blackwell Scientific. Quarterly. $335.00. ISSN 0021-8790.

Published for the British Ecological Society. Publishes "original research papers on any aspect of animal ecology."

Journal of Applied Ecology. v. 1– , 1964– . Oxford, UK: Blackwell Scientific. Quarterly. $335.00. ISSN 0021-8901.

Published for the British Ecological Society. Publishes "original research papers on most aspects of applied ecology. For the purposes of the Journal, this is defined as the application of ecological ideas, theories and methods to the use of biological resources in the widest sense."

Journal of Arid Environments. v. 1– , 1978– . London: Academic Press. 8/yr. $460.00. ISSN 0140-1963.

"The *Journal* will publish papers containing the results of original work, and review articles within the general field described by its title. It will be wide in scope, and will include physiological, ecological, anthropological, geological, and geographical studies related to arid environments."

Journal of Biological Rhythms. v. 1– , 19– . New York: The Guilford Press. Quarterly. $145.00. ISSN 0748-7304.

"Publishes original, full-length reports in English of empirical investigations into all aspects of biological rhythmicity."

Journal of Chemical Ecology. v. 1– , 1975– . New York: Plenum. Monthly. $695.00. ISSN 0098-0331.

The official journal of the International Society of Chemical Ecology. "Devoted to promoting an ecological understanding of the origin, function, and significance of natural chemicals that mediate interactions within and between organisms."

Journal of Comparative Psychology. v. 1– , 1983– . Washington, DC: American Psychological Association. Quarterly. $100.00. ISSN 0735-7036.

Continues, in part, *Journal of Comparative and Physiological Psychology.* "Publishes original research in the behavioral and cognitive abilities of different species (including humans) as they relate to evolution, ecology, adaptation, and development."

Journal of Ecology. v. 1– , 1913– . Oxford, UK: Blackwell Scientific. Quarterly. $335.00. ISSN 0022-0477.

Published for the British Ecological Society. Publishes "original research papers on all aspects of the ecology of plants (including algae) in both aquatic and terrestrial ecosystems."

Journal of Evolutionary Biology. v. 1– , 1988– . Basel, Switzerland: Birkhauser. Bimonthly. $464.00. ISSN 1010-061X.

Official journal of the European Society for Evolutionary Biology. "Publishes original empirical and theoretical research on biological evolution."

Journal of Experimental Marine Biology and Ecology. v. 1– , 1967– . Amsterdam: Elsevier. 20/yr. $1,843.00. ISSN 0022-0981.

"This journal provides a forum for work in the biochemistry, physiology, behaviour, and genetics of marine plants and animals in relation to their ecology; all levels of biological organization will be considered."

Journal of Experimental Psychology: Animal Behavior Processes. v. 1– , 1975– . Washington, DC: American Psychological Association. Quarterly. $109.00. ISSN 0097-7403.

Continues, in part, *Journal of Experimental Psychology.* "Publishes experimental and theoretical studies concerning all aspects of animal behavior processes. Studies of associative, nonassociative, cognitive, perceptual, and motivational processes are welcome."

Journal of Freshwater Ecology. v. 1– , 19– . Holmen, WI: Oikos Publishers. $48.00. ISSN 0270-5060.

"Intended to be a vehicle for the reasonably rapid dissemination of current limnological information."

Journal of Molecular Evolution. v. 1– , 1971– . New York: Springer
International. Monthly. $835.00. ISSN 0022-2844.

"This journal publishes articles in the following research fields: 1. Biogenic
evolution . . . ; 2. Evolution of informational macromolecules . . . ; 3.
Evolution of genetic control mechanisms; 4. evolution of enzyme systems and
their products; 5. evolution of macromolecular systems . . . ; 6. molecular bases for organismal evolution; 7. evolutionary aspects of molecular population
genetics."

Journal of the Experimental Analysis of Behavior. v. 1– , 1958– .
Bloomington, IN: Indiana University. Bimonthly. $100.00. ISSN
0022-5002.

Published for the Society for Experimental Analysis of Behavior. "Primarily for
the publication of experiments relating to the behavior of individual organisms."

Journal of Tropical Ecology. v. 1– , 1985– . New York: Cambridge
University Press. Quarterly. $165.00. ISSN 0266-4674.

"Contains original articles, review articles and short communications essential
for agriculturalists, environmentalists, foresters, wildlife biologists, developmental planners, and conservationists, covering: all tropical regions; all plant and
animal taxa; all tropical environments, both terrestrial and aquatic."

Journal of Wildlife Management. v. 1– , 19– . Bethesda, MD: The
Wildlife Society. Quarterly. $110.00. ISSN 0022-541X.

An official publication of the Wildlife Society. "Research papers dealing with
population dynamics, natural history, ecology, habitat use, genetics, physiology,
nutrition, systematics, modeling, research techniques, and reviews that develop
theory are published in *The Journal*."

Microbial Ecology: An International Journal. v. 1– , 19– . New York:
Sprinter International. Bimonthly. $278.00. ISSN 0095-3628.
Publishes "articles of original research and mini-review articles on those
areas of ecology involving microorganisms, including procaryotes, eucary-
otes, and viruses."

Molecular Biology and Evolution. v. 1– , 1983– . Chicago: University
of Chicago Press. Bimonthly. $330.00. ISSN 0737-4038.

Sponsored by The Society for Molecular Biology and Evolution. "Devoted to
the interdisciplinary science between molecular biology and evolutionary biology."

Oecologia. v. 1– , 1968– . New York: Springer Verlag. Monthly.
$2,468.00. ISSN 0029-8549.

Published in cooperation with the International Association for Ecology (Intecol). Publishes "original contributions and short communications dealing with the ecology of all organisms."

Oikos. v. 1– , 1948– . Copenhagen: Munksgaard. Semiquarterly. $404.00. ISSN 0030-1299.

Issued by the Nordic Society Oikos. "Theoretical as well as empirical work is welcome; however, theoretical papers should more than elaborate on previously published analyses, and empirical papers should test explicit hypotheses and/or theoretical predictions . . . . There is no bias as regards taxon, biome, or geographical region."

Origins of Life and Evolution of the Biosphere. v. 1– , 1968– . Dordrecht: Kluwer Academic. Bimonthly. $251.00. ISSN 0169-6149.

The journal of the International Society for the Study of the Origin of Life. "While any scientific study related to the origin of life has its place in this journal, the main interests revolve around experimental and theoretical studies: evolution of planetary atmosphere, prebiotic chemistry, biochemical evolution, and precambrian studies, to name only a few."

Researches on Population Ecology. v. 1– , 1952– . Kyoto: The Society of Population Ecology. Semiannual. $119.00. ISSN 0034-5466.

Publishes "original papers written in English dealing with various aspects of population ecology. The contributors, at least the senior author of an article, are confined to members of the Society of Population Ecology."

Sociobiology. v. 1– , 1976– . Chico, CA: California State University, Chico. Irregular. $96.00. ISSN 0361-6525.

"The serial is devoted to papers giving research results, review articles, or translations of classic papers on any aspect of the biology of social animals."

Soviet Journal of Ecology. v. 1– , 1970– . New York: Consultants Bureau. Bimonthly. $820.00. ISSN 0096-7807.

Translation of the Russian *Ekologiya.*

Vegetatio: The International Journal of Plant Ecology. v. 1– , 1949– . Dordrecht: Kluwer Academic. Monthly. $1,283.00. ISSN 0042-3106.

Publishes "original scientific papers dealing with the ecology of vascular plants and bryophytes in terrestrial, aquatic and wetland ecosystems. Papers reporting on descriptive, historical, and experimental studies of any aspect of plant population, physiological, community, ecosystem and landscape ecology as well as on theoretical ecology are within the scope of the journal."

Wetlands. v. 1– , 1981– . Lawrence, KS: Society of Wetland Scientists. Semiannual. $100 Library membership. ISSN 0277-5212.

The journal of the Society of Wetland Scientists. "Original articles dealing with freshwater, saltmarsh, or estuarine wetlands research from the viewpoint of any appropriate scientific discipline or from a management or regulatory viewpoint will be considered." Also included in the membership are the *SWS Newsletter* and a membership directory.

Wildlife Monographs. v. 1– , 1957– . Bethesda, MD: The Wildlife Society. Irregular. Price included with membership. ISSN 0084-0173.

A publication of The Wildlife Society. "*Wildlife Monographs* was begun in 1957 to provide for longer papers than those normally accepted for *The Journal of Wildlife Management*."

Wildlife Society Bulletin. v. 1– , 1973– . Bethesda, MD: The Wildlife Society. Quarterly. $80.00. ISSN 0091-7648.

"Manuscripts concerning all phases of management, law enforcement, economics, education, administration, philosophy, and contemporary problems related to wildlife are considered." Includes features listing recent publications of interest to Wildlife Society members.

Reviews of the Literature

Advances in Ecological Research. v. 1– , 1962– . New York: Academic Press. Price varies. ISSN 0065-2504.

The series aim is "to allow ecologists in general to remain aware not only of the advances that are made, but of the lacunae that remain in a subject that grows every [sic] more diverse."

Annual Review of Ecology and Systematics. v. 1– , 1970– . Palo Alto, CA: Annual Reviews. $47.00. ISSN 0066-4162.

The editors "invite qualified authors to contribute critical articles reviewing significant developments within each major discipline."

Ecological Studies: Analysis and Synthesis. v. 1– , 1970– . New York: Springer-Verlag. Irregular. Price varies. ISSN 0070-8356.

Each volume reviews an ecological topic in depth. Recent volumes include *Biodiversity and Ecosystem Function*, *Pinnipeds and El Niño*, and *Plankton Regulation Dynamics.*

Evolutionary Biology. v. 1– , 1967– . New York: Plenum Press. Annual. $85.00. ISSN 0071-3260.

Focuses on "critical reviews, commentaries, original papers, and controversies in evolutionary biology. The topics of the reviews range from anthropology to molecular evolution and from population biology to paleobiology."

Oxford Surveys in Evolutionary Biology. v. 1– , 1984– . New York: Oxford University Press. Annual. ISSN 0265-072X.

"The goal of this series, which reviews new theoretical ideas and frameworks, is to stimulate discussion and outline progress in evolutionary studies. It covers the entire field, and presents special features, such as reviews of books in areas of particular interest, essays in response to the publication of major works, and comments on previously published articles."

Perspectives in Ethology. v. 1– , 1973– . New York: Plenum. Irregular. ISSN 0738-4394.

Each volume contains articles organized around a general topic relating to animal behavior, such as animal awareness or the future of ethology.

Trends in Ecology and Evolution. v. 1– , 1986– . Barking, UK: Elsevier. Monthly. $490.00. ISSN 0169-5347.

A journal of "news, reviews and comments on current developments in ecology and evolutionary biology. It is not a vehicle for the publication of original research, hypotheses, synthesis or meta analyses."

Societies

American Society of Naturalists
c/o Dr. Barbara Bentley, State University of New York, Dept. of Ecology and Evolution, Stony Brook, NY 11794.

Founded: 1883. 700 members in 1993. Professional naturalists. Affiliated with the American Association for the Advancement of Science. Publications: sponsors *The American Naturalist* and publishes *Records of the American Society of Naturalists.*

Animal Behavior Society
c/o Dr. Janis W. Driscoll, Univ. of Colorado at Denver, Dept. of Psychology, C8173, PO Box 173364, Denver, CO 80217-3364.

Founded: 1964. 2,600 members in 1993. Professional society for the study of animal behavior. Closely associated with the Division of Animal Behavior of the American Society of Zoologists. Affiliated with the Association for the Study of Animal Behaviour (see below). Publications: publishes *Graduate Programs in Animal Behavior* and the *Newsletter* and co-publishes *Animal Behaviour*.

Association for the Study of Animal Behaviour
>   c/o Dr. C. K. Catchpole, Dept. of Biology, Royal Holloway and Bedford New Coll, Egham, Surrey TW20 OEX, England.

Founded: 1936. 1,000 members in 1992. A multinational association for the study of animal behavior. Affiliated with the Animal Behavior Society. Publications: publishes the *Newsletter* and co-publishes *Animal Behaviour*.

British Ecological Society.
>   Burlington House, Piccadilly, London W1V OLQ, England.

Founded: 1913. 4,700 members in 1992. Publications: *Functional Ecology, Journal of Animal Ecology, Journal of Applied Ecology*, and *Journal of Ecology*.

Ecological Society of America.
>   Arizona State University, Center for Environmental Studies, Tempe, AZ 85287.

Founded: 1915. 6,100 members in 1993. The largest ecological association in the United States. Affiliated with the American Institute of Biological Sciences. Publications: the *Bulletin, Ecological Applications, Ecological Mongraphs*, and *Ecology*. Also publishes a biennial *Directory*.

International Society of Behavioral Ecology
>   For information, contact Journal Fulfillment Department, Oxford University Press, 2001 Evans Rd., Cary, NC 27513.

Founded in 1986 to promote the field of behavioral ecology. Publications: *Behavioral Ecology*.

International Society for the Study of the Origin of Life
>   Mail Stop #245-1, NASA Ames Research Center, Moffet Field, CA 94035.

300 members. For scientists of all disciplines interested in studying the origin of life. Bestows the A. I. Oparin medal. Publications: *Origins of Life*, membership directory.

Society for the Study of Evolution
>   c/o Dr. Barbara Schall, Washington University, Dept. of Biology, 1 Brookings Dr., St. Louis, MO 63130.

Founded: 1946. 2,500 members in 1993. Biologists working in the area of organic evolution. Publications: *Evolution* and a membership directory.

Swedish Society Oikos (Svenska Foreningen Oikos)
    Dept. of Zoology, Box 561, S-751 22 Uppsala, Sweden.

Founded: 1949. 300 members. For researchers and students interested in improving ecological conditions in Sweden. Publications: *Oikos, Nordecol Newsletter*.

The Wildlife Society
    5410 Grosvenor L., Bethesda, MD 20814-2197.

Founded: 1937. 8,900 members in 1993. Society for wildlife biologists and conservationists. Publications: *Journal of Wildlife Management, Wildlife Monographs, Wildlife Society Bulletin*, and the newsletter *Wildlifer*. Also publishes annual *Membership Directory and Certification Registry*.

Textbooks

Aber, John D. and Jerry M. Melillo. *Terrestrial Ecosystems*. Philadelphia: Saunders College Publishing, 1991. 429 p. $28.00 (paper). ISBN 0030474434 (paper).

Alcock, John. *Animal Behavior: An Evolutionary Approach*. 5th ed. Sunderland, MA: Sinauer Associates, 1993. 640 p. $47.95. ISBN 0878930175.

Begon, Michael, John Harper, and Colin Townsend. *Ecology: Individuals, Populations, and Communities*, 2nd ed. Cambridge, MA: Blackwell Scientific Publications, 1990. 450 p. $49.95. ISBN 0865421110.

Begon, Michael and M. Mortimer. *Population Ecology: A Unified Study of Animals and Plants*, 2nd ed. Cambridge, MA: Blackwell Scientific Publications, 1986. 288 p. $32.95 (paper). ISBN 0632014431 (paper).

Brown, Jerram L. *The Evolution of Behavior*. New York: Norton, 1975. 761 p. $27.95. ISBN 039309295X.

Bulmer, Michael G. *Theoretical Evolutionary Ecology*. Sunderland, MA: Sinauer Associates, 1994. 416 p. $65.00, $35.00 (paper). ISBN 0878930795, 0878930787 (paper).

Camhi, Jeffrey M. *Neuroethology: Nerve Cells and the Natural Behavior of Animals*. Sunderland, MA: Sinauer Associates, 1984. 416 p. $44.50. ISBN 0878930752.

Caughley, Graeme and Anthony R. E. Sinclair. *Wildlife Ecology and Management*. Cambridge, MA: Blackwell Scientific, 1994. 337 p. $45.00 (paper). ISBN 0865421447 (paper).

Contains overview of ecology and fundamentals of wildlife management.

Cockburn, Andrew. *An Introduction to Evolutionary Ecology*. Boston: Blackwell Scientific Publications, 1991. 370 p. $39.95 (paper). ISBN 0632027290 (paper).

Advanced undergraduate or graduate students.

Colinvaux, Paul. *Ecology 2*, 2nd ed. New York: John Wiley and Sons, 1993. 688 p. ISBN 0471558605.

Undergraduate.

Cowen, Richard. *History of Life*, 2nd ed. Boston: Blackwell Scientific, 1995. 462 p. $34.95 (paper). ISBN 0865423547 (paper).

Encyclopedia of Environmental Biology. Edited by William A. Nierenberg. San Diego, CA: Academic Press, 1995. 3 v. $475.00. ISBN 0122267303.

Futuyma, Douglas J. *Evolutionary Biology*, 2nd ed. Cambridge, MA: Blackwell Scientific Publications, 1986. 600 p. $45.95. ISBN 0878931880.

Goodenough, Judith, Betty McGuire, and Robert A. Wallace. *Perspectives on Animal Behavior*. New York: John Wiley and Sons, 1993. 762 p. ISBN 0471536237.

Upper-level undergraduates, both in biology and psychology.

Gould, James L. *Ethology: The Mechanisms and Evolution of Behavior*. New York: Norton, 1982. 544 p. ISBN 0393014886.

Krebs, Charles J. *Ecology: The Experimental Analysis of Distribution and Abundance*, 3rd ed. New York: Harper and Row, 1985. 800 p. $54.50. ISBN 0060437782.

Advanced undergraduate and graduate.

Krebs, J. R. and N. B. Davies. *An Introduction to Behavioural Ecology*, 3rd ed. Cambridge, MA: Blackwell Scientific Publications, 1993. 432 p. $36.95 (paper). ISBN 0632035463 (paper).

Undergraduate.

Krebs, J. R. and N. B. Davies. *Behavioural Ecology: An Evolutionary Approach*, 3rd ed. Cambridge, MA: Blackwell Scientific Publications, 1991. 500 p. $46.95 (paper). ISBN 0632027029 (paper).
Advanced undergraduate and graduate students.

Louw, Gideon N. *Physiological Animal Ecology*. London, UK: Longman Scientific and Technical, 1993. 288 p. ISBN 0582059224 (paper).
This text was designed to "bridge the gap between physiology and ecology." For undergraduates.

Manning, Aubrey and Marian Stamp Dawkins. *An Introduction to Animal Behaviour*, 4th ed. New York: Cambridge University Press, 1992. 206 p. $69.95, $24.95 (paper). ISBN 0521417597, 0521427924.

Maynard Smith, John. *Evolutionary Genetics*. New York: Oxford University Press, 1989. 325 p. $35.00 (paper). ISBN 0198542151 (paper).
For advanced undergraduate and graduate students.

Maynard Smith, John. *The Theory of Evolution*, 3rd ed. New York: Cambridge University Press, 1993. 375 p. $11.95. ISBN 0521451280.
Describes the theory of evolution and the changes in our understanding of the theory over time.

McFarland, David. *Animal Behavior: Psychobiology, Ethology, and Evolution*, 2nd ed. London, UK: Longman Scientific and Technical, 1993. 585 p. ISBN 0582067219.
Undergraduates, both biology and psychology. More psychological than Alcock or Goodenough.

McKinney, Michael L. *Evolution of Life: Processes, Patterns, and Prospects*. Englewood Cliffs, NJ: Prentice Hall, 1993. $44.00 (paper). ISBN 0132929392.
For nonmajors.

Meffe, Gary K. and C. Ronald Carroll. *Principles of Conservation Biology*. Sunderland, MA: Sinauer Associates, 1994. 600 p. $46.95. ISBN 0878935193.

Mitsch, William J. and James G. Gosselink. *Wetlands*, 2nd ed. New York: Van Nostrand Reinhold, 1993. 722 p. $59.95. ISBN 0442008058.

Pianka, Eric R. *Evolutionary Ecology*. 4th ed. New York: Harper and Row, 1988. 468 p. $46.00. ISBN 0060452161.

Primack, Richard B. *Essentials of Conservation Biology*. Sunderland, MA: Sinauer Associates, 1993. 564 p. $28.95. ISBN 0878937226.

Putman, R. J. *Community Ecology*. New York: Chapman and Hall, 1994. 178 p. $70.00, $34.00 (paper). ISBN 0412544903, 0412545004 (paper).

Ricklefs, Robert E. *Ecology*, 3rd ed. New York: W. H. Freeman, 1990. 896 p. ISBN 0716720779.
Advanced undergraduates and graduates.

Ricklefs, Robert E. *The Economy of Nature: A Textbook in Basic Ecology*, 3rd ed. New York: W. H. Freeman, 1993. 576 p. ISBN 071672409X.
Undergraduate.

Ridley, Mark. *Evolution*. Cambridge, MA: Blackwell Scientific Publications, 1992. 670 p. $39.95. ISBN 0865422265.
Undergraduate.

Strickberger, Monroe W. *Evolution*. Boston: Jones and Bartlett Publishers, 1990. 579 p. $45.00. ISBN 0867201177.

Tivy, Joy. *Biogeography: A Study of Plants in the Ecosphere*, 3rd ed. New York: John Wiley, 1993. 452 p. $25.00 (paper). ISBN 0470220783.

Trivers, Robert. *Social Behavior*. Menlo Park, CA: Benjamin/Cummings, 1985. 462 p. ISBN 080538507X.

Wilson, Edward O. and William H. Bossert. *Primer of Population Biology*. Stamford, CT: Sinauer Associates, 1971. 192 p. $13.95. ISBN 0978939261.

Wittenberger, James F. *Animal Social Behavior*. Boston: Duxbury Press, 1981. 722 p. ISBN 0878722955.

Bibliography

McGraw-Hill Dictionary of Scientific and Technical Terms, 4th ed. Sybil P. Parker, Editor-in-Chief. New York: McGraw-Hill, 1989. 2088 p. ISBN 007045270.

10

Plant Biology

The study of botany has a long and distinguished history. This fact, coupled with the complexity of the subject as it has grown from descriptive botany to the molecular plant sciences, is reflected in its literature, producing a complicated and often confusing array of resources. For the purposes of this book, plant biology encompasses the literature of botany and the plant kingdom, including fungi. This chapter does not include agriculture, forestry, horticulture, or any of the applied areas of plant science, except biotechnology. The sources men-tioned here are not comprehensive, but they are recommended as starting points appropriate for the informed layperson, the student, teacher, or librarian.

Because divisions in the study of plant biology are often, necessarily, arbitrary, it is important to consult other chapters in this book for relevant materials that are annotated in the chapters on general sources, biochemistry, genetics, and microbiology (Chapters 3, 5, 7, and 8).

Abstracts, Bibliographies, and Indexes

Abstracts and indexes and their associated print or electronic databases are discussed in a separate chapter (Chapter 4), as are general sources useful for plant biologists. Basically, the databases that are important for botanists are the same as the notable indexing serials used by other biologists.

In addition to the general indexes, however, there are various other, more specialized resources and taxonomic indexes. These resources are listed in this section.

Bibliography of Systematic Mycology. v. 1– , 1943– . Wallingford, Oxon, UK: CAB International. Semi-annual. $68/yr. ISSN 0006-1573. In print, online, diskette, and CD-ROM.

Lists papers and books on all aspects of the taxonomy of fungi compiled from world literature. Each issue has an author and classified index, and book reviews of interests to botanists.

Current Advances in Plant Science (CAPS). v. 1– , 1972– . Oxford, England: Elsevier. Monthly. $1,065/yr. ISSN 0306-4484. Available in print, online, diskette, and CD-ROM.

Competitor to *Biological Abstracts* (p. 74, Chapter 4) but not as comprehensive.

Excerpta Botanica. v. 1– , 1959– . New York: VCH. Section A: Taxonomica et chorologia. 2 vols./yr. $524/yr. ISSN 0014-4037. Section B: Sociologica. Quarterly. $138/yr. ISSN 0014-4045.

Section A provides abstracts for articles on systematic botany, herbaria, gardens, etc. Section B is a world list of monographs dealing with plant geography and ecology.

Index of Fungi. v. 1– , 1940– . Wallingford, Oxon, UK: CAB International. Semi-annual. $84/yr. ISSN 0019-3895.

Lists of names of new genera, species, and varieties of fungi, new combinations and new names. Compiled from world literature. Supersedes Petrak's *Lists . . .* (p. 264).

Index of Mosses: A Catalog of the Names and Citations for New Taxa, Combinations, and Names for Mosses Published during the Years 1963 through 1989 with Citations of Previously Published Basionyms and Replaced Names Together with Lists of the Names of Authors of the Names and Lists of Names of Publications Used in the Citations. Compiled by Marshal R. Crosby et al. St. Louis: Missouri Botanical Garden, 1992. (Monographs in Systematic Botany from the Missouri Botanical Garden, v. 42). 646 p. ISSN 0161-1542.

The title provides a good description of the importance and convenience of this source.

Kew Record of Taxonomic Literature Relating to Vascular Plants, 1971– . London: Her Majesty's Stationery Office. Quarterly. $200/yr. ISSN 0307-2835.

Comprehensive publication of worldwide taxonomic literature of flowering plants, gymnosperms, and ferns. Systematic arrangement; also includes citations to phytogeography, floristics, nomenclature, chromosome surveys, chemotaxonomy, anatomy, reproductive biology, personnalia, etc. Cumulative indexes.

Review of Plant Pathology. v. 1– , 1922– . Wallingford, Oxon, UK: CAB International. Monthly. $554/yr. ISSN 0034-6438.

Although somewhat out of scope, this abstracting service is useful for journal articles, reports, conferences, and books dealing with diseases of crop plants and ornamental plants caused by fungi, bacteria, viruses, and mycoplasma-like organisms. Information on taxonomy, morphology, genetics, fungicides and antibiotics, physiology, and molecular biology. Includes books and occasional review articles.

Retrospective Sources

Bay, J. C. "Bibliographies of Botany. A Contribution toward a Biblio-theca Bibliographia." *Progressus Rei Botanicae* 3(2): 331–456, 1910.

A valuable source. Arranged by topic: methodology, periodicals and reviews, collective indexes to periodicals, general and comprehensive bibliographies, national bibliographies, morphology and anatomy, plant geography, libraries of institutions, booksellers' catalogs, etc. Many entries are annotated.

Botanical Abstracts. v. 1. 1–15, 1918–1926. Baltimore: Williams and Wilkins.

Monthly abstracting serial, international in scope. Continued by *Biological Abstracts*.

Botanisches Zentralblatt, Referiendes Organ fuer das Gesamtgebiet der Botanik. Im Auftrage der Deutschen Botanischen Gesellschaft. v. 1–179, 1880–1945. Jena: Fischer.

Abstracting publication in German.

Catalogue of Botanical Books in the Collection of Rachel McMasters Miller Hunt. v. 1–3. Pittsburgh: Hunt Botanical Library, 1958–61.

v. 1: Printed books, 1477–1700. v. 2, pt. 1: Introduction. v. 2, pt. 2: Printed books, 1701–1800. Informative historical introduction. Numerous illustrations; all books are annotated.

Flowering Plant Index of Illustration and Information. Garden Center of Greater Cleveland Staff. Boston: G. K. Hall, 1979. 3 v. $230.00. ISBN 0816103011.

Updates *Index Londinensis*. Provides source for colored illustrations of flowering plants. Cross-references for common and botanical names. *First Supplement*, 1982. 2 vols. $260.00. ISBN 0816104034.

Fungorum Libri Bibliothecae Joachim Schliemann. Edited by W. Uellner. Forestburgh, NY: Lubrecht & Cramer, 1976 reprint. (Books and Prints of Four Centuries Series). $42.00. ISBN 3768210758.

Complete bibliographic information with reference to descriptions in other bibliographies, arranged by author. Citations to biographical information is included when available.

Guide to the Literature of Botany: Being a Classified Selection of Botanical Works, Including Nearly 600 Titles Not Given in Pritzel's Thesaurus. Edited by Benjamin D. Jackson. Champaign, Il.: Koeltz reprint of the 1881 edition. 626 p. $64.00. ISBN 3874290697.

An essential companion to Pritzel (p. 265), arranged by subject.

Hall, E. C. *Printed Books: 1481–1900 in the Horticultural Society of New York.* New York: Horticultural Society of New York, 1970.

Records 3,000 titles with author, short title, publication place and date, number of volumes, and illustrator.

Hawksworth, D. L., and M. R. D. Seaward. *Lichenology in the British Isles, 1568–1975; An Historical and Bibliographic Survey.* Surrey, England: Richmond, 1977. 240 p. $40.25. ISBN 0916422321.

Comprehensive survey of lichenology in the British Isles, including 2,695 entries. Books, journal articles, thesis manuscripts, exsiccatae that were published through 1975 are included with some titles added during 1976. Collectors and the particular herbaria with which they are/were associated are listed, and a biographical index is included.

Henrey, B. *British Botanical and Horticultural Literature before 1800; Comprising a History and Bibliography of Botanical and Horticultural Books printed in England, Scotland and Ireland from the Earliest Times until 1800.* 3 vols. New York: Oxford University Press, 1975.

Comprehensive source; includes location of materials within the British Isles.

Index Kewensis Plantarum Phanerogamarum Nomina et Synonyma Omnium Generum et Specierum a Lennaeo usque ad Annum MDCCCLXXXV Complectens Nomine Recepto Auctore Patria Uni cuique Plantae Subjectis. Sumptibus beati Caroli Roberti Darwin ductu et consilio Josephi D. Hooker confecit B. Dayton Jackson . . . . Oxford, England: Clarendon, 1893–95. v. 1–2: "An enumeration of the genera and species of flowering plants from the time of Linnaeus to year 1885 inclusive together with their authors' names, the works in which they were first published, their native countries and their synonyms." *Supplementum* v. 1– , 1886– . Continued by *Kew Index for* . . . . Compiled by R. A. Davies and K. M. Lloyd. New York: Oxford University Press, 1987– .

This indispensable index is an alphabetical listing of plant names with bibliographic references to the place of first publication. In 1992 it became available, up through Supplement 16, on CD-ROM. The electronic version is published by Oxford University Press and edited at the Herbarium of the Royal Botanic Gardens, Kew. There are plans to update the compilation.

Index Londinensis to Illustrations of Flowering Plants, Ferns and Fern Allies; Begin an Amended and Enlarged Edition continued up to the End of the Year 1920 of Pritzel's Alphabetical Register of Representations of Flowering Plants and Ferns compiled from Botanical and Horticultural Publications of the 18th and 19th Centuries. 6 vols. plus 2 supplements. Originally published for the Royal Horticultural Society of London, 1929–31 by the Clarendon Press, 1941; 1979 reprint by Lubrecht & Cramer. $1,775.00. ISBN 3874291510.

Updated by *Index Kewensis* supplements (p. 262) and the *Flowering Plant Index* . . . (p. 261).

Index Muscorum. v. 1–5; 1959–1969 reprint. Champaign, IL: Koeltz. (Regnum Vegetabile, v. 17, 26, 33, 48, 65). $400 (set).

Alphabetical lists of the genera and subdivisions of genera of the Musci. Information includes name of genus, its author, place and date of publication.

Index to American Botanical Literature, 1886–1966. Compiled by the Torrey Botanical Club. 4 vol. Boston: G. K. Hall, 1969. $405/set.

Invaluable card index in book form arranged by author. Continued as a section in the *Bulletin of the Torrey Botanical Club* compiled by the staff of the New York Botanical Garden.

Index to Botanical Monographs: A Guide to Monographs and Taxonomic Papers relating to Phanerogams and Vascular Cryptogams Found Growing Wild in the British Isles. Compiled by Douglas H. Kent. London: Published for the Botanical Society of the British isles by Academic Press, 1967. 163 p.

Systematic arrangement for publications since 1800. Includes a list of abbreviations of the titles of periodicals.

Index to Plant Chromosome Numbers, 1956– . Variously published by the California Botanical Society, the University of North Carolina Press, and by the Missouri Botanical Garden for the International Bureau for Plant Taxonomy and Nomenclature under the auspices of the International Organization of Plant Biosystematists by its Committee for Plant Chromosome numbers.

The index is arranged alphabetically by family within broad groupings of algae, fungi, bryophytes, pteridophytes, spermatophytes. It is currently published in the

Monograph series of the Missouri Botanical Garden and is a continuation of the Index published in the Regnum Vegetabile series.

Index to Plant Distribution Maps in North American Periodicals Through 1972. Compiled by W. Louis Phillips. Boston: G. K. Hall, 1978. $125.00. ISBN 0816100098.

This index contains 28,500 entries arranged alphabetically by taxa, representing 268 periodicals published by societies, universities, museums, herbaria, botanical gardens, and arboreta.

International Bibliography of Vegetation Maps, 2nd ed. Edited by A. W. Kuchler. Lawrence: University of Kansas, Libraries, 1980. (Library series, no. 45, University of Kansas). 324 p.

Contains vegetation maps of North and South America, Europe, former USSR, Asia, Australia, Africa, and the world. Arrangement is geographic and then chronological. Data includes map title, date of preparation, color, scale, legend, author, publication information.

International Catalogue of Scientific Literature, 1901–1914. Section M: Botany, 3 vols. Johnson reprint of the 1902 edition.

Listing of the botanical literature for the date covered. For more information, see the Retrospective Tools section (p. 81) of Chapter 4, "Abstracts and Indexes."

Junk, W. *Bibliographia Botanica,* including supplement (1916). Berlin: Junk, 1909.

Bibliography of 6,891 botanical papers and books arranged by subject and then by author. There is a list of periodicals of importance.

Lindau, Gustav and P. Sydow. *Thesaurus Litteraturae Mycologicae et Lichenologicae.* 5 vols. Johnson reprint of the earlier edition. $275.00.

Mycological books and papers to 1930.

Nissen, Claus. *Die Botanische Buchillustration, Ihre Geschichte und Bibliographie,* v. 1–2. Stuttgart: Hierseman, 1951. Supplement, 1966.

Vol. 1: History. Vol. 2: Bibliography with indexes for titles, artists, plants, countries, and authors.

Petrak, F. *List of New Species and Varieties of Fungi, New Combinations and New Names Published, 1920–39.* Kew, England: Commonwealth Mycological Institute, 1950–57.

Mycological literature from 1922 to 1935. *Index of Fungi* supplements Petrak's *Lists.* . . .

Pfister, Donald H., Jean R. Boise and Maria A. Eifler. *A Bibliography of Taxonomic Mycological Literature, 1753-1821*. Published for the New York Botanical Garden in collaboration with the Mycological Society of America by J. Cramer, 1990. (Mycological Memoir, no. 17). 161 p. $55.00. ISBN 3443760074.

Fills in some gaps in the mycological taxonomic literature.

Plant Science Catalog; Botany Subject Index. Compiled by U.S. Dept. of Agriculture Library. v. 1–15. Boston: G. K. Hall, 1976 reprint of the 1958 edition. $1,635.00 (set). ISBN 0816105065 (set).

Pritzel, G. A. Thesaurus Literature Botanicae, 2nd ed. Champaign, IL: Koeltz, 1972 reprint of the 1877 edition. $120.00. ISBN 3874290352.

Most important retrospective source indexing early botanical works to 1870.

Rehder, A. *The Bradley Bibliography*. v. 1–5. Cambridge, MA: Harvard University, Arnold Arboretum.

"A guide to the literature of the woody plants of the world published before the beginning of the twentieth century."

Repertorium Commentationum a Societatibus Litterariis Editarum, T. 2: *Botanica et Mineralogia*. Edited by J. D. Reuss. Gottingae: Dieterich, 1803. (Burt Franklin reprint 1961).

Index to the publications of learned societies to 1800.

Saccardo, Pier A. *Sylloge Fungorum Omnium Hucusque Cognitorum*. v. 1–25. New York: Johnson Reprint. $2,070.00 (set). ISBN 0384528309 (set).

Genera and species to 1921.

Stafleu, Frans Antonie and Richard S. Cowan. *Taxonomic Literature: A Selective Guide to Botanical Publications and Collections with Dates, Commentaries and Types*, 2nd ed. (Regnum Vegetabile series). Utrecht: Bohn, Scheltema & Holkema, 1976– . 7 volumes and 2 supplements.

Excellent resource with comprehensive information given for each entry. There are indexes to titles and names.

Biography and History

Ainsworth, Geoffrey Clough. *Introduction to the History of Mycology.*
Cambridge: Cambridge University Press, 1976. 359 p. ISBN
0521210135. Indexes. Bibliography.

Detailed discussion of the development of mycology.

Ainsworth, Geoffrey Clough. *Introduction to the History of Plant Pathology.* Cambridge: Cambridge University Press, 1981. 315 p.
$89.95. ISBN 0521230322. Index. Bibliography.

Although slightly out of scope, this history is of general interest to the study of
botany as a discipline. Contains comprehensive bibliographical and biographical
references.

Arber, Agnes. *Herbals: Their Origin and Evolution; a Chapter in the
History of Botany 1470–1670,* 3rd ed. Cambridge: Cambridge
University Press, 1987. 358 p. $24.95. ISBN 0521338794. In-
dex. Bibliography. With an introduction by W. T. Stearn.

The classic history of herbals with appendixes including a chronological list of
the principal herbals, bibliography of references, and a subject index.

*Biographical Dictionary of Botanists Represented in the Hunt Institute Por-
trait Collection.* Hunt Botanical Library Carnegie-Mellon University.
Boston: G. K. Hall, 1972. 451 p. $30.00. ISBN 0816110239.

Biographical information and portraits for 11,000 botanists.

Biographical Notes upon Botanists. The New York Botanical Garden.
Compiled by J. H. Barnhart. Boston: G. K. Hall, 1965. 3 v. $370
(set). ISBN 0816106959.

Information on the life, academic history, obituary notices, location of
portraits, travels, and collections of botanists from the earliest times through
the 1940's.

Blunt, Wilfrid, with the assistance of William T. Stearn. *The Art of
Botanical Illustration,* new edition revised and enlarged.
Forestburgh, NY: Lubrecht and Cramer, 1994. 369 p. $59.00.
ISBN 1851491775.

A scholarly historical survey of botanical illustration from prehistoric to modern
times. Beautifully illustrated.

Coats, Alice M. *The Plant Hunters; being a History of the Horticultural
Pioneers, Their Quests and Their Discoveries from the Renaissance*

to the Twentieth Century. New York: McGraw-Hill, 1969. 400 p. ISBN 0070114757.

A history of 500 professional traveling gardeners.

Coffey, Timothy. *· The History and Folklore of North American Wildflowers.* New York: Facts on File. 384 p. $40.00. ISBN 0816026246.

Compendium of popular lore and practical uses of North American wildflowers from pre-colonial times to 20th century medicine.

Dictionary of British and Irish Botanists and Horticulturists (including Plant Collectors, Flower Painters, and Garden Designers), revised and completely updated edition.   Edited by Ray Desmond. London: Taylor & Francis, 1994. 825 p. $250.00. ISBN 0850668433.

Over 13,000 detailed entries make this biographical dictionary indispensable for anyone who wants to learn about the lives and achievements of past and present horticulturalists working in the United Kingdom and around the world. A key reference.

Desmond, Ray. *A Celebration of Flowers; Two Hundred Years of Curtis's Botanical Magazine.* Bentham-Moxon Trust, Royal Botanic Gardens, Kew, 1987. 206 p. $29.95. ISBN 0600550735. Color plates. Index.

The story of the oldest extant botanical magazine.

The Early Days of Yeast Genetics and Molecular Biology. Edited by Michael N. Hall and Patrick Linder. Plainview, NY: Cold Spring Harbor Laboratory Press, 1993. 477 p. $75.00. ISBN 0879693789. Indexes.

Reminiscences of early investigators whose pioneering studies before 1975 brought yeast biology to its current maturity. Identification methods for yeasts and other fungi.

Evans, Howard Ensign. *Pioneer Naturalists: The Discovery and Naming of North American Plants and Animals.* With drawings by Michael G. Kippenhan. New York: Holt, 1993. 294 p. $22.50. ISBN 0805023372.

". . . a book about eponyms and the stories they suggest" (from the introductory chapter).

Fifty Years of Botany; Golden Jubilee Volume of the Botanical Society of America. Edited by W. C. Steere. New York: McGraw-Hill, 1958. 638 p.

History of the Botanical Society of American highlighting exploration in the New World, botanical history, and a review of developments over fifty years.

Green, J. R. *A History of Botany, 1860 to 1900, being a Continuation of Sachs 'History of Botany', 1530–1860.* New York: Russell & Russell, 1909, reissued in 1967. 543 p.

A classic history of botany and a continuation of von Sachs monumental work. To update Green, see Morton (p. 269) or Weevers (p. 270).

Harshberger, John William. *Phytogeographic Survey of North America: A Consideration of the Phytogeography of the North American Continent, including Mexico, Central America and the West Indies, together with the Evolution of North American Plant Distribution.* Monticello, NY: Lubrecht and Cramer, 1958 reprint of the 1911 ed. 790 p., plus map of North American showing the phytogeographic regions. $75.00. ISBN 37682000035.

Part I: History and literature of the botanic works and explorations of the North American continent; Part II: Geographic, climatic and floristic survey; Part III: Geologic evolution, theoretic considerations and statistics of North American plants; Part IV: North American phytogeographic regions, formations associations.

Harvey, John Hooper. *Mediaeval Gardens*. London: B. T. Batsford, 1981. 199 p. ISBN 0713423951.

Detailed account of British gardens from 1,000 to 1,500 A.D.

Huntia; A Journal of Botanical History. v. 1– . Pittsburgh, PA: Hunt Institute for Botanical Documentation, 1964– . Irregular. Publication suspended 1965–1979. $50/yr. ISSN 0073-4071.

Publishes on all aspects of the history of botany.

Index Herbariorum. Part 2. Index of Collectors. Utrecht: Bohn, Scheltema & Holkema, 1983–88. (Regnum Vegetabile series, v. 2, 9, 86, 93, 109, 114, 117).

A list of collectors, including dates, collection specialty, location of vouchers, and sources.

Isely, Duane. *One Hundred and One Botanists*. Ames, IA: Iowa State University Press, 1994. 351 p. $32.95. ISBN 0813824982.

This collection of essays is arranged chronologically featuring such important botanists as Aristotle, Winona Hazel Welch, Henry Gleason, Asa Gray, and Konrad Gesner.

Keeney, Elisabeth B. *The Botanizers: Amateur Scientists in Nineteenth-Century America*. Chapel Hill: University of North Carolina Press, 1992. 206 p. ISBN 0807820466.

Engrossing history of amateur botanists active in the 1800's.

Meisel, Max. *A Bibliography of American Natural History: The Pioneer Century*.

See p. 24 in Chapter 3, "General Sources," for more information for this important guide to early American natural history.

Morton, A. G. *History of Botanical Science; An Account of the Development of Botany from Ancient Times to the Present Day*. New York: Academic, 1981. 474 p. $39.95 ISBN 0125083823.

A "modern" history of botany that traces the emergence of philosophical concepts.

Reveal, James L. *Gentle Conquest; the Botanical Discovery of North America with Illustrations from the Library of Congress*. Washington, DC: Starwood Publishing, 1992. (Library of Congress Classics). 160 p. $39.95. ISBN 1563730022. Bibliography. Index.

Beautifully illustrated account of the adventures of early botanical explorers in North America.

A Short History of Botany in the United States. Edited by Joseph Ewan. New York: Hafner, 1969. 174 p. $12.50. ISBN 0686378709.

Botany in the United States from 300 B.C. into the mid-twentieth century.

Stafleu, Fran Antonie. *Linnaeus and the Linnaeans: The Spreading of Their Ideas in Systematic Botany, 1735–1789*. Utrecht: Published for the International Association for Plant Taxonomy, 1971. (Regnum Vegetabile, v. 79). 386 p. Index. Bibliography.

Analysis of the impact of one of the most important botanists in the world.

Stevens, Peter F. *Development of Biological Systematics: Antoine-Laurent de Jussieu, Nature, and the Natural System*. New York: Columbia University Press, 1994. 616 p. $65.00. ISBN 0231064403.

An important botanist in the history of biological classification.

Von Sachs, Julius. *History of Botany (1530–1860)*, Rev. ed. Edited by I. B. Balfour. Translated by H. E. Garnsey. New York: Russell and

Russell, 1967 reprint of the 1890 ed. 568 p. $50.00. ISBN 084621797X.

Classic history of botany.

Weevers, Theodorus. *Fifty Years of Plant Physiology*. Netherlands: Scheltema & Holkema, 1949. 308 p.

Continuation of botanical history begun by Von Sachs and Green, with emphasis on European botanical literature.

Classification and Nomenclature

This section lists books concerning classification, nomenclature, and taxonomy as well as periodical sources for updating this kind of information. See the Abstracts, Bibliographies, and Indexes section (p. 259) for taxonomic indexes.

ATCC Names of Industrial Fungi. Rockville, MD: ATCC-Books, 1994. 280 p. $32.00.

Featured are valid names of 600 genera and 3,800 species, 250 industrial products of fungi, taxonomically preferred names with synonyms, a list of taxa from species to phylum, and a guide to fungal nomenclature.

Benson, Lyman. *Plant Classification*, 2nd ed. Lexington, MA: Heath, 1979. 901 p. ISBN 0669014893.

This text is appropriate for an introduction to the classification of living plants. Vocabulary, plant characteristics for identification, preparation and preservation of specimens, the basis for classification, and the association of species in natural vegetation are some of the topics discussed.

Brittonia. v. 1- , 1931- . Bronx, NY: New York Botanical Garden. Quarterly. $60/yr. ISSN 0007-196X.

Primary research papers concerned with systematic botany in a broad sense.

Brummit, R. K. *Vascular Plant Families and Genera: A Listing of the Genera of Vascular Plants of the World According to Their Families . . . with an Analysis of Relationships of the Flowering Plant Families According to Eight Systems of Classification*. Royal Botanic Garden, UK, distributed by Lubrecht and Cramer, 1992. 804 p. $75.00. ISBN 0947643435.

Crawford, Daniel J. *Plant Molecular Systematics: Macromolecular Approaches*. New York: Wiley, 1990. 388 p. ISBN 0471807605. References. Index.

This book discusses, and evaluates, the use of macromolecular (protein and DNA) data in plant systematics.

Cronquist, Arthur. *The Evolution and Classification of Flowering Plants*, 2nd ed. Bronx: New York Botanical Garden, 1988. 555 p. $42.00. ISBN 0893273325.

Based on Cronquist's *Integrated System* (below), this text provides a more compact of version of his system of classification including general information on taxonomy, species and speciation, and the origin and evolution of flowering plants.

Cronquist, Arthur. *An Integrated System of Classification of Flowering Plants*, corrected edition. Foreword by Armen Takhtajan. New York: Columbia University Press, 1993. 1,262 p. $180.00. ISBN 0231038801.

Indispensable for plant classification. There is information on the division, class, order, family, and basic features of almost 400 families of flowering plants.

Dictionary of Generic Names of Seed Plants. Edited by Tatiana Wielgorskaya; Armen Tahktajan, Consulting Editor. New York: Columbia University Press, 1994. 472 p. $60.00. ISBN 0231078927.

This dictionary follows the system devised by Armen Tahktajan to provide a comprehensive listing of all currently accepted generic names of seed plants, including synonyms.

Gibbs, R. Darnley. *Chemotaxonomy of Flowering Plants*. Montreal: McGill-Queen's University Press, 1974. 4 v. $160.00 (set). ISBN 0773500987.

A highly technical survey of plant chemistry pertaining to plant taxonomy. There are discussions on the history of chemotaxonomy, criteria used in taxonomy, plant constituents, families and orders of dicot and monocots, and a compilation of information on plant chemistry for all species investigated up to the publication date.

Infraspecific Classification of Wild and Cultivated Plants. Edited by B. T. Styles. New York: Oxford University Press for the Systematics Association, 1986. (The Systematics Association Special Volume, no. 29). 435 p. $85.00. ISBN 019857701X. Index.

This volumes reports the proceedings of an international symposium concentrating on the way in which the wild and cultivated variants of plant species are classified and named.

International Code of Botanical Nomenclature. Adopted by the Fifteenth
  International Botanical Congress, Tokyo, 1993. Edited by W.
  Greuter et al. Konigstein: Koeltz Scientific Books, 1994. (Regnum
  Vegetabile, v. 131). 328 p. $51.00. ISBN 3874292789.
Updated by Taxon or Mycotaxon.

Jeffrey, Charles. *An Introduction to Plant Taxonomy*, 2nd ed. Cambridge, England: Cambridge University Press, 1982. 154 p.
  $24.95. ISBN 0521287758. Index.
Fundamentals and process of classification, taxonomic hierarchy, naming of
plants, systems of classification, and an outline of plant classification.

Kartesz, John T. *A Synonymized Checklist of the Vascular Flora of the
  United States, Canada, and Greenland*, 2nd ed. Biota of North
  American Program of the North Carolina Botanical Garden. Portland,
  OR: Timber Press, 1994. 2 v. Vol. 1: Checklist; Vol. 2: Thesaurus.
  ISBN 0881922048 (set).
Names and synonyms in current use of native or naturalized plants found in
North America. This is one of the best sources for verifying names and locating
synonyms.

Little, Elbert L., Jr. *Checklist of United States Trees (Native and Naturalized)*. Washington, DC: Forest Service, U.S. Dept. of Agriculture,
  G.P.O., 1979. (Agriculture Handbook, no. 541). 375 p.
This checklist represents the official standard for tree names used by the Forest
Service. Information includes approved common and scientific names, synonyms, and geographic ranges of native and naturalized trees of the United States,
including Alaska but not Hawaii.

Molecular Systematics of Plants. Edited by Pamela S. Soltis et al. New
  York: Chapman and Hall, 1992. 434 p. $75.00. ISBN 04120223-
  11. Bibliography. Index.
The aim of this is: to summarize the achievements of plant molecular systematics; to illustrate the potential of molecular characters in addressing
phylogenetic and evolutionary questions; and to suggest the appropriate techniques for systematic inquiries.

Mycotaxon. An International Journal of Research on Taxonomy and
  Nomenclature of Fungi, including Lichens. v. 1– , 1974– . Ithaca,
  NY: Mycotaxon, Ltd. Quarterly. $60/vol. ISSN 0093-4666.
Devoted to all phases of mycological taxonomy and nomenclature. Papers may
be in French or in English.

National List of Scientific Plant Names. Washington, DC: U.S. Dept. of
Agriculture, Soil Conservation Service, 1982. 2 v. SCS-TP-159.

Vol. 1 lists plant names; vol. 2 provides synonymy and includes names and
symbols that have been incorrectly used, as well as directing users to the
accepted name. Information is also included for symbols for scientific names;
accepted names for genera, species, subspecies, and varieties; authors of plant
names; symbols for source manuals, family names; symbols for plant habits; and
symbols for regions of distribution.

*NCU-1. Family Names in Current Use for Vascular Plants, Bryophytes,
and Fungi.* Edited by Werner Greuter on behalf of the Special
Committee on Names in Current Use. Champaign, IL: Koeltz Scientific Books, for the International Association for Plant Taxonomy,
1993. (Regnum Vegetabile, v. 126). 95 p. $40.00. ISBN
1878762427.

This booklet lists family names found to be currently used for organisms treated
as plants under the *International Code of Botanical Nomenclature*.

*NCU-2. Names in Current Use in the Families Trichocomaceae, Cladoni-
aceae, Pinaceae, and Lemnaceae.* Edited by Werner Greuter on
behalf of the Special Committee on Names in Current Use.
Champaign, IL: Koeltz Scientific Books, for the International Association for Plant Taxonomy, 1993. 150 p. $43.50. ISBN
1878762443.

Lists of names currently used covered by the *International Code of Botanical
Nomenclature*.

NCU-3. Names in Current Use for Extant Plant Genera. Compiled and
edited by Werner Greuter et al. on behalf of the Special Committee
on Names in Current Use. Champaign, IL: Koeltz Scientific Books
for the International Association for Plant Taxonomy, 1993. (Regnum Vegetabile, v. 129). 1,464 p. $267.00. ISBN 1878762486.

A list of 28,041 generic names currently in use for extant algae, bryophytes,
ferns, flowering plants, and fungi.

Quicke, D. L. J. *Principles and Techniques of Contemporary Taxonomy.*
London: Chapman and Hall, 1993. 200 p. $50.00. ISBN
075140019X.

This general source provides rationale and techniques of systematics as applied
to all groups of organisms. It covers principles of nomenclature and classification, logic and practice of cladistics, and the techniques of good taxonomic
practice.

Radford, Albert E. et al. *Fundamentals of Plant Systematics*. New York: Harper & Row, 1986. 498 p. ISBN 0060453052.

This is more than taxonomic concepts, processes, and principles. There are also sections discussing the botanic library and taxonomic literature, the herbarium, botanic gardens and arboreta, and taxonomic resource information, such as data analysis.

Sivarajan. V. V. *Introduction to the Principles of Plant Taxonomy*, 2nd ed. New York: Cambridge University Press, 1991. 292 p. $79.95. ISBN 0521355877.

A readable exploration of current empirical and theoretical problems in plant taxonomy, phylogenetics and evolutionary systematics, concepts of taxa, characters employed in plant systematics, nomenclature and history.

Stace, Clive A. *Plant Taxonomy and Biosystematics*, 2nd ed. New York: Distributed in the United States by Routledge, Chapman and Hall, 1989. 264 p. $74.95. ISBN 0713129808.

Text on taxonomy suitable at the undergraduate level. The second edition pays particular attention to two major developments: the general availability of molecular biological techniques, and the application of the principles of cladistics.

Stuessy, Tod F. *Plant Taxonomy: The Systematics Evaluation of Comparative Data*. New York: Columbia University Press, 1990. 514 p. $63.00. ISBN 0231067844. Bibliography. Indexes.

Text introducing the philosophical and theoretical aspects of plant taxonomy. The first part of the book discusses the principles of taxonomy, and the second part focuses on taxonomic data, its types and handling.

Systematic Botany. v. 1– , 1976– . Notre Dame, IN: Department of Biology, St. Mary's College, Notre Dame for the American Society of Plant Taxonomists. Quarterly. $90/yr. ISSN 0363-6445.

Society journal publishing original articles pertinent to modern and traditional aspects of systematic botany. Papers longer than 50 printed pages and dealing with plant systematics, especially taxonomic monographs and revisions, appear in Systematic Botany Monographs, published by the Society. Vol. 40: Revision of Cymbopetalum and Porcelia (Annonaceae) by Nancy A. Murray. 1993. 121 p. $16.00.

Taxon. Journal of the International Association for Plant Taxonomy. v. 1– , 1951– . Berlin: International Bureau for Plant Taxonomy and Nomenclature. Quarterly. $108/yr. ISSN 0040-0262.

Devoted to systematic and evolutionary biology, containing current taxonomic information, with emphasis on botany.

Terrell, Edward E. *A Checklist of Names for 3,000 Vascular Plants of Economic Importance*. Washington, DC: U.S. Dept. of Agriculture, 1986. (Agriculture Handbook, no. 505). 241 p.

Alphabetic checklists for scientific and common names of 3,000 plants.

The Yeasts: A Taxonomic Study, 3rd rev. and enl. ed. Edited by N. J. W. Kreger-van Rij. Amsterdam: Elsevier Science, 1984. 1,082 p. $181.25. Glossary. Bibliography. Index.

Classic treatise on yeast taxonomy. Also, refer to Barnett (p. 288) in the Identification Manuals section of this chapter.

Dictionaries and Encyclopedias

Ainsworth and Bisby's Dictionary of the Fungi, 7th ed. E. L. Hawksworth, B. C. Sutton, G. C. Ainsworth. New York: French and European Publications, 1983. 445 p. $125.00. ISBN 0828851727.

Over 16,500 entries for fungi and lichens identifying status, systematics, number of species, distribution, and references to key publications.

Coombes, Allen J. *Dictionary of Plant Names*. Portland, OR: Timber Press, 1987. 210 p. $10.95. ISBN 0881920231.

Pronunciation, derivation, and meaning of botanical names, and their common name equivalents. A very complete dictionary at a bargain price.

The Concise Oxford Dictionary of Botany. Edited by Michael Allaby. New York: Oxford University Press, 1992. 442 p. $11.95 (paper). ISBN 0192860941 (paper).

Includes all areas of botany.

Dictionary of Gardening, Anthony Huxley, editor-in-chief. New York: Stockton Press, 1992. 4 v. $795.00 ISBN 1561590010. Also called *The New Royal Horticultural Society Dictionary of Gardening*.

This award-winning dictionary is useful, comprehensive, and includes over 50,000 plant entries with brief description, distribution, cultivation, and important species. Biographies of famous botanists and over 180 articles on aspects of plant biology are a major attraction of this important set. There is an index to this *Dictionary* that supplies over 60,000 names for ornamental and economic plants: *Index of Garden Plants,* published by Timber Press in 1994 (1,200 p., $59.95, ISBN 0881922463).

Dictionary of Plant Virology in Five Languages, English, Russian, German, French, and Spanish. Compiled by V. Bojnansky and A. Fargasova. New York: Elsevier, 1991. 472 p. $141.00. ISBN 0444987401.

This dictionary consists of two parts: a polyglot dictionary of general plant virology terms and a separate listing of common names of virus and virus diseases of higher plants.

Donnelly, Danielle J. and William E. Vidaver. *Glossary of Plant Tissue Culture.* Portland, OR: Dioscorides Press, 1988. 144 p. $22.95. ISBN 0931146127.

"A comprehensive guide for interpreting the current literature pertinent to plant cell and tissue culture," (preface) for about 1,500 terms.

Elsevier's Dictionary of Botany. Compiled by Paul Macura. New York: Elsevier, 1979–1982. 2 v. Vol. 1: Plant names in English, French, German, Latin, and Russian (1979) $169.25. ISBN 0444417877. Vol. 2: General terms in English, French, German, and Russian (1982) $169.25. ISBN 0444419772.

A multilingual dictionary of over 16,000 terms for botany and related disciplines.

Elsevier's Dictionary of Wild and Cultivated Plants in Latin, English, French, Spanish, Dutch, and German. Compiled by W. E. Clason. Amsterdam: Elsevier, 1989. 1,016 p. $498.00. ISBN 0444429778.

Latin and vernacular names of wild and cultivated European plants.

Encyclopedia of Herbs and Herbalism. Edited by Malcolm Stuart. New York: Grosset and Dunlap, 1979. 304 p. ISBN 0448154722.

An outstanding book discussing cultivation, preservation, uses, history, folklore, and chemistry of herbs. There is a reference section describing 420 of the most important herbs, which has been published separately as the *VNR Color Dictionary of Herbs and Herbalism* (New York: Van Nostrand Reinhold, 1982. 160 p. ISBN 0442283385). Also, see *Rodale's Illustrated Encyclopedia of Herbs* (p. 278).

Encyclopedia of Mushrooms. Edited by Colin Dickinson and John Lucas. New York: Crescent Books, 1983. 280 p. ISBN 0517374846. Color illustrations.

Written for the nonspecialist, this book discusses history, biology, habitat, identification, edible properties and utilization of mushrooms. Also, see Turner and Szczawinski, *Common Poisonous Plants and Mushrooms . . .* (p. 279).

Flowering Plants of the World, Updated ed. Consultant Editor, Vernon H. Heywood. New York: Oxford University Press, 1993. 336 p. $84.00. ISBN 0195210379. Glossary.

Introductory material discusses forms, structure, ecology, uses, and classification of the flowering plants of the world. Over 300 angiosperm families are described and illustrated, many in color, with information on distribution, significant features, classification, economic uses, and number of species and genera.

Gledhill, David. *The Names of Plants*, 2nd ed. New York: Cambridge University Press, 1989. 202 p. $49.95. ISBN 0521366682.

This book is arranged in two parts: a brief description of botanical nomenclature suitable for students and amateurs, and a glossary which translates the more descriptive scientific names into English.

Hortus III: A Concise Dictionary of Plants Cultivated in the United States and Canada. Initially compiled by Liberty H. Bailey and Ethel Z. Bailey, revised and expanded by the staff of the Liberty Hyde Bailey Hortorium. New York: Macmillan, 1976. 1,290 p. $150.00. ISBN 0025054708.

"Inventory of accurately described and named plants of ornamental economic importance." Updated by the *New York Botanical Garden Illustrated Encyclopedia of Horticulture*.

Jackson, Benjamin Dayton. *A Glossary of Botanic Terms with Their Derivation and Accent*, 4th ed. Revised and enlarged. Forestburgh, NY: Lubrecht and Cramer, 1986. 481 p. $35.00. ISBN 8121100054.

This unique glossary includes 25,000 entries providing derivation and author of terms, accent, and definitions. Especially of historical value.

Jaeger, Edmund C. *A Source-Book of Biological Names and Terms*, 3rd ed. Springfield, IL: C. C. Thomas, 1978. 306 p. $48.50. ISBN 0398009163.

Updates and expands on Jackson's work. There is a complete annotation for this (p. 36) in Chapter 3, "General Sources."

Longman Illustrated Dictionary of Botany, by Andrew Sugden. Harlow, England: Longman, 1989. 192 p. ISBN 0582556961.

Arranged by subject, this colorful dictionary defines and explains over 1,200 terms in the botanical sciences.

Mabberley, D. J. *The Plant Book: A Portable Dictionary of the Higher Plants*. New York: Cambridge University Press, 1987. 706 p. $44.95. ISBN 0521340608.

This updates Willis's *A Dictionary of the Flowering Plants and Ferns*" to include all accepted generic and family names of flowering plants and ferns, with selected common names. Entries provide information on number of species in each genus and the number of genera in each family, distribution, botanical details, and use.

Moerman, Daniel E. *American Medical Ethnobotany; A Reference Dictionary*. New York: Garland, 1977. (Garland Reference Library of Social Science Series, v. 34). 527 p. $51.00. ISBN 082409979.

A guide to native American medicinal uses of plants and the supporting literature. This valuable sources includes information for genera, indications, families, and cultures.

New York Botanical Garden Illustrated Encyclopedia of Horticulture. Compiled by Thomas H. Everett. New York: Garland, 1980–1982. 10 v. $1,070.00. ISBN 0824072227.

This beautifully illustrated set supplements *Hortus III* and is comprehensive, authoritative, and contains major botanical articles. Descriptions are provided for 20,000 species and varieties of plants including 10,000 photographs. Also, see *Dictionary of Gardening* (p. 275).

Rodale's Illustrated Encyclopedia of Herbs. Edited by Claire Kowalchik and William H. Hylton. Emmaus, PA: Rodale Press, 1987. 545 p. $24.95. ISBN 0878576991. Index.

More detailed than *Encyclopedia of Herbs and Herbalism* (p. 276), but with fewer examples, this encyclopedia provides articles on lotions, teas, dyes, crafts, and some spices.

Stearn, William T. *Botanical Latin: History, Grammar, Syntax, Terminology and Vocabulary*, 4th ed. Newton Abbot, England: David & Charles, 1992. 546 p. $45.00. ISBN 0715300520. Bibliographical references and index.

Authoritative, scholarly discussion of Latin used for the botanical description and naming of plants. Also see Jackson (p. 277) and Jaeger (p. 277).

Stearn, William T. *Stearn's Dictionary of Plant Names: A Handbook on the Origin and Meaning of the Botanical Names of Some Cultivated Plants*. London: Cassell Publishers, 1992. 363 p. ISBN 0304341495.

This book provides a reliable source of information on the significance of botanical names attached to cultivated plants. A revised edition of A. W. Smith's *A Gardener's Dictionary of Plant Names* (1972), this is a valuable tool for both gardeners and botanists.

Turner, Nancy J. and Adam F. Szczawinski. *Common Poisonous Plants and Mushrooms of North America*. Portland, OR: Timber Press, 1991. 324 p. $55.00. ISBN 0881921793. Color photographs.

This is an important reference source for poisonous plants found in homes, buildings, gardens, urban areas, and the wild. There are full descriptions of each plant with information on occurrence, toxicity, and treatment.

Directories

This list of directories is selective for plant biology. For directories with a broader scope, consult Chapter 3, "General Sources." For more detailed information about associations and societies, refer to the *Encyclopedia of Associations* (p. 39). Also, consult the Internet via Gopher or World Wide Web.

Agricultural Information Resource Centers: A World Directory 1990. Edited by Rita C. Fisher et al. Urbana, IL: International Association of Agricultural Librarians and Documentalists, 1990. 641 p. ISBN 0962405205.

There are listings for 2,531 international agricultural information resource centers, biological and botanical libraries.

American Society of Plant Taxonomists Membership Directory. The Society is located at the University of Georgia, Dept. of Botany, Athens, GA 30602. Founded: 1937. 1,200 members. Botanists and others interested in all phases of plant taxonomy. Bestows awards. Publications: *Directory, Newsletter, Systematic Botany*, and *Systematic Botany Monographs*. Annual meeting. The *Directory* is a benefit of membership and is available in print or on the "Biodiversity and Biological Collections at Harvard" Gopher server on the Internet. Information includes address, telephone, fax, e-mail, and taxonomic specialty for members.

Botanical Society of America. *Membership Directory and Handbook*.

Society offices are in Columbus, OH. All members receive the *Membership Directory* which includes society officers and committees, names, addresses, and research specialization for 3,000 members. There are periodical updates.

Compendium of Bryology: A World Listing of Herbaria, Collectors, Bryologists, and Current Research. Compiled by Dale H. Vitt et al. Forestburgh, NY: J. Cramer for the International Association of Bryologists, 1985. (Bryophytorum Bibliotheca, no. 30). 355 p. ISBN 3768214336.

Useful for its world listing of 471 herbaria, over 2,000 collectors, and 535 current researchers in bryology.

Hudson, Kenneth and Ann Nicholls. *The Directory of Museums and Living Displays*, 3rd ed. New York: Stockton Press, 1985. 1,047 p. $150.00. ISBN 094381876.

"Living displays" includes zoos, aquaria, botanical gardens, and living historical farms.

Index Herbariorum. Part I: The Herbaria of the World, 8th ed. Edited by Patricia K. Holmgren et al. Bronx, NY: New York Botanical Garden, published and distributed for the International Association for Plant Taxonomy, 1990. (Regnum Vegetabile, v. 120). 693 p. $70.00. ISBN 0893273589. Indexes.

Directory of the public herbaria of the world. For *Part II: Collectors*, see annotation (p. 268) in the Biography and History section.

International Directory of Botanical Gardens V, 5th ed. Compiled by Christine A. Heywood et al. Champaign, IL: Koeltz Scientific, 1990. 1,021 p. $ 162.00. ISBN 187876201X.

Arranged by country, this directory lists gardens with address, status, area, latitude and longitude, altitude, rainfall, taxa included, specialties, greenhouses, publications, accessibility, and names of the Director and other staff.

International Register of Specialists and Current Research in Plant Systematics. Compiled and edited by Robert W. Kiger et al. Pittsburgh, PA: Hunt Institute for Botanical Documentation, 1981. 346 p. ISBN 0913196398.

The directory lists over 1,500 scientists by name, institution, position, address, telephone number, projects, and specialties. A new edition is forthcoming in print and electronic formats.

Membership Directory. Association for Tropical Biology. Lawrence, KS: Allen Press, 1990. Annual. Issued as a supplement to the fourth quarterly issue of *Biotropica*. ISSN 0006-3606.

Information in the *Directory* includes name, location, address, e-mail address, voice and FAX telephone numbers.

Miller, Everitt L. and Jay S. Cohen. *The American Garden Guidebook: A Traveler's Guide to Extraordinary Beauty Along the Beaten Path*. New York: M. Evans, 1987. 293 p. ISBN 0871314991.

" . . . the finest botanical gardens, parks, and arboreta of the 28 states and four provinces of the Eastern half of the United States and Canada."

The Nature Catalog. Edited by Joel Makower et al. New York: Vintage Books, 1991. 327 p. $18.00. ISBN 0679733000.

Information that is useful for a wide range of people includes organizations, products, publications, computer software, photographs, short articles and essays, and supplemental readings.

The Naturalists' Directory and Almanac (International), 46th ed. and 116th yr. Compiled and edited by Ross H. and Mary E. Arnett. Gainesville, FL: Sandhill Crane Press, 1993. 308 p. $35.00. ISBN 187774316X.

"An index to contemporary naturalists and their special interests." Information includes name, address, phone number, and fields of interest.

North American Horticulture: A Reference Guide, 2nd ed. Compiled by the American Horticultural Society, Thomas M. Barrett, editor. New York: Macmillan, 1991. 427 p. $75.00. ISBN 002870012.

A very useful compendium of botanical gardens, arboreta, conservatories, and other public horticultural institutions, arranged by state. Listings also include historical and demonstration gardens, national horticultural organizations, plant societies, and cemeteries with notable landscaping.

The Plant Collections Directory. Edited by Jean H. Schumacher. Swarthmore, PA: AABGA, American Association of Botanical Gardens and Arboreta, 1988. 158 p. ISBN 0934843015. Indexes.

Complete information for important documented collections of living plants in Canada and the United States.

Plant Specialists Index: Index to Specialists in the Systematics of Plants and Fungi, Based on Data from Index Herbariorum (Herbaria, ed. 8). Prepared and edited by Patricia K. Holmgren and Noel H. Holmgren. Konigstein, Germany: Koeltz Scientific, 1992. (Regnum Vegetabile, v. 124). ISSN 0080-0694.

This index gives convenient access to the research and geographic specialties of staff listed in the 8th edition of *Index Herbariorum (Herbaria)* and is complementary to the *International Register of Specialists and Current Research in Plant Systematics.*

Plant Taxonomists Online (PTO).

Available on the University of New Mexico Gopher, this version will be updated as names are added. For more information, e-mail Jane Mygatt, PTO Editor: jmygatt@bootes.unm.edu. Also, see the *American Society of Plant Taxonomists Membership Directory* (p. 279).

Traveler's Guide to American Gardens. Edited by Mary Helen Ray and
    Robert P. Nicholls. Chapel Hill, NC: University of North Carolina
    Press, 1988. 375 p. ISBN 0807817872.

This is the second edition of *A Guide to Historic & Significant Gardens of America* (1982) offering details about gardens open to the public, including a ratings scale of "must see" and "superior."

Guides to Internet Resources

Electronic databases have come into their own as a source of information for museum or herbaria collections, updated floras, and taxonomic decisions. Many of the currently available electronic databases are on the Internet, accessible using Gopher, Veronica, or WAIS. At the moment, the best source of systematic botany databases is the Biodiversity and Biological Collections Gopher at Harvard University which includes the Harvard Herbaria Mint Type Specimen database, the Harvard *Gray Herbarium Card Index* of New World Plants, and the *Botanical Collectors Authority File*. Another source of taxonomic information based at Harvard is the TAXACOM server with services including a discussion list for taxonomists, a dial-access Bulletin Board Server (BBS), the electronic journal *Flora Online*, and the newsletter *Bean Bag* for legume systematists. Databases available through BBS include the American Society of Plant Taxonomists Membership/Research Specializations and selected Clinton Herbarium holdings. Another source of electronically based data is the Smithsonian Institution's Natural History Gopher Server which includes, among other databases: information on 88,000 of the most commonly requested plants in the *Type Register of the U.S. National Herbarium*, the *Index to Principal Collections in the U.S. National Herbarium*, the *Biological Conservation Newsletter*, the *Checklist of the Flora of the Guianas* (1992), and the *ASPT Newsletter* published quarterly for the American Society of Plant Taxonomists. Also, check out the World Wide Web for a subject listing for plant biology in the Biosciences category.

Guides to the Literature

B-P-H (Botanico Periodicum Huntianum). Pittsburgh, PA: Hunt Botanical
    Library, 1968. 1,063 p. $20.00. ISBN 091319610X. *B-P-H/S*

(Botanico Periodicum Huntianum/Supplementum). Edited by Gavin D. R. Bridson. Pittsburgh, PA: Hunt Institute for Botanical Documentation, Carnegie Mellon University, 1991. 1,068 p. $95.00. ISBN 0913196541. ". . . both a supplement to and a partial revision of *B-P-H* . . . ."

One of the most important guides to the botanical literature, this serials list and compendium of information for periodicals is extraordinarily useful. It provides titles, abbreviations, subtitles, places of publication, volumes, date of publication, location information, synonymous abbreviations, and abbreviations for synonymous titles.

Bridson, Gavin. *History of Natural History: Annotated Bibliography*. New York: Garland, 1994. (Garland Reference Library of the Humanities, v. 991; Bibliographies on the History of Science and Technology, v. 24). 740 p. ISBN 0824023196. $98.00 Indexes.

Historical sources for personnel, organizations, current awareness, biographies, natural history libraries, botanical bibliographies, and historical methods.

Davis, Elisabeth B. and Diane Schmidt. *Guide to Information Sources in the Botanical Sciences*, 2nd ed. Littleton, CO: Libraries Unlimited, 1995. 1350 p. $49.00. ISBN 1563080753.

Comprehensive guide to the botanical literature.

Guide to Reference Books, 10th ed. Edited by Eugene P. Sheehy. Chicago: American Library Association, 1986. 1,560 p. ISBN 0838903908. *Supplement* to the 10th edition, 1992.

The "standard" guide to reference sources in the arts and sciences.

Guide to Sources for Agricultural and Biological Research. Edited by J. Richard Blanchard and Lois Farrell. Sponsored by the U.S. National Agricultural Library. Berkeley: University of California Press, 1981. 735 p. ISBN 0520032268.

Covers all areas of the biological sciences with concentration on agriculture, horticulture, and other applied areas.

Handbooks and Databases

Databases and handbooks of specific botanical relevance are included in this section. For additional information on botanical databases, check out the following suggested readings:

"Database for Plant Species." *Nature.* v. 348: 275, 1990.

Designs for a Global Plant Species Information System, edited by F. A. Bisby, G. R. Russell, and R. J. Pankhurst. New York: Oxford University Press, 1994. (Systematics Association Special Volume No. 48). 350 p. $90.00. ISBN 1985777605.

"The Rise of the Botanical Database." *BioScience.* v. 43 (5): 274–279, 1993.

"Some Recent Computer-Based Developments in Plant Taxonomy." *Botanical Journal of the Linnean Society.* v. 106 (2): 121–128, 1991.

For materials applicable to general biology, biochemistry, molecular biology, genetics and biotechnology, and consult Chapters 3–7.

AATDB (An *Arabidopsis thaliana* Database). Harvard Medical School/ Massachusetts General Hospital. Available on the Internet via Gopher.

This database includes genetic maps, bibliographic references, information to 128 strains from the Nottingham seed collections, and sequence data. Other features: addresses of workers, genetic variations, scanned images of hybridization results, an expanded list of keywords, and information on characteristics of mutant alleles.

American Type Culture Collection Catalogues
    12301 Parklawn Dr., Rockville MD 20852.

The American Type Culture Collection (ATCC) is a nonprofit, private organization that was established to acquire, preserve, and distribute well-characterized biological cultures for the international research community. ATCC publishes catalogs and manuals and provides various services which are available in print, on disk, CD-ROM, and the Internet. Following are some of the catalog that are of interest to botanists:

Algae and Protozoa, 17 ed. 1991.

Lists over 1,000 strains, includes a special index for culture applications.

Filamentous Fungi, 18th ed. 1991.

Lists over 19,000 strains, includes characterization information, media formulae, and an index of culture applications.

Plant Tissues.

Listing of ATCC plant germplasm collection.

Yeast, 18th ed. 1990.

Lists over 4,000 strains.

Arabidopsis; An Atlas of Morphology and Development. Edited by John
   Bowman. New York: Springer-Verlag, 1993. 350 p. $69.00. ISBN
   0387940899.

This atlas provides documentation for *Arabidopsis thaliana*, the primary model
for research in plant genetics, by covering embryogenesis, vegetative growth,
root growth, reproductive structures, and host–pathogen interactions.

Bracegirdle, Brian and Philip H. Miles. *An Atlas of Plant Structure.* New
   York: Heinemann Educational Books, 1971. 2 v. Vol. 1: $27.95;
   ISBN 0435603124. Vol. 2: $22.50; ISBN 0435603140.

This unique set assists students in interpreting lab specimens by providing an
extensive array of photomicrographs and line drawings for bacteria, algae, fungi,
lichens, hepaticae, musci, and other plant issues.

Brown, Richard A. *A Guide to the Computerization of Plant Records.*
   Swarthmore, PA: American Association of Botanical Gardens and
   Arboreta, Computer Information Services Committee, 1988. 112 p.
   ISBN 0934843023. Glossary. Bibliography. Index.

This guide offers assistance for computerization of botanic garden records from
the planning stages to the final product.

CRC Handbook of Flowering. Edited by A. H. Halevy. Boca Raton: CRC,
   1985. 6 v. Vol. VI: 776 p. $420.00. ISBN 0849339162.

Comprehensive for specific data for control and regulation of flowering for over
5,000 species of plants.

Duke, James A. *Handbook of Biologically Active Phytochemicals and
   Their Activities.* Boca Raton, FL: CRC Press, 1992. 183 p. ISBN
   08493367708.

Compilation of data, including references, for 3,000 compounds. Reported
activities are listed for each chemical, with effective or inhibitory concentrations
or doses reported when available.

Genbank (Internet)

For information on genetics databases and handbooks, see pp. 146 and 151 of
Chapter 7, "Genetics."

Guidelines for Acquisition and Management of Biological Specimens.
   Lawrence, KS: Association of Systematics Collections, Museum of
   Natural History, University of Kansas, 1982. 42 p. ISBN
   0942924029.

"A Report of the Participants of a Conference on Voucher Specimen Management" which discusses methods of collection, identification, preparation, repository selection, legislation, publication policies, costs, fees, and funding.

Handbook of Natural Toxins. Edited by Richard J. Keeler and Anthony T. Tu. New York: Dekker, 1983. 8 v. Vol. 1: ISBN 0824718933, $195. Indexes.

Volume 1 discusses plant and fungal toxins. Toxins are grouped by effects on cardiovascular, pulmonary, reproductive, neurologic, gastrointestinal systems; carcinogenic properties; species interactions; and usefulness in medicine.

The Herbarium Handbook, rev. ed. Edited by Leonard Forman and Diane Bridson. Kew, England: Royal Botanic Gardens, 1989. 214 p. ISBN 0947643206. Bibliography. Index.

A definitive and authoritative source for curation and management of herbaria.

LiMB (Listing of Molecular Biology Databases)

See p. 116 in Chapter 6, "Molecular and Cellular Biology."

The Maize Handbook. Edited by Michael Freeling and Virginia Walbot. New York: Springer-Verlag, 1994. 759 p. $79.00. ISBN 0387978267.

This very useful laboratory guide covers development and morphology, cell biology, genetics, molecular biology, and cell culture, with detailed protocols.

Mycology Guidebook. Edited by Russell B. Stevens. Published under the auspices of the Mycology Guidebook Committee, Mycological Society of America. Seattle: University of Washington Press, 1981. 712 p. $50.00. ISBN 0295958413.

This indispensable handbook assembles information on field collecting, isolation techniques, culture maintenance, taxonomic groups, ecological groups, and fungi as biological tools.

Phytochemical Dictionary: A Handbook of Bioactive Compounds from Plants. Edited by Jeffrey B. Harborne and Herbert Baxter. New York: Taylor and Francis, 1993. 791 p. $350.00. ISBN 0850667364.

This dictionary covers 3,000 substances which are found in plants, including insect antifeedants, carcinogens, phytoalexins, etc. Information is provided for synonyms, structure, molecular weight and formula, natural occurrence, biological activity, and uses. Also, see *Phytochemical Dictionary of the Leguminosae* (p. 287).

Phytochemical Dictionary of the Leguminosae. Compiled by I. W.
Southon using the International Legume Database and Information
Service and the Chapman & Hall Chemical Database. New York:
Chapman & Hall, 1993. 1,200 p. $795.00. ISBN 0412397706.

Divided into two sections: 1. The plant section lists all legume species from
which chemical substances have been reported. Entries include accepted species
name/common names/synonyms, distribution, life form, botanical source, economic uses, type of compound, organs isolated from, and chemical citations; 2.
The chemical section gives chemical data on all identified phytochemicals occurring in the Leguminosae, including accepted and other chemical names, structure
diagram, CAS registry number, molecular formula/weight, use and biological
activities. Extensively indexed. Also, see *Phytochemical Dictionary* (p. 286).

Plant Molecular Biology LABFAX. Oxford: BIOS Scientific Publisher/Black-
well Scientific, 1993. 382 p. $45.00 (spiral bound). ISBN
1872748155.

A compendium of essential information and accurate data for plant anatomy,
nucleic acids, genetic index, transformation and expression vectors, PCR techniques, and the like.

Storage of Natural History Collections: Ideas and Practical Solutions.
Edited by Carolyn L. Rose and Amparo R. de Torres. Pittsburgh, PA:
Society for the Preservation of Natural History Collections, 1992.
346 p. ISBN 0963547607. Indexes.

This handbook presents 113 ideas and practical solutions for the storage of the
complete range of natural history collections. The book is organized according
to storage configurations to encourage the transfer and adaptation of methods.
Sections include: supports, covers, containers, environmental control, labels,
general guidelines. There are appendices that include a glossary, conversion
tables, materials and supplies, and suppliers and manufacturers.

Identification Manuals

There are many excellent field guides for the identification of plants, and this
section includes some of the best. For more information, investigate the well-
known field guide series that are annotated (p. 41) in Chapter 3, "General
Sources."

Ammirati, Joseph F., James A. Traquair, and Paul A. Horgen. *Poisonous
Mushrooms of the Northern United States and Canada.* Minneapo-

lis: University of Minnesota Press, 1985. 396 p. $75.00. ISBN
0816614075.

An outstanding, standard reference source for researchers, physicians, and amateurs.

Angier, Bradford. *Field Guide to Edible Wild Plants*. Harrisburg, PA:
Stackpole, 1974. 255 p. $14.95 (paper). ISBN 0811720187
(paper).

"A quick, all in color identifier of more than 100 edible wild foods growing free in the United States and Canada." Updated by Duke and Elias.

Barnett, James Arthur et al. *Yeasts: Characteristics and Identification*,
2nd ed. New York: Cambridge University Press, 1990. 1,002 p.
$250.00. ISBN 0521350565. Bibliography.

Commonly used procedures for yeast identification, including detailed descriptions for all known yeast species. This book is also an important source for yeast taxonomy.

Bremness, Lesley. *The Complete Book of Herbs*. New York: Viking
Studio Books, 1988. 288 p. ISBN 0670818941.

This book includes the most common garden herbs as well as a few naturalized "weeds" found in the United States and Great Britain.

Common Weeds of the United States. Prepared by the Agricultural
Research Service of the U.S. Dept. of Agriculture. New York:
Dover, 1971. 463 p. $7.50. ISBN 0486205045. A republication
of *Selected Weeds of the United States*, Washington, DC: USDA,
1970.

This valuable source, including 224 of the most important U.S. weeds, assists in weed identification for establishing control measures.

Duke, James A. *CRC Handbook of Medicinal Herbs*. Boca Raton, FL:
CRC Press, 1985. 696 p. $275.00. ISBN 0849336309.

This manual discusses over 365 species of folk medicinal herbs whose safety is, or has been, questioned by various Federal agencies. Information includes common and scientific name, description of use, medicinal applications, chemical content, and toxicity. This is a nice companion to Duke's *Handbook of Edible Weeds*, also published by CRC Press in 1992, 232 p., $19.95, ISBN 0849342252. Also, see Angier (above) and Elias (p. 289).

Elias, Thomas S. *The Complete Trees of North America: Field Guide and Natural History*. New York: Gramercy Pub. Co; distributed by Crown Publishers, 1987. Reprint of the 1980 ed. 948 p. ISBN 0517641046.

This book assists in identifying over 750 North American trees.

Elias, Thomas S. and Peter A. Kykeman. *Field Guide to North American Edible Wild Plants*. New York: Outdoor Life; distributed in New York by Van Nostrand Reinhold, 1982. 286 p. ISBN 0442222009.

An alternative to Angier and Duke.

Everard, Barbara. *Wild Flowers of the World*. Paintings by Barbara Everard; text by Brian D. Morley. New York: Avenel Books, 1988 reprint of the 1970 edition. 432 p.

Over 1,000 plants are included in this selection of wildflowers from around the world. There is information on plant nomenclature, morphology, classification, ecology, geography, and history.

Flora of North America, North of Mexico. Edited by Flora of North America Editorial Committee, Nancy R. Morin, convening editor. New York: Oxford University Press, 1993– v. 1, 1993. $75.00. ISBN 0195057139.

This projected 14 volume monumental work aims to survey and classify all the more than 20,000 plant species known to grow spontaneously from the Florida Keys to the Aleutian Islands. This authoritative set provides identification keys; distribution maps; summaries of habitat and geographical ranges; precise descriptions for families, genera, and species; chromosome numbers; pertinent synonymies; line drawings; endangered and threatened plants; selected references. The set will be updated by a computer database for taxonomic information housed at the Missouri Botanical Garden in St. Louis, MO.

Gray, Asa. *Gray's Manual of Botany: A Handbook of the Flowering Plants and Ferns of the Central and Northeastern United States and Canada*, 8th (Centennial) ed. Portland, OR: Dioscorides Press, 1987 reprint of the 1950 ed. 1,632 p. $57.95. ISBN 0931146097.

The standard descriptive manual for flowering plants and ferns.

Gray Herbarium Card Index. On microfiche and the Internet.

See the Guides to Internet Resources section (p. 282).

Grieve, Maud. *A Modern Herbal: The Medicinal, Culinary, Cosmetic and Economic Properties, Cultivation and Folk-Lore of Herbs, Grasses, Fungi, Shrubs and Trees, with all Their Modern Scientific Uses, with a New Service Index.* New York: Dover, 1971. 2 v. $8.95 each. ISBN 0486227987 (v. 1); 0486227995 (v. 2).

This encyclopedic treatment presents synonyms, common and scientific names, part used, habitat, descriptions, medicinal actions and uses, constituents, dose, poisons, history, and recipes. Complementary to Harris (below).

Harris, Ben Charles. *The Compleat Herbal: Being a Description of the Origins, the Lore, the Characteristics, the Types, and the Prescribed uses of All Common Medicinal Plants.* Barre, MA: Barre Publishers, 1972. 243 p. ISBN 0827172117.

Not as comprehensive as Grieve, but the information is a bit more contemporary. Also, see Duke (p. 288).

Hitchcock, Albert Spear. *Manual of the Grasses of the United States*, 2nd ed. New York: Dover, 1971 reprint. 2 v. (USDA Miscellaneous Publication No. 200). $9.95 each. ISBN 0486227170 (v. 1); 0486227189 (v. 2).

The classic manual and definitive encyclopedia of all grasses known to grow in the continental United States, excluding Alaska.

Lampe, Kenneth F. and Mary Ann McCann. *AMA Handbook of Poisonous and Injurious Plants.* Chicago: American Medical Association distributed by the Chicago Review Press, 1985. 432 p. $18.95 (paper). ISBN 0899701833.

This authoritative reference book is designed for health care professionals as a convenient field guide to the identification, diagnosis, and management of human intoxications from plants and mushrooms of the United States and Canada. Information includes botanical nomenclature, scientific and common name indexes, and sections on systemic plant poisoning, plant dermatitis, and mushroom poisoning. Each plant is completely described, including distribution, toxic part, toxin, symptoms, management, and reference to the literature.

Mitchell, Alan, illustrated by David More. *The Trees of North America.* New York: Facts on File, Inc., 1987. 208 p. Illus. Index.

All the common and frequently seen trees plus a wide selection of rare trees. Five hundred species are accurately illustrated together with 250 varieties with information on origin, history, distribution, and growth.

The Official World Wildlife Fund Guide to Endangered Species of North America. Edited by John R. Matthews and Charles J. Moseley.

Washington, DC: Beacham Pub., 1990– . 2 v. plus sup. $195.00.
ISBN 0933833172.

Endangered plant species are covered in volume 1, including maps, photographs, descriptions, habitat, historical and current range, conservation and recovery data, and bibliography for each species. For information on endangered U.S. plants, see *Flora of North America*. Also, see Chapter 9, on ecology, for environmental resources.

Phillips, Roger. *Mushrooms of North America*. Boston: Little, Brown,
1991. 319 p. $35.00. ISBN 0316706124.

Designed for laypeople, the bulk of this book consists of photographs and descriptions for over 1,000 mushroom species. Also, see Smith and Weber (below).

Rickett, Harold W. and W. Niles. *Wild Flowers of the United States*.
New York: New York Botanical Garden and McGraw-Hill, 1966–
1975. 6 v. plus complete index.

A superior, comprehensive work that is beautifully done and scientifically accurate.

Smith, Alexander H. and Nancy S. Weber. *The Mushroom Hunter's Field
Guide: All Color and Enlarged*. Ann Arbor: University of Michigan
Press, 1980. 316 p. $15.95. ISBN 0472856103.

A standard, accurate and dependable beginner's field guide to mushrooms.

Zomlefer, Wendy B. *Guide to Flowering Plant Families*. Chapel Hill, NC:
University of North Carolina Press, 1994. 424 p. $55.00, $27.50
(paper). ISBN 0807821608, 0807844705 (paper).

A handbook to 130 North American plant families, both temperate and tropical. Has glossary and numerous illustrations, including illustrations of dissected specimens.

Guides to Floras

Rather than enumerate all the floras that are available, in the interests of space flora will be identified by guides to location. Although some of these guides are dated, they still contain very important sources of information.

Blake, Sidney Fay and Alice Cary Atwood. *Geographic Guide to Floras of
the World. Pt. 1: Africa, Australia, North American, South America
and Islands of the Atlantic, Pacific, and Indian Oceans*. Reprint of
the 1942 ed. New York: Hafner, 1967. 336 p.

Updated by Frodin (below).

Blake, Sidney Fay. *Geographical Guide to Floras of the World. Pt. 2:*
    Western Europe: Finland, Sweden, Norway, Denmark, Iceland, Great
    Britain with Ireland, Netherlands, Belgium, Luxembourg, France,
    Spain, Portugal, Andorra, Monaco, Italy, San Marino, Switzerland.
    Reprint of the 1961 ed. Monticello, NY: Lubrecht and Cramer,
    1974. 742 p. $84.00. ISBN 3874290603.
Updated by Frodin (below).

Blake, Sidney Fay. *Guide to Popular Floras of the United States and*
    Alaska: An Annotated, Selected List of Nontechnical Works for the
    Identification of Flowers, Ferns and Trees. Washington, DC: Gov-
    ernment Printing Office, 1954. (U.S. Dept. of Agriculture Biblio-
    graphical Bulletin 23). 56 p.
A useful, short history of popular botanical works.

Frodin, D. G. *Guide to Standard Floras of the World.* New York:
    Cambridge University Press, 1985. 580 p. ISBN 0521236886.
"An annotated, geographically arranged systematic bibliography of the principal
floras, enumerations, checklists, and horological atlases of different areas."

Scientific and Technical Books and Serials in Print. 1978– . New York:
    Bowker. Annual. ISSN 0000-054X.
To find floras, look in the Subject section under "Botany" or "Plants."

Takhtajan, Armen. *Floristic Regions of the World*, translated by Theodore
    J. Crovello, under the editorship of Arthur Cronquist. Berkeley, CA:
    University of California Press, 1986. 522 p. $75.00. ISBN
    0520040279.
Geographic distribution of the plant world.

Methods and Techniques

Biotechnological Applications of Plant Cultures. Edited by Peter D.
    Shargool and That T. Ngo. Boca Raton, FL: CRC, 1994. (CRC
    Series of Current Topics in Plant Molecular Biology). 214 p. ISBN
    0849382629.
State-of-the-art reviews on current techniques in plant culture work, covering
four broad areas: production of secondary metabolites by plant cells; plant cell
transformation techniques; breeding and micropropagation techniques; and plant
cell and tissue bioreactor design.

Causton, David R. *An Introduction to Vegetation Analysis: Principles, Practice, and Interpretation*. Boston: Unwin Hyman, 1988. 342 p. ISBN 0045810249.

This book discusses principles, methods, and interpretation of field data. Also, see Greig-Smith (p. 294) for additional information.

Cell Culture and Somatic Cell Genetics of Plants. Vol. 1– . New York: Academic, 1984– . Vol. 8: Scale-up and Automation in Plant Propagation. 267 p. 1991. $105.00. ISBN 0127150080.

This continuing treatise provides key reference works, descriptions, and discussions of all aspects of modern plant cell and tissue culture techniques.

Collin, H. *Plant Cell Culture*. London: Chapman & Hall, 1994. (Introduction to Biotechniques). 160 p. $30.00. ISBN 182748473.

Overview of all the basic tissue culture techniques.

Collins, Christopher Herbert and Patricia M. Lyne. *Microbiological Methods*, 6th ed. Stoneham, MA: Butterworth-Heinemann, 1989. $49.95. ISBN 0750614285.

This manual also includes techniques for yeasts, molds, and pathogenic fungi, with information on laboratory safety and equipment, sterilization, culture media, staining, and mycological methods.

Electron Microscopy of Plant Cells. Edited by J. L. Hall and C. Hawes. San Diego: Academic Press, 1991. 466 p. $100.00. ISBN 0123188806. Illustrations.

Techniques for the preparation of botanical specimens, including basic principles, applications, and procedures.

Endress, R. *Plant Cell Biotechnology*. New York: Springer-Verlag, 1994. 353 p. $136.00. ISBN 0387569472.

"This book describes methods and techniques for gaining highly productive cells and plants to produce secondary metabolites, enabling the production of plant products on an industrial scale independent of environmental influences and natural resources" (from the back cover).

Glimn-Lacy, Janice and Peter B. Kaufman. *Botany Illustrated; Introduction to Plants, Major Groups, Flowering Plant Families*. New York: Van Nostrand, 1984. 146 p. $34.95. ISBN 0442229690. Ill. Bibl. Index.

Methods of illustration for flowering plants. There is also an introduction to botanical details and a section on major flowering plant families. Other books of interest: Holmgren and Angell (below), West (p. 297), and Wunderlich (p. 298).

Greig-Smith, Peter. *Quantitative Plant Ecology*, 3rd ed. Berkeley, CA: University of California University Press, 1983. (Studies in Ecology, v. 9). 355 p. $39.95. ISBN 0520050800.

This book is still in print and valuable because it fills a niche providing guidance on the most profitable means of obtaining and handling ecological data.

Handbook of Phycological Methods: Ecological Field Methods, Macroalgae. Edited by Mark S. Littler and Diane S. Littler. New York: Cambridge University Press, 1985. $89.95. ISBN 0521249155.

Still valuable.

Harborne, Jeffrey B. *Phytochemical Methods: A Guide to Modern Techniques of Plant Analysis*, 2nd ed. New York: Chapman and Hall, 1984. 288 p. $39.50 (paper). ISBN 0412343304.

Phytochemical techniques and lists of addresses for purchasing rare chemicals and chromatographic equipment.

Hill, Stephen A. *Methods in Plant Virology*. Oxford: Blackwell, 1985. (Methods in Plant Pathology, v. 1). 167 p. $41.95. ISBN 06320-09950. Index.

Published for the British Society for Plant Pathology, this handbook present basic techniques, rationale, required materials, procedures and protocols, interpretation of results, and references to the literature.

Holmgren, N. H. and B. Angell. *Botanical Illustration: Preparation for Publication*. Bronx, NY: New York Botanical Garden, 1986. 74 p. $10.00 (paper). ISBN 0893272728.

An authoritative "how to." Also, see Glimn-Lacy (p. 293), West (p. 297), and Wunderlich (p. 298).

Kearns, Carol Ann and David William Inouye. *Techniques for Pollination Biologists*. Niwot, CO: University Press of Colorado, 1993. 583 p. $37.50. ISBN 0870812793. Glossary. Bibliography.

Practical information for experimental field studies with recipes of how to deal with whole plants, flowers, gynoecia, pollen, nectar, bees, birds, bats, butterflies, beetles, etc. Appendices provide equipment and chemical vendors.

Journal of Tissue Culture Methods. Vol. 1– . Columbia, MD: Tissue Culture Association, 1975– . Quarterly. $80/yr. ISSN 0271-8057. Formerly *TCA Manual*.

Procedures for plants and animals.

MacFarlane, Ruth B. Alford. *Collecting and Preserving Plants for Science and Pleasure*. New York: Arco, 1985. 184 p. $13.95. ISBN 0668060093.

Procedures for collecting, preserving, identifying, labeling, mounting and packaging, storage and display, city botany, the herbarium, and ornamental uses of plants.

Maintenance of Microorganisms and Cultured Cells: A Manual of Laboratory Methods. Edited by B. Kirsop and A. Doyle, 2nd ed. San Diego, CA: Academic Press, 1991. 308 p. ISBN 0124103510.

Preservation methodology for bacterial, yeasts and other fungi, algae, protozoa, and animal and plant cell cultures.

The Maize Handbook

See annotation (p. 286) in the Handbooks and Databases section.

Methods in Arabidopsis Research. Edited by Csaba Koncz, Nam-Hai Chua, and Jeff Schell. River Edge, NJ: World Scientific Publishing, 1992. 482 p. ISBN 981029045. Index.

Methods for investigating the Arabidopsis genome as a model system to explore the physiological, cellular, and biochemical properties of plants.

Methods in Comparative Plant Ecology: A Laboratory Manual. Edited G. A. F. Henfry and J. P. Grime. London: Chapman & Hall, 1993. 252 p. (ring bound). ISBN 041246662303.

Ninety diagnostic techniques for investigating the effects on plants of contrasting ecology. This book is a companion to the text *Comparative Plant Ecology: A Functional Approach to Common British Species*. Edited by J. P. Grime et al., Chapman & Hall, 1988.

Methods in Enzymology

This outstanding and useful series is annotated (p. 57) in Chapter 3, "General Sources."

Methods in Plant Biochemistry. v. 1– . San Diego: Academic Press, 1989. Each volume has a distinctive title; price varies.

Highly acclaimed series providing comprehensive, practical information on analytical techniques for a particular family of plant compounds. Vol. 9: *Enzymes of Secondary Metabolism* (1993).

Methods in Plant Molecular Biology and Biotechnology. Edited by Bernard R. Blick and John E. Thompson. Boca Raton, FL: CRC Press, 1993. 360 p. $70.00. ISBN 0849351642.

Comprehensive handbook covering recombinant DNA technology, methods for the production and analysis of plant mutants, relevant computer software, DNA mapping and analysis of DNA polymorphism, and detection and characterization of plant pathogens.

Modern Methods of Plant Analysis. v. 1– . New York: Springer-Verlag, 1985– . Irregular. Price varies.

This is a highly regarded, valuable series that devotes an entire volume to one topic. Vol. 13: *Plant Toxin Analysis*, 1992, 389 p., $229.00.

Modern Phytochemical Methods; Proceedings of the 30th Meeting of the Phytochemical Society of North America, Quebec City, Canada, August 11–15, 1990. Edited by Nikolaus H. Fischer, et al. New York: Plenum, 1991. (Recent Advances in Phytochemistry, v. 25). 418 p. $85.00. ISBN 0306439255.

More Gene Manipulations in Fungi. Edited by J. W. Bennett and Linda L. Lasure. San Diego: Academic Press, 1991. 470 p. $80.00. ISBN 0120886421. Bibliography. Index.

Sequel to *Gene Manipulations in Fungi* (1985). State-of-the-art methods and techniques for molecular mycology.

Phytochemical Analysis: An International Journal of Plant Biochemical Techniques. v. 1– . Chichester, UK: Wiley, 1990– . Bimonthly. $365/yr. ISSN 0958-0344.

Review journal for biochemical methods.

Plant Cell and Tissue Culture. Edited by Jeffrey W. Pollard and John M. Walker. Clifton, NJ: Humana Press, 1990. (Methods in Molecular Biology, v. 6). 597 p. ISBN 0896031611.

Protocols for a broad range of basic techniques and culture conditions for plant cells.

Plant Cell and Tissue Culture. Edited by Angela Stafford and Graham Warren. Milton Keynes, England: Open University Press, 1991. (Biotechnology Series). 251 p. $45.00 (paper). ISBN 0335158234.

Approved by the Institute of Biology in London, this manual examines theory, processes, and applications.

Plant Cell Biology: A Practical Approach. Edited by N. Harris and K. J. Oparka. New York: IRL at Oxford University Press, 1994. 329 p. $45.00. ISBN 0199633983.

Both established techniques and new advances provide a comprehensive guide to methods for studying characterization of growth and differentiation through structural, biochemical, and molecular analysis. This is a natural companion to *Plant Molecular Biology: A Practical Approach* (below).

Plant Genome Analysis

A review of molecular techniques, annotated (p. 307) in the "Reviews of the Literature" section.

Plant Genomes: Methods for Genetic and Physical Mapping. Edited by Jacques S.Beckmann and Thomas C. Osborn. Dordrecht: Kluwer Academic, 1992. 250 p. ISBN 0792316304.

Introduction to methods written for new investigators in this area.

Plant Molecular Biology: A Practical Approach. Edited by C. H. Shaw. Washington, DC: IRL Press, 1988. 313 p. $45.00. ISBN 18552210575.

Techniques for plant molecular biology and genetic engineering. Also, see *Plant Cell Biology: A Practical Approach* (p. 296).

Plant Molecular Biology Manual. Edited by Stanton B. Gelvin and Robert A. Schilperoort. Dordrecht: Kluwer Academic, 1988. (ring binder). ISBN 9024736331. *Supplements* 1– and updates, 1991– .

Routine, basic techniques convenient for use in the laboratory.

Plant Tissue Culture Manual: Fundamentals and Applications. Edited by K. Lindsey. Dordrecht: Kluwer Academic, 1991. (ring binder). ISBN 0792311159. *Supplement* 1– , 1992– .

Step-by-step protocols for basic techniques. Companion to *Plant Molecular Biology Manual.*

Smith, Roberta H. *Plant Tissue Culture: Techniques and Experiments.* San Diego: Academic Press, 1992. 171 p. $34.95 (spiral bound). ISBN0126503400. Index.

Laboratory exercises demonstrating major concepts, appropriate for students in secondary and higher education.

Technique; A Journal of Methods in Cell and Molecular Biology. v. 1– , 1989– . Philadelphia, PA: Saunders. Bimonthly. $210/yr. ISSN 1043-4658.

Useful for botanists; provides detailed protocols for laboratory use.

West, Keith. *How to Draw Plants.* New York: Watson-Guptill, 1989. 152 p. $16.95. ISBN 0823023567.

Other useful books by West are *How to Draw and Paint Wild Flowers* (1993) and *Painting Plant Portraits: A Step-by-Step Guide* (1991). Also, check out Glimn-Lacy (p. 293), Holmgren and Angell (p. 294), and Wunderlich (below).

Womersley, J. S. *Plant Collecting and Herbarium Development: A Manual*. Rome: Food and Agriculture Organization of the United Nations, 1981. (FAO Plant Production and Protection Paper no. 33). 148 p. $17.00. ISBN 9251011443.

Practical information for the beginner, including specialized techniques for field collection, preservation, identification, organization of the herbarium processing collections, curation, and services. Also, see *The Herbarium Handbook* (p. 286).

Wunderlich, Eleanor B. *Botanical Illustration in Watercolor*. New York: Watson-Guptill Pubs., 1991. $32.50. ISBN 082305291.

Also, see Glimn-Lacy (p. 293), Holmgren and Angell (p. 294), and West (p. 297).

Periodicals

American Journal of Botany. v. 1– , 1914– . Ames, IA: Botanical Society of America. Monthly. $155/yr. ISSN 0002-9122.

"Devoted to all branches of plant sciences." This is a leading botanical journal and an official publication of the Botanical Society of America. Abstracts of the annual meeting, poster sessions, and symposium talks are published as a supplement to the journal.

Annals of Botany (London). v. 1– , 1887– . London: Academic Press. 12 times/yr. $384/yr. ISSN 0305-7364.

An international journal publishing in all areas of botany.

Annals of the Missouri Botanical Garden. v. 1– , 1914– . St. Louis, MO: Missouri Botanical Garden. Quarterly. $125/yr. ISSN 0026-6493.

This journal publishes original papers in systematic botany, primarily from authors connected with the Missouri Botanical Garden.

Botanical Journal of the Linnean Society. v. 1– , 1855– . New York: Academic. Monthly. $618/yr. ISSN 0024-4074.

Published for the Linnean Society of London, the main focus of this journal is systematic as it relates to all areas of botany.

Brittonia. v. 1– , 1931– . New York: New York Botanical Garden. Quarterly. $60/yr. ISSN 0007-196X.

One of the most important journals for systematic botany.

Bryologist. v. 1– , 1898– . Buffalo, NY: American Bryological and Lichenological Society. Quarterly. $55/yr. ISSN 0007-2745.

"Devoted to the study of bryophytes and lichens."

Bulletin of the Torrey Botanical Club. v. 1– , 1870– . New York: New York Botanical Garden. Quarterly. $45/yr. ISSN 0040-9618.

An important journal of long standing. Each issue contains reports of original research; review papers and papers on local flora, field trips, obituaries, book reviews, notes; and the "Index to American Botanical Literature" compiled from the resources of the New York Botanical Garden Library. This last part is currently under review and may not be continued because its compilation is time consuming and available electronically elsewhere.

Economic Botany. v. 1– , 1947– . New York: New York Botanical Garden. Quarterly. $65/yr. ISSN 0013-0001.

Published for the Society for Economic Botany and is "Devoted to past, present, and future uses of plants by man," with emphasis on uses of plants.

Environmental and Experimental Botany. v. 1– , 1961– . Elmsford, NY: Pergamon. Quarterly. $424/yr. ISSN 0098-8472.

Publishes papers in all fields of experimental and environmental botany.

Experimental Mycology. v. 1– , 1977– . San Diego: Academic Press. Quarterly. $128/yr. ISSN 0147-5975.

Devoted to experimental investigations relating structure and function to growth, reproduction, morphogenesis, and differentiation of fungi and their allies.

International Journal of Plant Sciences. 1993– . (Formerly *Botanical Gazette*, 1886–1992). Chicago: University of Chicago Press. Qquarterly. $135/yr. ISSN 1058-5893.

Well-considered journal for all areas of original research.

Journal of Bryology. v. 7– , 1972– . England: W. S. Maney & Son, Ltd. Twice annually. $155/yr. ISSN 0373-6687.

Formerly *Transactions of the British Bryological Society*, and published for the Society. This journal focuses on mosses and liverworts.

Journal of Experimental Botany, for the Pure and Applied Aspects of Physiology, Biochemistry, Molecular Biology and Biophysics of Plants. v. 1– , 1950– . Oxford, England: Oxford University Press. Monthly. $515/yr. ISSN 0022-0957.

Published for the Society for Experimental Botany and designated as an official journal of the Federation of European Societies of Plant Physiology.

Journal of Phycology. v. 1– , 1965– . Lawrence, KS: Phycological Society of America. Bimonthly. $235/yr. ISSN 0022-3646.

Original research on algae.

Journal of Plant Physiology. v. 115– , 1984– . New York: G. Fischer. Monthly. $932/yr. ISSN 0044-328X.

Formerly *Zeitschrift fuer Pflanzenphysiologie,* 1909–1983. Incorporating *Biochemie und Physiologie der Pflanzen BPP* in 1993, this international journal publishes in plant physiology, biochemistry and molecular biology of plants, in English.

Mycologia. v. 1– , 1909– . New York: New York Botanical Garden. Bimonthly. $90/yr. ISSN 0027-5514.

Official publication of the Mycological Society of America, issuing original research articles on all aspects of the fungi.

Mycological Research. The International Journal of Fungal Biology. v. 1– , 1896– . Cambridge, England: Cambridge University Press. Monthly. $360.00/yr. ISSN 0953-7562.

Formerly *Transactions of the British Mycological Society.* Publishes original research articles in all areas of mycology, notes and brief articles, as well as book reviews.

Mycotaxon. v. 1– , 1974– . Ithaca, NY: Mycotaxon, Ltd. Irregular, quarterly issues. $60.00/yr. ISSN 0093-4666.

"An international journal designed to expedite publication of research on taxonomy and nomenclature of fungi and lichens." Papers may be in French or English with summaries in any language.

New Phytologist. v. 1– , 1902– . New York: Cambridge University Press. Monthly. $625.00/yr. ISSN 0028-646X.

"An International Journal of the Plant Sciences." Articles include commissioned and submitted reviews, research papers, and book reviews on all aspects of the plant sciences.

Photosynthesis Research. v. 1– , 1980– . Dordrecht: Kluwer.
　　Monthly. $748.00/yr. ISSN 0166-8595.

An international journal publishing papers dealing with both basic and applied aspects of photosynthesis.

Phycologia. v. 1– , 1961– . Lawrence, KS: International Phycological
　　Society. Bimonthly. $352.00/yr. ISSN 0031-8884.

Dedicated to the promotion of phycology and to the encouragement of international cooperation among phycologists and phycological institutes. Contributions may be in English or French.

Physiologia Plantarum. v. 1– , 1948– . Copenhagen: Munksgaard.
　　Monthly. $553.00/yr. ISSN 0031-9317.

Official journal of the Federation of European Societies of Plant Physiology, published by the Scandinavian Society for Plant Physiology. The journal publishes papers in English on all aspects of experimental plant biology. Two different types of review articles, "Minireviews" and "What's New in Plant Physiology," are considered for publication.

Phytochemistry. v. 1– , 1962– . Elmsford, NY: Pergamon. Semi-
　　monthly. $1,432.00/yr. ISSN 0031-9422.

An international journal of plant biochemistry, and an official organ of the Phytochemical Society of Europe and the Phytochemical Society of North America. This journal publishes research on all aspects of pure and applied plant biochemistry. The journal issues review articles, as well as papers on growth and metabolism, ecological biochemistry, biosynthesis, cell culture and biotechnology, chemotaxonomy, plant chemistry, and short reports. Book reviews are published regularly.

Phytopathology. v. 1– , 1911– . St. Paul, MN: American Phytopatho-
　　logical Society. Monthly. $230.00/yr. ISSN 0031-949X.

The official journal of the American Phytopathological Society. Membership in the Society is not a prerequisite for publication, but page charges, which are mandatory, are higher for nonmembers than for members. Papers are accepted in the following categories: 1) original research on some aspect of plant pathology; 2) obituaries; 3) letters to the Editor. The journal also provides reports of the officers, meetings, and news of the Society.

Plant and Cell Physiology. v. 1– , 1960– . Kyoto, Japan: Japanese
　　Society of Plant Physiologists. Monthly. $180.00/yr. ISSN
　　0032-0781.

An international journal, text in English, for physiology, biochemistry, molecular biology, and cell biology of plants and microorganisms. Contributors must hold membership in the Japanese Society of Plant Physiologists.

Plant and Soil. v. 1– , 1948– . Dordrecht: Kluwer. Semimonthly. $1,865.00/yr. ISSN 0032-079X.

Issued under the auspices of the Royal Netherlands Society of Agricultural Science. "An international journal on plant-soil relationships . . . ."

Plant Cell. v. 1– , 1989– . Rockville, MD: American Society of Plant Physiologists. Monthly. $825.00/yr. ISSN 1040-4651.

This journal features original research articles specializing in plant cell biology and plant tissue culture.

Plant, Cell & Environment. v. 1– , 1978– . London: Blackwell. Monthly. $950.00/yr. ISSN 0140-7791.

Original papers in any field of the physiology of green plants, including plant biochemistry, molecular biology, biophysics, cell physiology, whole plant physiology, crop physiology and physiological ecology, together with structural, genetical, pathological and micrometeorological aspects as related to plant function.

Plant Cell Reports. v. 1– , 1981– . New York: Springer-Verlag. Monthly. $576.00/yr. ISSN 0721-7714.

This journal presents original research results dealing with new advances concerning all aspects of research and technology in plant cell science, plant cell culture and molecular biology including biochemistry, genetics, cytology, physiology, phytopathology, plant regeneration, genetic manipulations and nucleic acid research.

Plant Growth Regulation. v. 1– , 1982– . Dordrecht: Kluwer. Bimonthly. $245.00/yr. ISSN 0167-6903.

Offers original research dealing with plant regulators and plant hormones.

Plant Journal. v. 1– , 1991– . Oxford, UK: Blackwell Scientific. Monthly. $675.00/yr. ISSN 0960-7412.

Rapid publication of original work on plant molecular sciences. Features include review articles, research papers, short communications, technical advances, and gene and mutant directories.

Plant Molecular Biology. v. 1– . Dordrecht: Kluwer, 1981-. Semimonthly. $1,275.00/yr. ISSN 0167-4412.

An international journal on fundamental research and genetic engineering published in cooperation with the International Society for Plant Molecular Biology. This journal provides a "rapid publication outlet for all types of research concerned and connected with plant molecular biology, biochemistry and plant molecular genetics." In 1995, the section *Molecular Breeding* became an independent journal publishing papers on all aspects of research focused on applications of plant molecular biology benefiting the seed industry, farmers, and the processing industries, for the environment as well as the consumer.

Plant Physiology. v. 1– , 1926– . Rockville, MD: American Society of Plant Physiologists. Monthly. $825.00/yr. ISSN 0032-0889.

Devoted to physiology, biochemistry, cellular and molecular biology, biophysics, and environmental biology of plants. Supplements to the journal record abstracts of papers presented at the annual meeting of the society.

Plant Science. v. 38– , 1985- . Limerick, Ireland: Elsevier Scientific. Semimonthly. $1,962.00/yr. ISSN 0168-9452.

Formerly *Plant Science Letters*, 1973–1984. Original work in any area of experimental plant biology is welcome. Areas covered include cell and tissue culture, cell organelles, cellular and molecular genetics, enzymology, inorganic metabolism, molecular biology, nucleic acids, photosynthesis, phytochemistry, pigments, regulation of growth and development, ultrastructure, viruses, and water relations.

Plant Systematics and Evolution/Entwicklungsgeschichte und Systematik der Pflanzen. v. 1– , 1851– . New York: Springer-Verlag. Semimonthly. $1,375.00/yr. ISSN 03782697.

A continuation of *Osterreichische Botanische Zeitschrift*. Original papers on the morphology and systematics of plants in the widest sense.

Planta. v. 1– , 1925– . New York: Springer-Verlag. Monthly. $1,915.00/yr. ISSN 0032-0935.

This international journal publishes original articles in all aspects of plant biology, particularly in molecular and cell biology, ultrastructure, biochemistry, metabolism, growth, development and morphogenesis, ecological and environmental physiology, biotechnology and plant–microorganism interactions.

Planta Medica. v. 1– , 1953– . Stuttgart, W. Germany: Thieme. Bimonthly. $278.00/yr. ISSN 0032-0943.

This journal of medicinal plant research publishes original articles and reports and is the official organ of the Society for Medicinal Plant Research. Articles deal with pharmacology, toxicology, medicinal applications, biochemistry, physiology, *in vitro* cultures, biological activities, and phytochemistry. Letters, phytochemical notes, and book reviews are featured.

Systematic Botany. v. 1– , 1976– . Kent, OH: American Society of Plant Taxonomists. Quarterly. $90.00/yr. ISSN 0363-6445.

This journal accepts original articles pertinent to modern and traditional aspects of systematic botany, including theory as well as applications. Papers longer than 50 printed pages and dealing with plant systematics, especially taxonomic monographs and revisions, appear in *Systematic Botany Monographs* published by ASPT.

Taxon. v. 1– , 1951– . Utrecht, Netherlands: International Bureau for Plant Taxonomy and Nomenclature. Quarterly. $108.00/yr. ISSN 0040-0262.

Published for the International Association for Plant Taxonomy. The journal is "devoted to systematic and evolutionary biology with emphasis on botany." Issues contain original articles, methods and techniques, nomenclature, proposals to conserve or reject, news, "Index Herbariorum," book reviews, and announcements.

Vegetatio, International Journal of Plant Ecology. v. 1– , 1949– . Dordrecht: Kluwer. Monthly. $1,215.00/yr. ISSN 0042-3106.

Original articles in the fields of geobotany and vegetation science, specifically in the areas of vegetation descriptions, population analyses of plants in the context of plant communities, and functional aspects of plant communities.

Reviews of the Literature

Advances in Botanical Research. v. 1– , 1963– . London, New York: Academic. Irregular. $96.00/yr. (for the year 1991). ISSN 0065-2296.

This well-established series traditionally surveys progress across the whole spectrum of botanical studies.

Advances in Bryology. v. 1– , 1981– . Forestburgh, NY: Lubrecht & Cramer. Biennial. v. 4: Bryophyte Systematics, 1991, 264 p., $112.00, ISBN 3443520022.

The official publication of the International Association of Bryologists. This series aims to present authoritative and current reviews, essays, and summary syntheses of the different fields of bryology written by leaders in the area.

Advances in Economic Botany. v. 1– , 1984– . Bronx, NY: New York
    Botanical Garden. Irregular. Priced individually. ISSN 0741-8280.
    v. 9: *Non-Timber Products from Tropical Forests; Evaluation of a
    Conservation and Development Strategy.* 176 p. $18.95. ISBN
    0893273767.

This series was established to provide an outlet for monographs and symposia
on all subjects in the field of economic botany.

Advances in Plant Cell Biochemistry and Biotechnology. v. 1– , 1992– .
    London: JAI Press. 300 p. $75.00. ISBN 1559383577.

This series was launched to promote current awareness of the progress being
made in the understanding and utilization of plant materials. Aimed at a
research and professional audience.

Annual Review of Phytopathology. v. 1– , 1963– . Palo Alto, CA:
    Annual Reviews. Annual. $46.00/yr. ISSN 00664286.

This is a highly respected series devoted to review articles, by individual
authors, dealing with plant pests and diseases, their history, nature, effects,
ecology, and control.

Annual Review of Plant Physiology and Plant Molecular Biology. v. 1– ,
    1950– . Palo Alto, CA: Annual Reviews. Annual. Formerly *Annual
    Review of Plant Physiology.* v. 1–38. $44.00/yr. ISSN 1040-
    2519.

This series is addressed to the advanced student doing research in plant physiology and plant biochemistry. This series, according to *Science Citation Index*,
has an Impact Factor ranking of #1 out of 120 journals in the botany category,
an excellent indication of its importance.

Botanical Monographs. v. 1– , 1961– . Oxford: Blackwell. Infrequent.
    Priced individually.

Each volume provides a specific review for a particular topic. Previous volumes include *The Biology of Dinoflagellates, Molecular Biology of Plant
Development, Plant Cell Culture Technology.*

Botanical Review. v. 1– , 1935– . New York: New York Botanical
    Garden. Quarterly. $65.00/yr. ISSN 0006-8101.

Authors are chosen primarily by invitation, but unsolicited manuscripts are also
considered. This journal bills itself as "interpreting botanical progress" and
serves the function of synthesizing the state of knowledge of the botanical
sciences for a broad spectrum of botanists.

Critical Reviews in Plant Sciences. v. 1– , 1983– . Boca Raton, FL:
      CRC Press. Irregular. $225.00/yr. ISSN 0735-2689.
This review series focuses on critical reviews for photosynthesis, nitrogen fixation, genetic engineering, and other areas of plant science.

Current Topics in Plant Molecular Biology. v. 1– , 1992– . Boca Raton,
      FL: CRC Press. Irregular. Price varies.
Each book in the series is designed to reflect different areas of research. Vol.
1: *Plant Biotechnology and Development.* Edited by Peter M. Gresshoff (1992);
Vol. 2: *Plant Responses to the Environment.* Edited by Peter M. Gresshoff
(1993).

Ecophysiology of Photosynthesis. Edited by Ernst-Detlef Schulze and
      Martyn M. Caldwell. New York: Springer-Verlag, 1994. (Ecological
      Studies 100). 576 p. $199.00. ISBN 0387559523.
Dedicated to Otto Ludwig Lange, this volume reviews 30 years of research in
the ecophysiology of photosynthesis at several levels of integration, from the
molecular level to canopy, ecosystem and global scales. Subject and species
indexes.

Encyclopedia of Plant Physiology, new series. v. 1– , 1975– .
      New York: Springer-Verlag. Irregular. Price varies.
This important monographic review series surveys all aspects of the botanical
sciences, devoting several volumes, as needed, to cover any particular topic.
The new series is in English, and continues the older series, in German, *Handbuch der Pflanzenphysiologie*, vol. 1–18, 1955–1967. Recent volumes in this
excellent series include: *Photosynthesis*, v. 1–3, edited by A. Trebst et al.,
1977–1986.

Handbook of Vegetation Science. Edited by Reinhold Tuxen, continued
      by H. Lieth. v. 1– , 1974– . The Hague: Junk. Price varies.
      Later published by Kluwer Academic.
This multivolume reference work reviews vegetation science in all its many
aspects. Recent volumes are *Vegetation Mapping*, 1988; *Fungi in Vegetation
Science.* Edited by W. Winterhoff, 1992.

The Molecular and Cellular Biology of the Yeast Saccharomyces. Edited
      by James R. Broach, John R. Pringle, and Elizabeth W. Jones.
      Plainview, NY: Cold Spring Harbor Laboratory, 1991–1995. (Cold
      Spring Harbor Monograph Series, v. 21 A, B, C). $285 (3 vols.).
      ISBN 087969355X (v. 1).

Vol. 1: *Genome Dynamics, Protein Synthesis, and Energetics*; Vol. 2: *Gene Expression*; Vol. 3: *Cell Cycle and Cell Biology*. A complete review of the molecular and cellular biology of *Saccharomyces* and a valuable reference for anyone using it as an experimental organism.

The Mycota: A Comprehensive Treatise on Fungi as Experimental Systems for Basic and Applied Research. Edited by K. Esser and P. A. Lemke. New York: Springer-Verlag, 1994– . Projected 7-vol. set. $200.00 (v. 1). ISBN 0387577815 (v. 1).

Leading research specialists have contributed to this treatise on the fungi covering: Vol. 1: Growth, Differentiation and Sexuality; Vol. 2: Genetics and Biotechnology; Vol. 3: Biochemistry and Molecular Biology; Vol. 4: Environmental and Microbial Relationships; Vol. 5: Plant Relationships; Vol. 6: Animal and Human Relationships; Vol. 7: Systematics and Cell Structure.

Oxford Surveys of Plant Molecular and Cell Biology. v. 1– , 1984– . New York: Oxford University Press. Annual. Price varies. ISSN 0264-861X.

This series contains review articles delineating current progress, news, and views, in the field of plant molecular and cell biology.

Plant Genome Analysis. Edited by Peter M. Gresshoff. Boca Raton, FL: CRC, 1994. (A CRC series of current topics in plant molecular biology). 247 p. ISBN 0849382645.

Edited by a well-known authority in the field, this book provides a review of molecular techniques for plant genome mapping and analysis. An added attraction is a set of three articles by patent lawyers on the patent process, legislative initiatives pertaining to biotechnology, and patenting DNA.

Plant Molecular Biology. Edited by Diter von Wettstein and Nam-Hai Chua. New York: Plenum Press, 1987. (NATO ASI series. Series A, Life Sciences, v. 140). 697 p. ISBN 030642696X. Updated by *Plant Molecular Biology 2*. Edited by R. G. Herrman and B. A. Larkins. New York: Plenum Press, 1991. (NATO ASI series. Series A, Life Sciences, v. 212). 766 p. ISBN 0306440245.

Extensive reviews of development in plant molecular biology; includes bibliographical references to the literature. Also, see *10 Years Plant Molecular Biology* (p. 308).

Progress in Botany. v. 36– , 1974– . New York: Springer-Verlag. Irregular. Price varies. ISSN 0340-4773. Continues *Fortschritte der Botanik*. v. 1–35, 1931–1973.

The basic mission of this series is to report on all areas of botany, rotating subjects every two to three years. The 1993 volume includes reviews on structural botany, taxonomy, geobotany, plant physiology, genetics, and floral ecology.

Progress in Phycological Research. v. 1– , 1982– . New York: Elsevier Science. Irregular. ISSN 0167-8574.

Excellent contribution to the algal literature relevant to physiologists, plant scientists, and ecologists.

Recent Advances in Phytochemistry. Proceedings of the Phytochemical Society of North America. Vol. 1–4, 1968–1972, New York: Appleton Century Crofts; Vol. 5–8, 1972–1974, New York: Academic Press; Vol. 9– ,1975– , New York: Plenum. Irregular. Price varies. ISSN 0079-9920.

Volume 27 *Phytochemical Potential of Tropical Plants*, 1993, contains reviews of the 32nd annual meeting of the Society.

10 Years Plant Molecular Biology. Edited by Robbert A. Schilperoort and Leon Dure. Dordrecht: Kluwer, 1992 reprint from *Plant and Soil*. v. 19, no. 1. $50.00. ISBN 0792314808.

State-of-the-art papers in genetic engineering, gene expression and its manipulation, microbe and insect interactions with plants, transposable elements and gene tagging, plant and organ development, and other topics in plant molecular biology.

Topics in Photosynthesis. v. 1–12, 1976–1992. Amsterdam, New York: Elsevier. Price varies. ISSN 0378-6099.

Although this series has apparently stopped, each volume was instituted to assist in keeping up with progress and development in particular fields of photosynthesis. Vol. 11: *Current Topics in Photosynthesis*; Vol. 12: *Crop Photosynthesis: Spatial and Temporal Determinants*.

Trends in Photosynthesis Research. Edited by J. Barber, et al. Newark, NJ: Intercept Ltd./Springer-Verlag, 1992. 426 p. $61.00. ISBN 0946707375.

Lectures from an advanced course held in Spain in Sept. 1990 that gives an excellent representation of many active areas of photosynthesis research.

Trends in . . . . Series published by Elsevier. There is no title specifically for the plant sciences, but there may be articles in the following series that are of interest to the botanist: *Trends in Biochemical Sciences, Trends in Biotechnology, Trends in Cell Biology, Trends in Ecology & Evolution, Trends in Genetics, Trends in Microbiology.*

Societies

Consult the multiVolume *Encyclopedia of Associations* or the *Research Centers Directory* for more information. Current information about meetings, programs, and job opportunities may be picked on the Internet, accessible by Gopher or World Wide Web.

American Association of Botanical Gardens and Arboreta (AABGA)
786 Church Rd., Wayne, PA 19087, (215) 688-1120.

Founded: 1940. 1800 members. Directors and staffs of botanical gardens, arboreta, institutions maintaining or conducting horticultural courses, and others. Bestows awards. Publications: *American Association of Botanical Gardens and Arboreta-Newsletter, Public Garden*, internship directory, technical information papers, etc. Annual conference.

American Botanical Council (ABC)
PO Box 201660, Austin, TX 78720, (512) 331-8868.

Founded: 1988. Nonmembership. Seeks to gather and disseminate information on herbs, medicinal plants, and herbal research; increase public awareness and professional knowledge of the historical role and current potential of plants in medicine; promote understanding regarding the importance of preserving native plant populations in temperate and tropical zones. Conducts educational and public outreach programs, operates speakers' bureau. Publications: *Herbal-Gram*.

American Bryological and Lichenological Society (ABLS)
c/o Robert S. Egan, University of Nebraska at Omaha, Dept. of Biology, Omaha, NE 68182-0072, (402) 554-2491.

Founded: 1898. 506 members. Professional botanists, botany teachers, and hobbyists interested in the study of mosses, liverworts, and lichens. Maintains moss, lichen, and hepatic exchange clubs. Publications: *Bryologist, Evansia*, membership directory. Annual meeting.

American Herb Association (AHA)
PO Box 1673, Nevada City, CA 95959, (916) 626-5046.

Founded: 1981. 1000 members. Enthusiasts of herbs and herbal products. seeks to increase public knowledge and provide up-to-date information on herbs. Created a network to exchange data and resources among members nationwide. Conducts laboratory research projects and maintains herb garden. Operates library on herb gardening, botanical art, and culinary and medicinal herbs. Publications: *American Herb Association Quarterly Newsletter*; also publishes list of herb books.

American Phytopathological Society (APS)
    3340 Pilot Knob Rd., St. Paul, MN 55121, (612) 454-7250.

Founded: 1908. 4474 members. Professional educators, researchers, and other interested in the study and control of plant diseases. Maintains 46 committees. Publications: *Biological and Cultural Tests, Molecular Plant-Microbe Interactions, Phytopathology, Plant Disease: An International Journal of Applied Plant Pathology*, etc. Annual meeting.

American Society of Plant Physiologists (ASPP)
    15501 Monona Dr., Rockville, MD 20855, (301) 251-0560.

Founded: 1924. 5200 members. Professional society of plant physiologists, plant biochemists, and other plant scientists engaged in research and teaching. Presents annual and biennial awards for outstanding research in plant physiology. Offers placement service for members, conducts educational programs. Publications: *ASPP Newsletter, Plant Cell, Plant Physiology* (program and abstracts of papers presented at the annual conference is issued as supplement to this journal). Annual meeting.

American Society of Plant Taxonomists (ASPT)
    University of Georgia, Dept. of Botany, Athens, GA 30602,
    (404) 542-1802.

Founded: 1937. 1200 members. Botanists and others interested in all phases of plant taxonomy. Bestows Cooley Award for best paper presented at the annual meeting and Asa Gray Award for outstanding service in the field of botanical systematics. Publications: *American Society of Plant Taxonomists-Directory, Systematic Botany*, newsletter. Annual meeting.

Botanical Society of America (BSA)
    c/o Christopher Haufler, University of Kansas, Dept. of Botany,
    Haworth Hall, Lawrence, KS 66045-2106, (913) 864-4301.

Founded: 1906. 2600 members. Professional society of botanists and others interested in plant science. Conducts special research programs, bestows awards. Publications: *American Journal of Botany, Membership Directory, Career Bulletin, Guide to Graduate Study in Botany for the U.S. and Canada, Plant Science Bulletin*. Annual Conference with symposium.

Botanical Society of the British Isles (BSBI)
    Dept. of Botany, Natural History Museum, Cromwell Rd., London
    SW7 5BD, England.

Founded: 1836. 2700 members. Amateur and professional botanists in England and the Republic of Ireland. Promotes the study of British and Irish flowering plants and ferns. Organizes field visits; conducts surveys and symposia.

Formerly: Botanical Society of London. Publications: *BSBI Abstracts, BSBI News, Watsonia.* Periodic exhibition meetings and conferences.

British Lichen Society (BLS)
c/o T. H. Moxham, University of Bath, Dept. of Plant Sciences, Bath, Avon BA2 7AY, England.

Founded: 1958. 526 members. Professional academic, and amateur lichenologists involved in research. Organizes lecture meetings, presents awards. Maintains 6000 volume library of lichenological reprints and books. Publications: *BLS Bulletin, The Lichenologist, Membership List.* Annual convention.

British Phycological Society (BPS)
c/o Dr. E. J. Cox, University of Sheffield, Dept. of Animal and Plant Sciences, Sheffield, S. Yorkshire S10 2TN, England.

Founded: 1952. 685 members. Scientists, students, and other interested persons organized to further phycology, the study of algae and seaweed. Conducts conservation activities, field excursions, and workshops. Assists in mapping the geographical distributions of seaweeds of the British Isles. Publications: *BPS Newsletter, British Phycological Journal, Seaweeds of the British Isles.* Annual scientific meetings.

Council on Botanical and Horticultural Libraries (CBHL)
c/o John F. Reed, New York Botanical Garden, Bronx, NY 10458, (718) 220-8728.

Founded: 1970. 250 members. Libraries and collections in botanical or horticultural materials; librarians, bibliographers, booksellers, publishers, researchers, and administrators. Purpose is to initiate and improve communication between persons and institutions concerned with the development, maintenance, and use of botanical and horticultural libraries. Assists in the organization and coordination of activities of benefit and interest to these libraries; advises members of exhibitions, acquisitions, location of rare books. Circulates members' lists of duplicate or wanted titles. Conducts workshops on conservation and preservation, and sponsors talks on literature, etc. Publications: *Directory of Member Libraries,* membership lists, newsletter. Convention: Annual conference and business meeting is usually held in the spring.

Herb Society of America (HSA)
9019 Kirtland Chardon Rd., Mentor, OH 44060, (216) 256-0514.

Founded: 1933. 2100 members. Regional and state groups. Scientists, educators and others interested in botanical and horticultural research on herbs and culinary, economic, decorative, fragrant, and historic use of herbs. Maintains herb gardens in arboreta and other public sites. Establishes and maintains gardens for the blind. Planned and funded the National Herb Garden which was donated to the National Arboretum in Washington, DC. Operates library on botany and horticulture, conducts slide shows and lectures, maintains speakers' bureau, bestows awards. Publications: *Herbalist*, membership directory, newsletter. Triennial conventions in Reading, PA.

International Association for Plant Physiology (IAPP)
    CSIRO Div. of Food Processing, PO Box 52, North Ryde, NSW
    2113, Australia.

Founded: 1964. National and regional societies of plant physiologist representing 10,000 individual members in 40 countries. Promotes plant physiology internationally and encourages the formation of national societies, especially in developing countries. Sponsors exchange among plant physiologists on topics of regional significance in an effort to encourage cooperation between scientists in developed countries and their colleagues in developing countries. Conducts regional seminars. Publications: *International Directory of Plant Physiologists*, newsletter. Annual conference with symposium and exhibits.

International Association for Plant Taxonomy (IAPT)
    Botanischer Garten und Botanisches Museum Berlin-Dahlem,
    Konigin-Luise-Strasse 6-8, W-1000 Berlin 33, Germany.

Founded: 1950. 2900 members. Coordinates work related to plant taxonomy and international codification of plant names. Bestows awards. Formerly: Commission on the Nomenclature of Plants. Publications: *Regnum Vegetabile, Taxon*.

International Association for Vegetation Science (IAVS)
    Wilhelm-Weber-Strasse 2, W-3400 Gottingen, Germany.

Founded: 1930. 800 members. Vegetation scientists in 51 countries including botanists and ecologists. Fosters contacts among vegetation scientists worldwide; promotes development of vegetation science. Works to define and clarify problems regarding taxonomy and nomenclature of plant communities. Cooperates with other societies for the protection of nature. Formerly: International Society for Vegetation Science. Publications: *Report of the International Symposium, Journal of Vegetation Science*. Annual convention with symposium and excursion.

International Association of Botanic Gardens (IABG)
    Botanic Garden of Adelaide, N. Terrace, Adelaide, SA 5000,
    Australia.

Founded: 1954. 798 members. Botanic gardens, arboreta, and similar institutions maintaining scientific collections of plants; members of their staffs.

Encourages international information exchange and cooperation between botanic gardens, arboreta, and related institutes; promotes the study of taxonomy of cultivated plants and of the introduction and acclimatization of plants; fosters the conservation and preservation of rare and endangered plants. Publications: *International Directory of Botanical Gardens.* Quadrennial conference.

International Board for Plant Genetic Resources (IBPGR)
    c/o Food and Agriculture Organization of the U.N., Via delle Sette Chiese 142, I-00145 Rome, Italy.

Founded: 1974. Promotes and coordinates plant genetic resource activities through 8 regional groups. Organizes seminars, symposia, and training programs. Publications: *Annual Report, Geneflow, Newsletter.* Annual board meeting.

International Commission for the Nomenclature of Cultivated Plants (ICNCP)
    c/o CRZ, PO Box 32, NL-6700 AA Wageningen, Netherlands.

Founded: 1955. Cultivated plant taxonomists in 14 countries representing the fields of agriculture, forestry, and horticulture. Publications: *International Code of Nomenclature for Cultivated Plants.* Quinquennial convention.

International Mycological Associations (IMA)
    c/o Dr. C. P. Kurtzman, National Center Agricultural Utilization Research, 1815 N. University St., Peoria, IL 61604, (309) 685-4011.

Founded: 1971. National and international societies representing 20,000 mycologists from 80 countries. Promotes the study of mycology in all its aspects. Constitutes the section for general mycology within the International Union of Biological Sciences. Publications: *IMA News, International Mycological Directory.* Conventions: Congress every 4–6 years.

International Photosynthesis Committee (IPC)
    c/o Dr. N. Murata, National Institute for Basic Biology Myodaiji, Okazaki 444, Japan.

Founded: 1968. Organizes Triennial International Congress on Photosynthesis. Publications: *Congress Directory, Congress Proceedings.*

International Phycological Society (IPS)
    c/o Gerald T. Boalch, The Laboratory, Citadel Hill, Plymouth, Devon PL1 2PB, England.

Founded: 1960. 900 members. Scientists working to develop phycology. Promotes international cooperation among individual phycologists and phycological organizations. Disseminates information, sponsors competitions and bestows awards. Publications: *Phycologia*. Triennial congress with exhibits.

International Plant Biotech Network (IPBNet)
    c/o TCCP, Colorado State University, Dept. of Biology, Ft. Collins,
    CO 80523, (303) 491-6996.
Founded: 1985. 1500 members. Scientist and other individuals interested in plant biotechnology as a means of improving crops. Functions as the informal networking unit of the Tissue Culture for Crops Projects. Seeks to establish working partnerships among researchers worldwide. Offers technical assistance to laboratories; provides consulting service. Publications: *International Plant Biotechnology network Newsletter, IPBNet Directory*. Biennial conference.

International Society for Plant Molecular Biology (ISPMB)
    University of Georgia, Biochemistry Dept., Athens, GA 30602,
    (404) 542-3239.
Founded: 1982. 1850 members. Scientists whose research involves the molecular biology of plants. Coordinates exchange of information. Publications: *Directory of Members, Plant Molecular Biology, Plant Molecular Biology Reporter*. Triennial International Congress of Plant Molecular Biology.

International Society of Plant Morphologists (ISPM)
    Dept. of Botany, University of Delhi, Delhi 110 007, India.
Founded: 1950. 805 members. Individuals in 26 countries interested in plant morphology and its allied sciences. Purpose is to promote international cooperation among botanists. Disseminates information in the fields of plant morphology, anatomy, embryology, and histochemistry. Publications: *Phytomorphology, Yearbook*.

Mycological Society of America (MSA)
    c/o Donald H. Pfister, Harvard University Herbaria, 22 Divinity Ave.,
    Cambridge, MA 02138, (617) 495-2368.
Founded: 1931. 1400 members. Researchers, industrial and medical mycologists, plant pathologists, students, and others interested in the study of fungi through research, teaching, and industrial applications. Offers 2 graduate fellowships in mycology to MSA members. Publications: *MSA Directory, Mycologia, Mycologia Memoirs, Mycology Guidebook*, career brochure, newsletter. Convention: annual conference.

National Wildflower Research Center (NWRC)
    2600 F. M. 973 North, Austin, TX 78725-4201, (512) 929-3600.

Founded: 1982. 18,000 members. Seeks to stimulate and conduct research on wildflowers and other native American flora. Acts as a clearinghouse of information on native plants, and promotes their conservation and use in landscaping. Conducts experiments, demonstration plants and research on the propagation and cultivation of wildflowers and other native plants. Develops education programs; maintains speakers' bureau and library; conducts seminars and workshops. Publications: *Wildflower.* No conventions or meetings.

North American Mycological Association (NAMA)
    3556 Oakwood St., Ann Arbor, MI 48104-5213, (313) 971-2552.

Founded: 1959. 1600 members consisting of regional, state, and local groups. Amateur and professional mycologists, students, and botanists. Promotes amateur mycology; sponsors field trips and taxonomic and mycological seminars. Publications: *McIlvainea*, newsletter, *Directory.* Annual convention.

Organization for Flora Neotropica (OFN)
    New York Botanical Garden, Bronx, NY 10458-5126, (718) 220-8742.

Founded: 1964. 150 members. Established by the United Nations Educational, Scientific and Cultural Organization. Representatives from countries and organizations actively concerned with the taxonomy of neotropical flora, and interested individuals. Objectives are to plan the general outline of flora of the Neotropics; assist botanists in obtaining the facilities needed for their work; establish collaboration with botanical institutions; promote the training of promising young plant taxonomists; advance botanical exploration; strengthen herbaria that house important tropical collections; promoted the protection of natural tropical vegetation; aid in the establishment and protection of biological reserves. Conducts research on plants of the New World. Publications: *Flora Neotropica Monographs.* Annual board and commission meeting.

Phycological Society of America (PSA)
    c/o Dr. Robert Sheath, Memorial University, Dept. of Botany, St. John's, NF, Canada A1B 3X9.

Founded: 1946. 1200 members. Educators, researchers, and others interested in the pure, applied, or avocational study and utilization algae. Maintains speakers' bureau, sponsors competitions, bestows awards. Publications: *Journal of Phycology, Phycological Newsletter*, membership directory. Annual meeting with symposia and exhibits.

Phytochemical Society of Europe (PSE)
    Div. of Biosphere Sciences, King's College London, University of London, Campden Hill Rd., London W8 7AH, England.

Founded: 1957. 500 members. Scientists in 17 countries working in the field of plant chemistry. Acts as a forum for specialists in plant chemistry, biochemistry, and biotechnology who are interested in applying their research findings to agriculture and industry. Publications: *Proceedings of the Phytochemical Society of Europe, Phytochemistry*.

Phytochemical Society of North America (PSNA)
    c/o Dr. Helen M. Haberman, Goucher College, Dept. of Biological
    Sciences, Towson, MD 21204, (301) 337-6303.

Founded: 1960. 405 members. Membership comprises primarily research scientists interested in all aspects of the chemistry of plants. Purpose is to promote phytochemical research and communication. Bestows awards, maintains archives. Publications: *Directory*, newsletter, *Recent Advances in Phytochemistry*. Annual conference with symposium and exhibits.

Plant Growth Regulator Society of America (PGRSA)
    c/o C. David Fritz, Rhône-Poulenc Agriculture Co., Alexander Dr.,
    Research Triangle Park, NC 27709, (919) 549-2408.

Founded: 1973. 850 members. Scientists concerned with plant growth regulation. Seeks to foster a better understanding of the processes of plant growth and development. Promotes research; provides a forum for scientists from diverse disciplines to exchange information. Publications: *PGRSA Quarterly*, newsletter, *Membership Directory*, books on techniques, etc. Annual convention.

Society for Economic Botany (SEB)
    University of Missouri, American Archaeology Division, 103
    Swallow Hall, Columbia, MO 65211, (314) 882-3038.

Founded: 1959. 800 members. Botanists, anthropologists, pharmacologists, and others interested in scientific studies of useful plants. Seeks to develop interdisciplinary channels of communication among groups concerned with past, present, and future uses of plants. Publications: *Economic Botany, Membership Directory*, newsletter. Annual conference.

Society for Medicinal Plant Research (Gesellschaft fur Arzneipflanzen-
    forschung-GA)
    c/o ASYA Pharma AG, Weismullerstrasse 45, W-6000 Frankfurt am
    Main 1, Germany.

Founded 1953. 1300 members. Scientists in 70 countries who promote medicinal plant research. Organized to serve as an international focal point for such interests as pharmacognosy, pharmacology, phytochemistry, plant biochemistry and physiology, chemistry of natural products, plant cell culture, and application of medicinal plants in medicine. Publications: *Newsletter, Planta Medica*. Annual symposium.

Torrey Botanical Club
c/o Dr. H. D. Hammond, New York Botanical Garden, Bronx, NY 10458, (718) 220-8987.

Founded 1860. Botanists and others interested in botany and in collecting and disseminating information on all phases of plant science. International membership, but active membership is concentrated in New York City area. Publications: *Bulletin of the Torrey Botanical Club, Memoirs of the Torrey Botanical Club.* Meets 1st and 3rd Tuesdays, October through December and March through May.

Texts and Treatises

Taxonomic textbooks are listed and annotated in the Classification and Nomenclature section of this chapter.

Arabidopsis. Edited by Elliot M. Meyerowitz and Chris R. Somerville. Plainview, NY: Cold Spring Harbor Press, 1994. (Monograph 27). 1,300 p. $175.00. ISBN 0879694289.

Bell, Adrian D., with line drawings by Alan Bryan. *Plant Form. An Illustrated Guide to Flowering Plant Morphology.* New York: Oxford University Press, 1991. 341 p. $75.00. ISBN 0198542798. 351 color and black-and-white illustrations.

The Biochemistry of Plants: A Comprehensive Treatise. Edited by P. K. Stumpf and E. E. Conn. New York: Academic Press, 1980. Currently up to v. 16: *Intermediary Nitrogen Metabolism*, 1990. $149.00. ISBN 0126754160.

Bold, Harold C. et al. *Morphology of Plants and Fungi*, 5th ed. New York: Harper & Row, 1987. 912 p. $74.00. ISBN 0060408391.

Capon, Brian. *Botany for Gardeners: An Introduction and Guide.* Portland, OR: Timber Press, 1990. 220 p. $17.95 (paper). ISBN 0881921637.

Fahn, A. *Plant Anatomy*, 4th ed. New York: Pergamon, 1990. 588 p. $99.00. ISBN 0080374905.

Fosket, Donald E. *Plant Growth and Development: A Molecular Approach.* San Diego: Academic Press, 1994. 580 p. $49.95. ISBN 0122624300.

Galston, Arthur W. *Life Processes of Plants: Mechanisms for Survival.* New York: Freeman, 1993. (Scientific American Library). 246 p. $32.95. ISBN 0716750449.

Goodwin, Trevor Walworth and E. I. Mercer. *Introduction to Plant Biochemistry*, 2nd ed. New York: Pergamon, 1988, corrected reprint. 677 p. ISBN 0080249221.

Grierson, D. and S. N. Covey. *Plant Molecular Biology*, 2nd ed. New York: Chapman & Hall, 1994. 320 p. $75.00. ISBN 0412043416.

Griffin, David H. *Fungal Physiology*, 2nd ed. New York: Wiley-Liss, 1994. $69.95. ISBN 0471595861.

Hall, D. O. and K. K. Rao. *Photosynthesis*, 5th ed. New York: Cambridge University Press, 1994. 150 p. $39.95 (paper). ISBN 052143222.

Ingold, Cecil Terrence and H. J. Hudson. *The Biology of Fungi*, 6th ed. New York: Chapman & Hall, 1993. 224 p. $84.00. ISBN 0412490404.

Johri, B. M. et al. *Comparative Embryology of Angiosperms*. New York: Springer-Verlag, 1992. 1,120 p. $300.00. ISBN 3540536337.

Lawlor, David W. *Photosynthesis: Molecular, Physiological, and Environmental Processes*, 2nd ed. Essex, England: Longman, 1993. $49.95. 318 p. ISBN 0582086574.

Long, P. E. *Fungi: A New Synthesis*. Oxford, England: Blackwell Scientific, 1993. 400 p. $55.00. ISBN 0632018011.

Lyndon, R. F. *Plant Development: The Cellular Basis*. New York: Chapman & Hall, 1990. (Topics in Plant Physiology Series 3). 320 p. $27.50. ISBN 0045810338.

Molecular Biology of Photosynthesis. Edited by Govindjee. Dordrecht: Kluwer, 1989; reprinted from *Photosynthesis Research*. v. 16–19, 1988–89. 815 p. $275.00. ISBN 0792300971.

North American Terrestrial Vegetation. Edited by Michael G. Barbour and William Dwight Billings. New York: Cambridge University Press, 1990. 444 p. $37.95 (paper). ISBN 0521386780.

Photosynthesis: Photoreactions to Plant Productivity. Edited by Yash Pal Abrol et al. Dordrecht: Kluwer, 1993. 607 p. ISBN 0792319435.

Plant Biochemistry and Molecular Biology. Edited by Peter J. Lea and Richard C. Leegood. New York: Wiley, 1993. 312 p. ISBN 0471938955.

Plant Physiology: A Treatise. Edited by F. C. Steward. New York: Academic Press, 1959–1991. 10 v. in 15.

The Plant Viruses. New York: Plenum Press, 1985– . Multivolume set, a subseries within *The Viruses*. Edited by Heinz Fraenkel-Conrat.

Raven, Peter H. *Biology of Plants*, 5th ed. New York: Worth, 1992. 791 p. $59.95. ISBN 0879015322.

Romberger, J. A. et al. *Plant Structure: Function and Development: A Treatise on Anatomy and Vegetative Development, with Special Reference to Woody Plants*. Berlin: Springer, 1993. 550 p. $100.00. ISBN 3540563059.

Rose, A. H. and J. S. Harrison. *Yeasts*, 2nd ed. San Diego: Academic Press, 1987– . 4 Vols. v. 1: *Biology of Yeast*, $107.00, ISBN 0125964110; v. 2: *Yeasts and the Environment*, $105.00, ISBN 0125964129; v. 3: *Metabolism and Physiology of Yeasts*, $130.00, ISBN 0125964137; v. 4: *Yeast Organelles*, $159.00, ISBN 0125964145; v. 5: *Yeast Technology*, $120.00, ISBN 0125964145; v. 6: *Yeast Genetics*, $99.95, ISBN 0125964145.

Rudall, Paula. *Anatomy of Flowering Plants: An Introduction to Structure and Development*, 2nd ed. New York: Cambridge University Press, 1992. 110 p. $19.95 (paper). ISBN 0521421543.

Schofield, W. B. *Introduction to Bryology*. New York: Macmillan, 1985. 431 p. $68.00. ISBN 002949660.

Silvertown, Jonathan W. and Jonathan Lovett Doust. *Introduction to Plant Population Biology*, 3rd ed. Boston: Blackwell Scientific, 1993. 210 p. $39.95. ISBN 0632029730.

Singh, Ram J. *Plant Cytogenetics*. Boca Raton, FL: CRC Press, 1993. 448 p. $79.95. ISBN 084938656X.

Stern, Kingsley Rowland. *Introductory Plant Biology*, 4th ed. Dubuque, IA: W. C. Brown, 1988. 498 p. ISBN 06970512845.

Takhtajan, Armen. *Evolutionary Trends in Flowering Plants*. New York: Columbia University Press, 1991. 241 p. $44.00. ISBN 0231073-283.

Wainwright, M. *An Introduction to Fungal Biotechnology*. New York: Wiley, 1992. (Wiley Biotechnology series). 202 p. $34.95 (paper). ISBN 047193528X.

11

Anatomy and Physiology

The subjects of anatomy and physiology make such convenient and logical associates that they are combined as a package in this chapter, along with the currently hot topics of neurobiology and endocrinology. The fifth edition of the *McGraw-Hill Dictionary of Scientific and Technical Terms* defines "anatomy" as a "branch of morphology dealing with the structure of animals and plants." The same source interprets "physiology" as the "study of the basic activities that occur in cells and tissues of living organisms by using physical and chemical methods." This chapter includes only *human* anatomy and physiology: plants and animals are discussed in Chapters 10, "Plant Biology," and 13, "Zoology." Although a few medical titles are included, emphasis in this chapter is on the biological sciences rather than the behavioral or clinical.

There is overlap between this chapter and Chapters 5, "Biochemistry and Biophysics," and 6, "Molecular and Cellular Biology," so don't neglect to broaden the search to these other chapters if the molecular aspects, for example, of a particular discipline are important. Another caution: this chapter is complete only if it is complemented by resources available on the Internet. As biologists gradually make the transition from print to the electronic information highway, it is essential to explore cyberspace for possibilities that already exist, and constantly emerge, in the form of electronic news, discussion groups, and databanks on Gopher and the World Wide Web.

Abstracts, Indexes, and Bibliographies

Consult Chapter 4, "Abstracts and Indexes," for a discussion of general indexing resources for anatomy and physiology, chiefly *Biological Abstracts* (p. 74), *Index Medicus* (p. 78), and various *Exerpta Medica* sections (p. 77). Several specialized bibliographies and indexes are noted below.

Berichte uber die gesamte Physiologie und experimentelle Pharmakologie.
    (Berichte uber die gesamte Biologie. Abt. B). Berlin: 1920–1969.

References to articles and books, usually with abstracts. Not as comprehensive as *Biological Abstracts.*

Bibliographia Physiologica. Zurich: Concilium Bibliographicum, 1893/94–
    1926.

Includes books and reports.

Calcium and Calcified Tissue Abstracts. v. 1– , 1969– . Bethesda, MD:
    Cambridge Scientific Abstracts. Quarterly. $495/yr. ISSN 1069-
    5540.

Covers all areas of anatomy and metabolism dealing with calcium, including bone metabolism, tooth development, nerve transmission, and others. Also available as part of the **Life Sciences Collection** (See Chapter 4, p. 78).

CSA Neurosciences Abstracts. v. 1– , 1982– . Bethesda, MD: Cambridge Scientific Abstracts. Monthly. $740/yr. ISSN 0141-7711.

Provides abstracts for articles in all areas of the neurosciences, both basic and applied. A special topics section features a two-year bibliography covering a different hot topic each issue. Also available as part of the **Life Sciences Collection** (see Chapter 4, p. 78).

Depth Studies: Illustrated Anatomies from Vesalius to Vicq d'Azyr; Exhibition Held at the Smart Museum of Art, March 17 to June 7, 1992. Chicago: Smart Museum of Art, 1992. 23 leaves.

This exhibition catalog is listed as a point of information for history of science scholars. The catalog was issued in conjunction with the Centennial Conference Imaging the Body: Art and Science in Modern Culture, April 1–4 (from the cover); it is further noted in the catalog that "All books on display were loaned by the Department of Special Collections, University of Chicago Library."

Haller, Albrecht von. *Bibliotheca Anatomica*. New York: Olms, 1969. 2 v. Reprint of the edition originally published in Zurich by Orell, Gessner, Fuessli, et Socc., 1774–1777.

One of the most important works ever published for the history and bibliography of anatomy.

Krogman, Wilton Marion. *A Bibliography of Human Morphology*, 1914–1939. Chicago: University of Chicago Press, 1941. (The University of Chicago Publications in Anthropology. Physical Anthropology Series). 385 p.

Although somewhat out of scope, this bibliography may be useful for anatomists searching for journal articles in the period covered. Arrangement is by broad subject.

Medical Reference Works, 1679–1966; A Selected Bibliography. Edited by John B. Blake and Charles Roos. Chicago: Medical Library Association, 1967. (Medical Library Association. Publication no. 3). 343 p. *Supplements 1–3*. Chicago: Medical Library Association, 1970–1975.

This is an extremely valuable bibliography, especially useful for retrospective materials. Complete bibliographic information is arranged by subject, and then by form, with indexes.

Physiological Abstracts. v. 1–22. London: Physiological Society (Great Britain and Ireland), 1916–1937.

References with abstracts to articles and books; prepared through v. 9 (1924/25) in cooperation with the American Physiological Society.

Russell, Kenneth Fitzpatrick. *British Anatomy, 1525–1800: A Bibliography of Works Published in Britain, America, and on the Continent*, 2nd ed. Winchester, Hampshire, UK: St. Paul's Bibliographies, 1987. 245 p. ISBN 0906795338.

Annotated bibliography covering human anatomy books by British authors published in Britain, America, and on the European Continent in all languages and editions. It also includes the works of Continental authors translated into English or published in Britain in their original language.

Zentralblatt für Physiologie. v. 1–34. Leipzig: Organ der Deutschen physiologischen Gesellschaft (from v. 19), 1887–1921.

Abstracts of articles, although some books, theses, and proceedings of societies are included.

Atlases

A wide variety of atlases are listed, from the classic to the recently trendy, in an effort to provide examples of illustrations of the human body geared to wide group of student, medical, or lay person audiences. See also the Visible Human Project, discussed (p. 327) under **Anatomy Via the Internet**.

Anderson, Paul D. *Human Anatomy and Physiology Coloring Workbook and Study Guide*. Boston, MA: Jones and Bartlett, 1990. (Jones and Bartlett Series in Biology). 288 p. ISBN 0867201452.

A good example of the popular "coloring book"–type study guides appropriate at the undergraduate, or uninitiated, level.

Colborn, Gene L. and John E. Skandalakis. *Clinical Gross Anatomy: A Guide for Dissection, Study, and Review*. Pearl River, NY: Parthenon, 1993. 581 p. $28.00. ISBN 1850705224.

A dissection handbook.

Crocker, Mark. *The Body Atlas*. New York: Oxford University Press, 1991. 64 p. ISBN 0195208455.

The text and maps describe the main working parts of the body, including the nervous, digestive, blood transport, lymphatic, and reproductive systems; skeleton; muscles; skin; and glands. This example is included as a useful atlas suitable at the juvenile level and appropriate for school or public libraries.

Despopoulos, Agamemnon and Stefan Silbernagl, with 156 color plates by Wolf Rudiger Gay and Astried Rothenburger. Translations by Joy

Wieser. *Color Atlas of Physiology*, 4th rev. and enl. ed. New York: Thieme, 1991. 369 p. ISBN 0865773823. Index and bibliography.

Beautifully reproduced color plates.

Feneis, Heinz. *Pocket Atlas of Human Anatomy: Based on the International Nomenclature*. Translation of *Anatomisches Bildworterbuch der Internationalen Nomenclatur* by Hans E. Kaiser. New York: Thieme, 1985. (Thieme Flexibook Series). 183 p. $25.95. ISBN 0865772061.

Human anatomy that can be conveniently transported.

Grant, John Charles Boileau. *Grant's Atlas of Anatomy*, 9th ed. By Anne M. R. Agur. Baltimore: Williams & Wilkins, 1991. About 650 p. $89.00. ISBN 0683037013, 068303703X (paper).

A classic atlas worthy of its reputation.

Gray, Henry. *Anatomy of the Human Body*, 30th ed. On spine: Gray's Anatomy. Edited by Carmine D. Clemente. Philadelphia: Lea & Febiger, 1985. 676 p. $89.50. ISBN 081210644X

Probably the most famous of all the anatomies, and justifiably so; the first edition of this standard atlas was published in 1858.

Hendelman, Walter J. *Student's Atlas of Neuroanatomy*. Philadelphia: Saunders, 1994. 236 p. $29.00 (paper). ISBN 0721654282. Bibliography; index.

The aim of this book is to assist the student in understanding the functional anatomy of the human central nervous system. Black and white illustrations (some color) and diagrams.

Krstic, Radivoj V. *Human Microscopic Anatomy: An Atlas for Students of Medicine and Biology*. New York: Springer-Verlag, 1991. 616 p. $69.00. ISBN 0387536663.

Detailed line drawings of organ systems based on standard histological sections with associated descriptive text that concisely summarizes the structure and function of each system. This book follows the author's earlier volume, *Ultrastructure of the Mammalian Cell: An Atlas*, published by Springer-Verlag in 1979.

Lillie, John H. and Brent A. Bauer. *Sectional Anatomy of the Head and Neck; A Detailed Atlas*. New York: Oxford University Press, 1994. 213 p. $55.00. ISBN 0195042972. Black and white illustrations and radiographs.

The goal is to present an accurate, detailed series of illustrations, with commentary, depicting the structural relationships of the head and neck.

Mackenna, B. R. and R. Callander. *Illustrated Physiology*, 5th ed.
Edinburgh; New York: Churchill Livingstone, 1990. 325 p. $29.00.
ISBN 0443040958. Rev. ed. of *Illustrated Physiology* by Ann B.
McNaught and Robin Callander, 4th ed., 1983.

This is an illustrated atlas of human physiology, useful, also, as a course outline
or syllabus.

Martini, Frederic H. and Michael J. Timmons, et al. *Human Anatomy*.
Englewood Cliffs, NJ: Prentice Hall, 1995. 832 p. $72.00. ISBN
0134441346.

Numerous illustrations, both black/white and color, with accompanying text
provide an introduction to human anatomy. The format is larger than usual
which allows larger illustrations than most anatomy books. While this book
does serve as a text, it also doubles as an atlas especially written with the
student in mind.

Netter, Frank H. *Atlas of Human Anatomy*. Summit, NJ: Ciba-Geigy,
1989. (Ciba Collection of Medical Illustration Series). $86.00. 514
plates, 36 p. ISBN 0914168185.

This famous author/illustrator provides an outstanding atlas of gross anatomy
which follows the nomenclature adapted by the Eleventh International Congress
of Anatomists in 1980. References and an index are included.

Netter, Frank H. *The Ciba Collection of Medical Illustrations: A Compilation of Pathological and Anatomical Paintings*. Summit, NJ: Ciba
Pharmaceutical, 1959–1991. Vol. 1–8, in 11 vols., including rev.
eds. in some vols. Vol. 8– published by Ciba-Geigy.

The standard collection of medical illustrations by the best medical illustra-
tor/artist of all time.

Rohen, Johannes W. and Chihiro Yokochi; with the collaboration of Lynn
Romrell. *Color Atlas of Anatomy: A Photographic Study of the
Human Body*, 3rd ed. New York: Igaku-Shoin, 1993. 484 p.
$85.00. ISBN 0896402282. Index.

Photographs of actual anatomic specimens add a spacial dimension, an effect that
is so often lacking, thus assisting students in the dissection course, and surgeons
in the operating room to see anatomical relationships exactly as they appear.
The volume is arranged by body region of macroscopic anatomy, and includes
schematic drawings, CT-scans, and MR-images of main tributaries of nerves and
vessels, muscles; and nomenclature.

Twietmeyer, Alan and Thomas McCracken. *Coloring Guide to Regional
Human Anatomy*, 2nd ed. Philadelphia: Lea & Febiger, 1992. 214
p. ISBN 0812115260. Index.

Atlas for regional anatomy.

Wolf-Heidegger's Atlas of Human Anatomy, 4th completely rev. ed.
     Edited by Hans Frick, Benno Kummer, and Reinhard Putz.
     Farmington, CT: Karger, 1990. 600 p. 593 color illustrations.
     $100.00. ISBN 3805542895.

Designed especially for the medical student, this volume documents the macroscopic anatomy of the human body and can be used in conjunction with any anatomy textbook. Legends to the illustrations conform to the 6th edition of Nomina Anatomica 1989; there is a detailed index locating illustrations by Latin, English, and German key words.

Dictionaries and Encyclopedias

Barber, Colin, B. H. Brown, and R. H. Smallwood. *Dictionary of Physiological Measurement*. Lancaster, England; Boston: MTP, 1984. 145 p. $15.00 (paper). ISBN 085200737X.

Dictionary of human physiology and physiological measurement.

Choulant, Johann Ludwig. *Geschichte und Bibliographie . . . .*
See this entry (p. 329) in the Histories section.

Dobson, Jessie. *Anatomical Eponyms; Being a Biographical Dictionary of Those Anatomists Whose Names Have Become Incorporated into Anatomical Nomenclature, with Definitions of the Structures to Which Their Names Have Been Attached and References to the Works in Which They Are Described*, 2nd ed. 235 p. Edinburgh: Livingstone, 1962. 235 p.

The title says it all.

Donath, Tibor. *Anatomical Dictionary: With Nomenclature and Explanatory Notes*; English ed. edited by G. N. C. Crawford. Oxford; New York: Pergamon, 1969. 634 p. $257.00. ISBN 0080123988.

Although this dictionary is old, it is in print and still valuable.

Encyclopedia of Neuroscience. Edited by George Adelman. Boston: Birkhauser, 1987. 2 v. (A Pro Scientia Viva Title). $195.00. ISBN 0817633359 (set).

Neuroscience Year, supplement to the *Encyclopedia of Neuroscience*, was published in 1989. The second updated edition of this prestigious encyclopedia will be published by Elsevier Science as a CD-ROM multimedia product and is scheduled for release in 1995.

Field, Ephraim Joshua and R. J. Harrison. *Anatomical Terms: Their Origins and Derivation*, 3rd ed., rev. and enl. Cambridge: Heffer, 1968. 212 p. ISBN 0852700016.

Especially useful for its historical information.

Nomina Anatomica. Authorized by the Twelfth International Congress of Anatomists in London, 1985. Together with *Nomina Histologica*, 3rd ed. *Nomina Embryologica*, 3rd ed. Revised and Prepared by Subcommittees of the International Anatomical Nomenclature Committee, 6th ed. Edinburgh; New York: Churchill Livingstone, 1989. 1 v. ISBN 0443040850.

The authoritative, standard nomenclature for anatomy, embryology, and histology.

Shaw, Diane L. *Glossary of Anatomy and Physiology*. Springhouse, PA: Springhouse, 1992. Reprint of *Anatomy and Physiology Glossary*. Thorofare, NJ: Slack, 1990. 420 p. $25.00. ISBN 0874344220.

Stedman's Anatomy and Physiology Words. Baltimore: Williams & Wilkins, 1992. 450 p. $28.00. ISBN 0683079417.

This new nomenclatural dictionary is based on, and extracted from, *Stedman's Medical Dictionary*, 25th ed., 1990.

Guides to Internet Resources

Consult the BIOSCI/bionet Frequently Asked Questions (FAQ) for descriptions and uses of various BIOSCI/bionet newsgroups. BIOSCI/bionet FAQ is posted the first of each month with a list of changes that occurred during the preceding month. This document is available for anonymous FTP from net.bio.net in pub/BIOSCI/biosci.FAQ or by e-mail to biosci@net.bio.net. Neuroscience (bionet.neuroscience) is an example of a newsgroup listed in the BIOSCI/bionet FAQ that is relevant to this chapter. Also, check the World Wide Web under "Biosciences" for information about neurobiology.

The American Physiological Society Gopher has a Gopher server with categories listing administration and membership, announcements and meeting notices, employment opportunities, information for authors, public affairs activities, publications, committee and section reports, and the like.

Anatomy Via the Internet. Bethesda, MD: National Library of Medicine.

Access to anatomy via the Internet is provided by the Visible Human Project and the related Visible Embryo Project. Together, these programs will create a national standard for anatomical images and will "transparently link the print library of functional-physiological knowledge with the image library of structural-anatomical knowledge into one unified resource of medical information," says Michael J. Ackerman of the National Library of Medicine (NLM). Currently (1994), investigators can log on to a demonstration via Mosaic on World Wide Web; the National Library of Medicine plans to release the entire dataset over the Internet when both sets of images are complete, probably late 1994. Check the NLM's WWW homepage (http://www.nlm.nih.gov/) for updates.

Neurosciences Internet Resource Guide. Compiled by Steve Bonario and Sheryl Cormicle, School of Information and Library Studies, University of Michigan. Dec. 1993, version 1.0. Available on the Internet from the University of Michigan Gopher.

This work in progress is a guide to free, Internet-accessible resources helpful to neuroscientists of all types. Indexes to resources are by alphabetical listing; by keyword for subject, format, and resource type; and by resources. Examples of keyword listings: addresses, atlases, data sets, directories, educational software, FAQ, gophers, images, journals, listservs, MRI, neural networks, neuroanatomy, neurobiology, neurophysiology, newsgroups, photographs, usenet groups, etc. Check for updated versions of the **Guide** in the University of Michigan Clearinghouse.

Handbooks

Handbook of Olfaction and Gustation. Edited by Richard L. Doty. New York: Marcel Dekker, 1994. 906 p. $225.00. ISBN 0824792521.

This reference book provides data on olfaction, gustation, and other chemosensory systems by examining the anatomy, biochemistry, physiology, and psychophysics of the senses of smell and taste. It includes over 4,400 citations to the literature, more than 275 tables, equations, drawings, and photographs useful for both the researcher and the clinician.

Handbook of Physiology, 1977– . Bethesda, MD: American Physiological Society. Irregular. Price varies per section; several volumes/ section.

Comprehensive and authoritative, the APS organizes this review series with contributions from scientists all over the world. Billed as "A spectrum of physiological knowledge and concepts," this series presents significant research results, extensive bibliographic references, and detailed indexes, making these volumes indispensable. The series is divided into eleven sections that are updated frequently: Section 1: *Nervous System*; Section 2: *Cardiovascular System*; Section 3: *Respiratory System*; Section 4: *Adaptation to the Environment* (out of print); Section 5: *Adipose Tissue* (out of print); Section 6: *Gastrointestinal System*;

Section 7: *Endocrinology*; Section 8: *Renal Physiology*; Section 9: *Reactions to Environmental Agents*; Section 10: *Skeletal Muscle*; Section 11: *Cell and General Physiology*.

Kirkpatrick, C. T. *Illustrated Handbook of Medical Physiology*. New York: Wiley, 1991. (A Wiley Phoenix Publication). 548 p. ISBN 047191455X.

Numerous line drawings help explain the mechanisms of the human body, each chapter describing a separate organ or system and providing an account of the relevant functional anatomy and histology.

Philo, Ronald and John H. Linner. *Guide to Human Anatomy*. Philadelphia: Saunders, 1985. 335 p. $30.50 (paper). ISBN 0721612032.

Can serve as a guide, or as a syllabus for a course in human anatomy.

Rayman, Rebecca. *The Body in Brief: Essentials for Healthcare*. El Paso, TX: Skidmore-Roth, 1993. 392 p. $26.95. ISBN 0944132766.

Essentially written for a nursing course in anatomy and physiology.

Thibodeau, Gary A. and Kevin T. Patton. *Quick Reference to Anatomy and Physiology*. St. Louis: Mosby, 1993. 60 p. ISBN 0801675308.

Histories

The American Association of Anatomists, 1888-1937: Essays on the History of Anatomy in America and a Report on the Membership—Past and Present. Edited by John E. Pauly, with the assistance of the editorial committee. Baltimore: Williams & Wilkins, 1987. 292 p. (paper). ISBN 0683068008.

Choulant, Johann Ludwig. (*Geschichte und Bibliographie der Anatomischen Abbildung*). *History and Bibliography of Anatomic Illustrations*; translated and annotated by Mortimer Frank; further essays by Fielding H. Garrison, Mortimer Frank, Edward C. Streeter, with a new historical essay by Charles Singer and bibliography of Mortimer Frank by J. Christian Bay. New York: Hafner, 1962. 435 p. Bibliographies and Index.

Originally published in German in Leipzig, 1852. Contributions and essays by major historians of science.

Circulation of the Blood: Men and Ideas. Edited by Alfred P. Fishman

and Dickinson W. Richards. Bethesda, MD: APS, 1982. 879 p.
$56.00. ISBN 0195206991.

A study of the origins, discovery, and progress of great ideas in this branch of
science.

Corner, George Washington. *Anatomical Texts of the Earlier Middle
Ages: A Study in the Transmission of Culture, with a Revised Latin
Text of Anatomia Cophonis and Translations of Four Texts*. New
York: AMS Press, 1977. Reprint of the 1927 ed. published by the
Carnegie Institution of Washington, which was issued as no. 364 of
its Publication. 112 p. ISBN 0404132502.

Dobson, Jessie. *Anatomical Eponyms*.

Refer to this entry (p. 326) in the Dictionaries and Encyclopedias section.

Endocrinology: People and Ideas. Edited by S. M.McCann. Bethesda,
MD: American Physiology Society, 1988. 484 p. $75.00. ISBN
0195207181.

Principal ideas and developments in endocrinology from Aristotle to the most
recent discoveries.

Fenn, Wallace O. *History of the American Physiological Society: The
Third Quarter Century, 1937–1962*. Washington. American
Physiological Society. 1963. 182 p. Illustrations, portraits.

See the earlier two volumes, *History of the American Physiological Society* (p.
331).

Finger, Stanley. *Origins of Neuroscience; A History of Explorations into
Brain Function*. New York: Oxford University Press, 1994. 462 p.
$75.00. ISBN 0195065034.

Over 350 illustrations help to trace the development of the history of science
relevant to brain function.

Foster, Michael. *Lectures on the History of Physiology During the 16th,
17th, and 18th Centuries*. Cambridge, England: Cambridge University Press, 1901. (Cambridge Biological Series). New York:
Readex Microprint, 1973. 4 cards (Landmarks of Science). 310 p.

A classic history by one of the founders of modern physiology.

Franklin, Kenneth J. *A Short History of Physiology*, 2nd ed. London;
New York: Staples Press, 1949. 147 p.

Although this book deals with the history of *animal* physiology, it shows the
evolution of the experimental method and discusses some of the most important

physiologists from the ancient world through the nineteenth century. There are portraits for 16 scientists and an index of personal names.

History of the American Physiological Society: The First Century, 1887–1987. Edited by John R. Brobeck, Orr E. Reynolds, and Toby A. Appel. Bethesda, MD: The Society, distributed for the Society by Williams & Wilkins, 1987. 533 p. $45.00. ISBN 0683010670.

History of the beginnings of a major scientific and medical society.

History of the American Physiological Society Semicentennial, 1887–1937. Baltimore, MD: American Physiological Society, 1938. 228 p. Indexes.

Includes portraits and biographical sketches of the original members and of the presidents of the society.

Hunter, William. *Hunter's Lectures of Anatomy*. Amsterdam, New York: Elsevier, 1972. 1 v. ISBN 0444409106.

Facsimile of two notebooks containing lecture notes from a course given by W. Hunter in Manchester, England, beginning Jan. 20, 1752. The lecture notes were taken by C. White of Manchester, and the notebooks have been prepared for publication by N. Dowd.

Hunterian Museum (Glasgow). *Catalogue of the Anatomical and Pathological Preparations of Dr. William Hunter*: in the Hunterian Museum, University of Glasgow, catalogue prepared by John H. Teacher. Glasgow: James MacLehose, 1900. 2 v. Bibliography and index.

Catalogs of anatomical and pathological biological specimens prepared by Dr. Hunter.

Knight, Bernard. *Discovering the Human Body: How Pioneers of Medicine Solved the Mysteries of the Body's Structure and Function*. New York: Lippincott & Crowell, 1980. 192 p. $25.00. ISBN 0690019289. Index.

Membrane Transport: People and Ideas. Edited by Daniel C. Tosteson. Bethesda, MD: APS, 1989. 420 p. $65.00. ISBN 0195207734.

A collection of personal accounts by investigators who have been at the forefront of research in membrane transport.

O'Connor, W. J. *Founders of British Physiology: A Biographical Dictionary 1820–1885*. Manchester: Manchester University Press, 1988. 278 p. $25.00. ISBN 0719025370.

Dictionary of nineteenth century British human physiologists, arranged by name. Includes bibliographies and index.

Persaud, T. V. N. *Early History of Human Anatomy: From Antiquity to the Beginning of the Modern Era.* Springfield, IL: Thomas, 1984. 200 p. ISBN 0398050384. Bibliography and index.

The author portrays the early history of anatomy in relation to the practice of medicine by charting the achievements and changing concepts from ancient times to the beginning of the scientific era, symbolized by Vesalius and his masterpiece, *De Humani Corporis Fabrica.*

Punt, H. *Bernard Siegfried Albinus (1697–1770), On "Human Nature": Anatomical and Physiological Ideas in Eighteenth Century Leiden.* Amsterdam: B. M. Israel, 1983. 223 p. ISBN 9060780884.

"Addendum: transcriptions and translations of the lecture notes on Albinus' physiology:" p. 135–183. Text in English and Latin with a summary in Dutch. Bibliography, p. 204–215.

Renal Physiology: People and Ideas. Edited by Carl W. Gottschalk, Robert W. Berliner, and Gerhard H. Giebisch. Bethesda, MD: APS, 1987. 520 p. $84.50. ISBN 0195207025.

Written by well-known physiologists, each chapter offers a unique, inside perspective on the historical record of the historical record of the discipline.

Stevens, Leonard A. *Explorers of the Brain.* New York: Knopf, 1971. 348 p. ISBN 0394429680.

History of anatomy including a substantial seventeen-page bibliography.

Tenney, S. Marsh. "The Father of American Physiology," *News in Physiological Sciences* 9: 43–44, Feb. 1994.

The life and work of Robley Dunglison.

Todd, Edwin M. *The Neuroanatomy of Leonardo da Vinci.* Park Ridge, IL: American Association of Neurological Surgeons, 1991. Reprint of the 1983 ed. 192 p. $75.00. ISBN 1879284057.

Focused history of one of the greatest scientists and artists of all times.

Women Physiologists; An Anniversary Celebration of Their Contributions to British Physiology. Edited by Lynn Bindman, Alison Brading, and Tilli Tansey. Brookfield, VT: Ashgate Publishing Co., distributor for Portland Press, 1993. 164 p. $15.00 (paper). ISBN 1855780496.

This book focuses on a group of women who have made significant contributions to physiology.

Methods and Techniques

Other ideas for laboratory work and aids may be found in the Guides to Internet Resources and Handbooks sections (pp. 327 and 328, respectively).

Culturing Nerve Cells. Edited by Gary Banker and Kimberly Goslin. Cambridge, MA: MIT Press, 1991. 468 p. $57.00. ISBN 0262023202.

Invaluable resource, with hands-on advice, for growing neurons in culture.

Electrophysiology; A Practical Approach. Edited by D. I. Wallis. New York: IRL Press, 1993. 320 p. $56.00. ISBN 0199633487.

As the subtitle suggests, this volume provides a practical introduction to the methods used for studying single cells and complex neural tissues. Detailed protocols and advice for handling and culturing neural tissues and cells, and for mathematical models of neuronal behavior, are also included.

Handbook of Endocrine Research Techniques. Edited by Flora de Pablo, Colin G. Scanes, and Bruce D. Weintraub. San Diego: Academic Press, 1993. 599 p. $99.00. ISBN 0122099206.

Written by experts in the field, this volume synthesizes in a single source up-to-date methods and strategies useful in endocrinological research. General concepts, detailed protocols, and extensive references to the original literature are provided.

Journal of Neuroscience Methods.

See the annotation (p. 340) in the Periodicals section.

Methods in Neurosciences. v. 1– , 1989– . San Diego, CA: Academic Press. Irregular. Each volume priced separately. Hard cover, or spiral bound.

This series, companion to *NeuroProtocols* also published by Academic Press, presents contemporary techniques significant to a particular branch of the neurosciences. Vol. 22: *Neurobiology of Steroids*, 1994. 542 p. $99.00. ISBN 012185292X.

Methods in Physiology Series. v. 1– , 1994– . Bethesda, MD: American Physiological Society. Irregular. Price varies.

This series describes experimental techniques in cellular, molecular and general physiology. Each book is edited by experts in the field and covers theory and history behind the methods, critical commentary, major applications with examples, and limitations and extensions of each technique. Vol. 1: *Membrane Protein Structure; Experimental Approaches*, 1994. 395 p. $65.00. ISBN 0196071123. Vol. 2: *Fractal Physiology*, 1994. 400 p. $55.00. ISBN 0195080130.

Microelectrode Techniques; The Plymouth Workshop Handbook, 2nd ed. Edited by D. Ogden. Cambridge, England: The Company of Biologists, 1994. 450 p. $38.00. ISBN 0948601493.

The aim is to provide an introduction to each technique and its instrumentation, and to discuss ideas and techniques of analysis and current practice covered by this famous and popular workshop.

NeuroProtocols; A Companion to Methods in Neurosciences. v. 1– , 1992– . San Diego, CA: Academic Press. 6 issues/yr. $148.00. ISSN 1058-6741.

Provides methodology of central significance to the neurosciences.

Neuroscience Protocols. Edited F. G. Wouterlood. Amsterdam: Elsevier, 1993. Modules 1–2, ring binder set: $317.25. Module 3, 1994: $117.75. Also available on a subscription basis.

Another neuroscience laboratory manual, this time from Elsevier. An electronic version is in preparation. While this set and the better established series from Academic Press are in competition, the two may also be used in conjunction with each other for supplemental or background information.

Quantitative Methods in Neuroanatomy. Edited by Michael G. Steward. Chichester; New York: Wiley, 1992. 349 p. $129.95. ISBN 0471933082.

Techniques for autoradiography, imaging, and neuroanatomical histology of the nervous system.

Sharif, N. A. *Molecular Imaging in Neuroscience*. Oxford: IRL Press, 1993. (The Practical Approach Series). 245 p. $39.00 (paper). ISBN 0199633800. Index.

As the series promises, this book provides practical aspects for detailed protocols and powerful new techniques for neuropharmacology, neuroanatomy, neurogenetics, and neuropathology.

A Source Book of Practical Experiments in Physiology Requiring Minimal Equipment. Prepared by the International Union of Physiological Sciences Commission on Teaching Physiology. Teaneck, NJ: World Scientific, 1991. 193 p. $38.00. ISBN 9810205708.

A laboratory manual for teaching human physiology suitable for developing countries.

Stamford, J. A. *Monitoring Neuronal Activity*. Oxford: IRL Press, 1992. (The Practical Approach Series). 294 p. (paper). $39.00. ISBN 019963243X. Index.

This volume draws together many different methods of monitoring activity of nerve cells. Detailed protocols are available, along with monitoring tips.

Tortora, Gerard J. *Anatomy and Physiology Laboratory Manual*, 4th ed. New York: Macmillan, 1994. 588 p. (paper). ISBN 0024210137.

A very successful laboratory manual for anatomy and physiology appropriate at the undergraduate level.

Periodicals

For convenience, journals are divided into subject sections for anatomy, endocrinology, neurobiology, and physiology. These divisions are, in some sense, arbitrary, and because titles are not duplicated between divisions, it may be necessary to consult several subject divisions when tracking down a particular journal. Because there are so many current journals in these areas, only the most prominent were selected for inclusion, chosen for their relative importance and high impact.

Anatomy

Acta Anatomica, including supplements. v. 1– , 1945– . Basel, Switzerland: Karger. Monthly. $1,272/yr. ISSN 0001-5180.

This international forum for experimental and theoretical work presents information on morphology at all levels of organization, from subcellular to macroscopy, with emphasis on humans and higher vertebrates.

Anatomical Record. v. 1– , 1906– . New York: Wiley/Liss. Monthly. $1,896/yr. ISSN 0003-276X.

Official publication of the American Association of Anatomists, issuing research reports, review articles, trends, special communications, and commentaries/letters to the editor concerning broad research interests in anatomy. The journal also publishes Abstracts and Proceedings of the Association's annual meeting.

Anatomy and Embryology. v. 1– , 1892– . Heidelberg: Springer-Verlag. Monthly. $1,498/yr. ISSN 0340-2061.

A leading European anatomy journal. Former titles: *Journal of Anatomy and Embryology; Zeitschrift fuer Anatomie und Entwicklungsgeschichte.*

Annals of Anatomy. v. 1– , 1886– . Jena, Germany: Fischer. Bi-monthly. $355/yr. ISSN 0940-9602.

This continuation of *Anatomischer Anzeiger* is a general anatomical journal.

Developmental Dynamics. v. 193– , 1992– . New York: Wiley/Liss.
    Monthly. $1,135/yr. ISSN 1058-8388.

Continues *American Journal of Anatomy.* An official publication of the
American Association of Anatomists, this journal provides a focus for
communication among developmental biologists studying the emergence of form
during human and animal development.

Journal of Anatomy. v. 1– , 1866– . New York: Cambridge University
    Press. Bimonthly. $630/yr. ISSN 0021-8782.

Presents articles and reviews covering normal human and comparative anatomy,
including applied anatomy, physical anthropology, neurology, endocrinology,
embryology.

Journal of Morphology. v. 1– , 1887– . New York: Wiley/Liss. Month-
    ly. $1,064/yr. ISSN 0362-2525.

Publishes original papers in cytology, protozoology, embryology, and general
morphology.

Endocrinology and Metabolism

Endocrinology. v. 1– , 1917– . Bethesda, MD: Endocrine Society.
    Monthly. $290/yr. ISSN 0013-7227.

Includes papers describing results of original research in the fields of endocrinology and metabolism for nonprimate biochemical and physiological studies.
Work on material of primate origin is not excluded.

General and Comparative Endocrinology. v. 1– , 1961– . San Diego,
    CA: Academic. Monthly. $736/yr. ISSN 0016-6480.

Published under the auspices of the Division of Comparative Endocrinology of
the American Society of Zoologists, and the European Society for Comparative
Endocrinology. Includes articles based on studies on cellular mechanisms of
hormone action, and on functional, developmental, and evolutionary aspects of
vertebrate and invertebrate endocrine systems.

Hormones and Behavior. v. 1– , 1969– . San Diego, CA: Academic.
    Quarterly. $210/yr. ISSN 0018-506X.

Publishes a broad range of original articles dealing with behavioral systems
known to be hormonally influenced. Scope extends from evolutionary significance of hormone/behavior relations to those dealing with cellular and molecular
mechanisms of hormonal actions on neural tissues and other tissues relevant to
behavior.

Journal of Endocrinology. v. 1– , 1939– . Bristol, England: Journal of Endocrinology Ltd. Monthly. $470/yr. ISSN 0022-0795.

Original research on all aspects of the nature and functions of endocrine systems.

Journal of Molecular Endocrinology. v. 1– , 1988– . Bristol, England: Journal of Endocrinology Ltd. Bimonthly. $280/yr. ISSN 0952-5041.

Original research papers, rapid communications, short reviews, and commentaries are accepted on molecular and cellular aspects of endocrine and related systems.

Journal of Neuroendocrinology. v. 1– , 1989– . Cambridge, MA: Blackwell Scientific. Bimonthly. $354/yr. ISSN 0953-8194.

This is an international journal acting as a focus for the newest ideas, knowledge, and technology in the neurosciences which are contributing to the rapid growth of neuroendocrinology. Deals with manuscripts for both vertebrate and invertebrate systems.

Journal of Steroid Biochemistry and Molecular Biology. v. 37– , 1990– . New York: Elsevier (Pergamon). Semimonthly. $1,820/yr. ISSN 0960-0760.

Continues *Journal of Steroid Biochemistry.* The journal is devoted to new experimental or theoretical developments in areas related to steroids. Original papers, mini-reviews, proceedings of selected meetings, and rapid communications are included.

Metabolism: Clinical and Experimental. v. 1– , 1952– . Philadelphia, PA: Saunders. Monthly. $219/yr. ISSN 0026-0495.

Reports on research into the metabolic aspects of nutrition, endocrines, genetics, dystrophies, diabetes, and gout.

Molecular and Cellular Endocrinology. v. 1– , 1974– . Limerick, Ireland: Elsevier. Semimonthly. $1,708/yr. ISSN 0303-7207.

Publishes on all aspects related to the biochemical effects, synthesis and secretions of extracellular signals (hormones, neurotransmitters, etc.), and cellular regulatory mechanisms involved in hormonal control.

Molecular Endocrinology. v. 1– , 1987– . Bethesda, MD: Endocrine Society. Monthly. $215/yr. ISSN 0888-8809.

Describes results of original research on the mechanistic studies of the effects of hormones and related substances on molecular biology, and genetic regulation of nonprimate and primate cells.

Neuropeptides. v. 1– , 1980– . Edinburgh, Scotland: Churchill Living-
stone/Longman Group. Monthly. $700/yr. ISSN 0143-4179.

The aim of this journal is the rapid publication of original research and review
articles dealing with the structure, distribution, actions, and functions of peptides
in the central and peripheral nervous systems.

Regulatory Peptides. v. 1– , 1980– . Amsterdam: Elsevier. Semimonthly. $1,427/yr. ISSN 0167-0115.

Provides a medium for the rapid publication of interdisciplinary studies on the
physiology and pathology of peptides of the gut, endocrine, and nervous systems
which regulate cell or tissue function.

Steroids. v. 1– , 1963– . Stoneham, MA: Butterworths/Heinemann.
Monthly. $460/yr. ISSN 0039-128X.

Accepts papers on basic endocrinology and on the organic, biochemical, physiological, pharmacological, and clinical phases of steroids.

Neurobiology

Biological Cybernetics. v. 1– , 1975– . New York: Springer-Verlag.
Monthly. $1,498/yr. ISSN 0340-1200.

Continues *Kybernetik*. Foremost in computational neuroscience, this journal provides an interdisciplinary medium for the exchange of experimental and theoretical information in quantitative analysis of behavior; quantitative physiological
studies of information processing; computational studies of perceptual and motor
information; biologically relevant studies in artificial intelligence, robotics,
information theory; and mathematical models of information processing, control,
and communication in organisms, including mechanisms of genetic expression
and development. Also, see *Neural Computation* (p. 340).

Brain. v. 1– , 1878/79– . New York: Oxford University Press. Bimonthly. $240/yr. ISSN 0006-8950.

Publishes papers on neurology and related clinical disciplines, and on basic
neuroscience, including molecular and cellular biology, and neuropsychology
when they have a neurological orientation and are relevant to the understanding
of human disease.

Brain Research. v. 1– , 1966– . Amsterdam: Elsevier. With *Brain Research Reviews, Cognitive Brain Research, Developmental Brain Research, Molecular Brain Research, Discussions in Neuroscience*. Frequency varies. $9,500/yr for all sections. ISSN varies with section.

"International multidisciplinary journal devoted to fundamental research in the
brain sciences." One of the leaders in the field and worth the expense.

Brain Research Bulletin. v. 1– , 1976– . Oxford, England: Elsevier (Pergamon). Monthly. $850/yr. ISSN 0361-9230.

Incorporated *Journal of Electrophysiological Techniques.* Publishes papers on all aspects of the nervous system including brief communications with describe a new method, techniques, or apparatus, and results of experiments.

Cellular and Molecular Neurobiology. v. 1– , 1981– . New York: Plenum. Bimonthly. $330/yr. ISSN 0272-4340.

Original research articles are accepted that are concerned with the analysis of neuronal and brain function at the cellular or subcellular levels.

European Journal of Neuroscience. v. 1– , 1989– . New York: Oxford University Press. Monthly. $595/yr. ISSN 0953-816X.

Published on behalf of the European Neuroscience Association, this journal has a broad scope ranging from the behavioral to the molecular, with a European focus but with a world-wide orientation. Supplements to the journal consist of the abstracts of the Annual Meeting of the European Neuroscience Association.

Glia. v. 1– , 1988– . New York: Wiley/Liss. Monthly. $700/yr. ISSN 0894-1491.

"A journal of glial and neuroglial research."

Journal of Chemical Neuroanatomy. v. 1– , 1988– . New York: Wiley. Bimonthly. $455/yr. ISSN 0891-0618.

Presents scientific reports relating to functional and biochemical aspects of the nervous system with its microanatomical organization. The scope covers micro-anatomical, biochemical, pharmacological, and behavioral approaches.

Journal of Comparative Neurology. v. 1– , 1891– . New York: Wiley/ Liss. Weekly. $6,475/yr. ISSN 0021-9967.

Publishes papers on the anatomy and physiology of the nervous system, not including clinical neurology, neuropathology, psychiatry, and introspective psychology unless these bear on the anatomy and physiology of the nervous system. Preference is given to papers which deal descriptively or experimentally with the nervous system, its structure, growth, and function.

Journal of Neurobiology. v. 1– , 1969– . New York: Wiley. Monthly. $825/yr. ISSN 0022-3034.

High-quality contributions in all areas of neurobiology are solicited, but there is emphasis on cellular, genetic, and molecular analyses of neurodevelopment and the ontogeny of behavior.

Journal of Neurochemistry. v. 1– , 1956– . New York: Raven. Monthly. $1,225/yr. ISSN 0022-3042.

Official journal of the International Society for Neurochemistry, devoted to the molecular, chemical, and cellular biology of the nervous system.

Journal of Neurophysiology. v. 1– , 1938– . Bethesda, MD: American Physiological Society. Monthly. $435/yr. ISSN 435/yr. ISSN 0022-3077.

Highest quality science on the function of the nervous system. Includes theoretical studies and rapid communications.

Journal of Neuroscience. v. 1– , 1981– . New York: Oxford University Press. Monthly. $765/yr. ISSN 0270-6474.

Official journal of the Society for Neuroscience. A leader in the field, this journal publishes broad, multidisciplinary science articles for effective coverage from molecular and cellular neurobiology to behavioral and system neuroscience.

Journal of Neuroscience Methods. v. 1– , 1979– . Amsterdam: Elsevier. Semimonthly. $1,176/yr. ISSN 0165-0270.

Publishes research papers and a limited number of broad and critical reviews dealing with new methods or significant developments of recognized methods, used to investigate the organization and fine structure, biochemistry, molecular biology, histo- and cytochemistry, physiology, biophysics, and pharmacology of receptors, neurones, synapses and glial cells, in the nervous systems of man, vertebrates, and invertebrates, or applicable to the clinical and behavioral science, tissue culture, neurocommunications, biocybernetics, or computer software.

Journal of Neuroscience Research. v. 1– , 1975– . New York: Wiley/ Liss. Semimonthly. $1,833/yr. ISSN 0360-4012.

Basic reports in molecular, cellular, and subcellular areas of the neurosciences, including clinical studies that emphasize fundamental and molecular aspects of nervous system dysfunction. The journal features full-length papers, rapid communications, and mini-reviews on selected areas.

Journal of Physiology (Paris); An Integrative Neuroscience Journal. v. 1– , 1889– . Paris: Elsevier. Bimonthly. $330/yr. ISSN 0928-4257.

Formerly *Journal de Physiologie* until 1992, this journal covers all aspects of neurobiology relevant to behavior and cognition and the integrative functions of the brain, with focus on functional imaging, development and plasticity, cellular neurobiology, systems, behavioral and neuromuscular physiology, endocrinology, and cognition.

Neural Computation. v. 1– , 1989– . Cambridge, MA: MIT Press. Bimonthly. $180/yr. ISSN 0899-7667.

Disseminates important, multidisciplinary research results and reviews on the interplay between experimental data, computational models, and theoretical analysis at all levels of organization in the brain, from the molecular to the systems levels.

Neurobiology of Learning and Memory. v. 63– , 1995 . San Diego: Academic Press. Bimonthly. $395/yr. ISSN 1074-7427. Formerly *Behavioral and Neural Biology.*

Publishes original research articles dealing with neural and behavioral plasticity, at all levels from the molecular to behavior. Also includes short communications, minireviews, and commentaries.

Neuron. v. 1– , 1988– . Cambridge, MA: Cell Press. Monthly. $350/yr. ISSN 0896-6273.

Publishes reports of novel results in any area of experimental neuroscience, in the form of research articles and mini-reviews. Some issues are accompanied by supplements titled *Cell Neuron.*

Neurochemistry International. v. 1– , 1980– . Oxford, England: Pergamon (Elsevier). Monthly. $595/yr. ISSN 0197-0186.

Publishes articles on the cellular and molecular aspects of neurochemistry in the form of original and rapid research communications; critical reviews and commentaries are included.

Neuroscience. v. 1– , 1976– . New York: Elsevier. Semimonthly. $3,305/yr. ISSN 0306-4522.

"An International Journal under the editorial direction of the International Brain Research Organization." The journal publishes papers describing the results of original research on any aspect of the scientific study of the nervous system.

Neuroscience Letters. v. 1– , 1975– . Limerick, Ireland: Elsevier. Fortnightly. $2,943/yr. ISSN 0304-3940.

Rapid publication of short, complete reports, but not preliminary communications, in all areas in the fields of neuroanatomy, neurochemistry, neuroendocrinology, neuropharmacology, neurophysiology, neurotoxicology, molecular neurobiology, behavioral sciences, biocybernetics, and clinical neurobiology. The overriding criteria for publication are novelty and interest to a multidisciplinary audience.

Synapse. v. 1– , 1987– . New York: Wiley/Liss. Monthly. $865/yr. ISSN 0887-4476.

Accepts articles on all aspects of synaptic structure and function, including neurotransmitters, neuropeptides, neuromodulators, receptors, gap junctions, metabolism, plasticity, circuitry, mathematical modeling, ion channels, patch recording, single-unit recording, development, behavior, pathology, toxicology, and so forth.

Physiology

American Journal of Physiology. v. 1– , 1898– . Bethesda, MD: American Physiological Association. Monthly. All 7 sections: $1,601/yr.

The consolidated *American Journal of Physiology* is billed as "the most comprehensive body of research covering the full spectrum of physiology." Its sections are: *Cell Physiology,* ISSN 0363-6143, the cutting edge of cell physiology research; *Endocrinology and Metabolism,* ISSN 0193-1849, original investigation on endocrine and metabolic systems on levels of organization, human and animal; *Gastrointestinal and Liver Physiology,* ISSN 0193-1857, papers on digestion, secretion absorption, metabolism, motility, microbiology and colonization, growth and development, and neurobiology; *Heart and Circulatory Physiology,* ISSN 0363-6135, original research on the heart, blood vessels, and lymphatics; *Lung Cellular and Molecular Physiology,* ISSN 1040-0605, deals with molecular, cellular, and morphological aspects of normal and abnormal function and response of cells and components of the respiratory system; *Regulatory, Integrative and Comparative Physiology,* ISSN 0363-6119, innovative articles illuminating physiological processes at all levels of biological organization; *Renal, Fluid and Electrolyte Physiology*, ISSN 0363-6127, information on kidney and urinary tract physiology, epithelial cell biology, and control of body fluid volume and composition.

Chemical Senses. v. 5– , 1980– . New York: Oxford University Press. Bimonthly. $270/yr. ISSN 0379-864X.

Formerly *Chemical Senses and Flavour*, providing an international forum for chemoreception research at morphological, biochemical, physiological, and psychophysical levels. Covers development and specific application of new methods.

European Journal of Applied Physiology and Occupational Physiology. v. 1– , 1928– . New York: Springer-Verlag. Monthly. $1,543/yr. ISSN 0301-5548.

Emphasis is on environmental and work physiology, encompassing a broad spectrum of animal experimentation seen as clearly relevant to the human condition.

European Journal of Physiology.

See *Pflueger's Archiv* (p. 343).

Experimental Physiology. v. 75– , 1990– . New York: Cambridge University Press. Bimonthly. $252/yr. ISSN 0958-0670.

Continues *Quarterly Journal of Experimental Physiology and Cognate Medical Sciences.* Includes research papers on all aspect of experimental physiology from molecular to animal studies.

Journal of Applied Physiology. v. 1– , 1948– . Bethesda, MD: American Physiological Society. Monthly. $528/yr. ISSN 8750-7587.

A leader in its field, this journal publishes original papers dealing with normal or abnormal function in respiratory physiology, nonrespiratory functions of the lungs, environmental physiology, temperature regulation, exercise physiology, and interdependence.

Journal of Cellular Physiology. v. 1– , 1932– . New York: Wiley/Liss. Monthly. $1,896/yr. ISSN 0021-9541.

Devoted to the publication of research papers concerned with physiology and pathology at the cellular level: the biochemical and biophysical mechanisms concerned in the regulation of cellular growth, differentiation, and function.

Journal of Developmental Physiology. v. 1– , 1979– . New York: Oxford University Press. Monthly. $457/yr. ISSN 0141-9846.

The journal publishes papers describing the results of original research on any aspect (anatomical, biochemical, endocrine, pharmacological, physiological, etc.) of the scientific study of the pregnancy, the fetus, or the neonate of man or experimental animals. Research papers, short communications, and solicited short reviews are acceptable.

Journal of General Physiology. v. 1– , 1918– . New York: Rockefeller University Press for the Society of General Physiologists. Monthly. $210/yr. ISSN 0022-1295.

Publishes articles concerned with the mechanisms of broad physiological significance covering research of prime importance for cellular and molecular physiology.

Journal of Physiology. v. 1– , 1878– . Cambridge, England: Cambridge University Press for the Physiological Society, London. Semimonthly. $1,575/yr. ISSN 0022-3751.

Covers physiological research in vertebrates: respiration, circulation, excretion, reproduction, digestion and homeostasis, with emphasis on neurophysiology and muscle contraction. Some issues report the proceedings of the scientific meetings of the Society.

Pflueger's Archiv; European Journal of Physiology. v. 1– , 1968– .

New York: Springer-Verlag. Semimonthly. $1,541/yr. ISSN 0031-6768.

Results of original research considered likely to further the physiological sciences in their broadest sense. Purely clinical papers will be excluded. The journal is subdivided into four sections: heart, circulation, respiration and blood, environmental and exercise physiology; transport processes, metabolism and endocrinology, kidney, gastrointestinal tract, and exocrine glands; excitable tissues and central nervous physiology; molecular and cellular physiology.

The Physiologist. v. 1– , 1957– . Bethesda, MD. American Physiological Society. Bimonthly. $37/yr. ISSN 0031-9376.

The newsletter of the American Physiological Society features articles on Society affairs, announcements, and articles of interest to physiologists, in general. Supplements appear irregularly.

Respiration Physiology. v. 1– , 1965– . Amsterdam: Elsevier. Monthly. $947/yr. ISSN 0034-5687.

Original articles for research in respiratory, pulmonary, and circulatory physiology.

Reviews of the Literature

Anatomy

Advances in Anatomy, Embryology, and Cell Biology. v. 1– , 1891– . Heidelberg: Springer-Verlag. Price varies. ISSN 0301-5556.

Reviews and critical articles covering the entire field of normal anatomy (cytology, histology, cyto- and histochemistry, electron microscopy, macroscopy, experimental morphology and embryology, and comparative anatomy.

Endocrinology and Metabolism

Endocrine Reviews. v. 1– , 1980– . Bethesda, MD: Endocrine Society. Bimonthly. $150/yr. ISSN 0163-769X.

Features in-depth review articles on both experimental and clinical endocrinology and metabolism.

Frontiers of Hormone Research. v. 1– , 1972– . Basel, Switzerland: Karger. Irregular. Vol. 19: $164.00. ISSN 0301-3073.

Focuses on areas of endocrinology undergoing active investigation by consolidating findings from both experimental and clinical work.

Frontiers in Neuroendocrinology. v. 1– , 1969– . San Diego, CA:
    Academic Press. Quarterly. $215/yr. ISSN 0091-3022.
Review articles for the broad field of brain–endocrine interactions.

Recent Progress in Hormone Research. v. 1– , 1947– . San Diego,
    CA: Academic Press. Irregular. Price varies. ISSN 0079-9963.
Proceedings of the Laurentian Hormone Conference. Vol. 49, 1994, discusses
multiple aspects of endocrine-related research, including neuroendocrinology,
pancreatic islet cell function, growth factors, novel humoral signals, second
messenger system, androgen action, and androgen-dependent diseases.

Trends in Endocrinology and Metabolism. v. 1– , 1989– . New York:
    Elsevier. Monthly. $400/yr. ISSN 1043-2760.
Review journal providing reviews of the literature, meeting reports, techniques,
viewpoints, book reviews, job trends, and calendar.

Vitamins and Hormones; Advances in Research and Applications. v. 1–
    1943– . San Diego, CA: Academic Press. Irregular. Price varies.
    ISSN 0083-6729.
Quality reviews of the literature of interest to endocrinologists and biochemists.
Vol. 48, 1994, $80.00.

Neurobiology

Annual Review of Neuroscience. v. 1– , 1978– . Palo Alto, CA: Annual
    Reviews. Annual. $47.00. ISSN 0147-006X.
A leader in review literature, this well-known and respected review annual provides systematic, periodic examination of scholarly advances in the selected field
through critical, authoritative reviews.

Critical Reviews in Neurobiology. v. 1– , 1985– . Boca Raton, FL:
    CRC. Quarterly. $245/yr. ISSN 0892-0915.
Presents comprehensive reviews, analyses, and integration of recently developed
substantive observations and information of processes involving the nervous
system.

Current Opinion in Neurobiology. v. 1– , 1991– . London: Current
    Biology. Bimonthly. $428/yr. ISSN 0959-4388.
Provides: views from experts on current advances in neurobiology; selections
annotated by experts of the most interesting papers; comprehensive bibliographic
listings of papers. For the purposes of this journal, neurobiology is divided into
several sections, each one reviewed annually. According to *Public-Access*

Computer Systems News 5(3), 1994, this journal will be available electronically early in 1995 through the OCLC Electronic Journals Online service.

Handbook of Membrane Channels: Molecular and Cellular Physiology. Edited by Camillo Peracchia. San Diego, CA: Academic Press, 1994. 591 p. $120.00. ISBN 0125506406.

A useful discussion on the state-of-the-art of channel research for molecular biologists, biophysicists, physiologists, neuroscientists, and other concerned biologists. Thirty- six chapters provide a thorough account of new directions at the molecular level.

International Review of Neurobiology. v. 1– , 1959– . San Diego, CA: Academic Press. Irregular. ISSN 0074-7742. Price varies: 1994, $99.00.

This review covers the whole field of neurobiology to include work within a basic science as well as in neurology and psychiatry. The aim is to enable active researchers in neurobiology, neurochemistry, neuroanatomy, neuropharmacology, neurophysiology, psychopharmacology, etc., to give an account of the latest advances in their field.

Neuroscience and Biobehavioral Reviews. v. 1– , 1977– . Oxford, England: Pergamon (Elsevier). Quarterly. $520/yr. ISSN 0149-7634.

Publishes original, major reviews of the literature in anatomy, biochemistry, embryology, endocrinology, genetics, pharmacology, physiology, and all aspects of biological sciences with relevance to the nervous system or the investigation of behavior.

Progress in Brain Research. v. 1– , 1963– . Amsterdam: Elsevier. Irregular. Priced individually. ISSN 0079-6123.

Each volume reviews a particular topic. Vol. 101, 1994: *Biological Function of Gangliosides*, proceedings of Nobel Symposium 83.

Progress in Neurobiology. v. 1– , 1973– . Oxford, England: Elsevier (Pergamon). Monthly. $1,185/yr. ISSN 0301-0082.

Reviews advances in the field of neurobiology, with coverage of all relevant disciplines.

Seminars in the Neurosciences. v. 1– , 1989– . San Diego, CA: Academic Press. Bimonthly. ISSN 1044-5765. $185.00.

Each issue is devoted to an important topic and edited by a guest editor, an acknowledged expert in the field. The articles review various aspects of the chosen topic, discussing the latest advances, current results and models, and specific implications.

Studies of Brain Function. v. 1– , 1977– . New York: Springer-Verlag. Irregular. Price varies. ISSN 0172-5742.

A series of monographs, each reviewing a particular area in the neurosciences. Vol. 19, 1993: *Synaptic Modifications and Memory; An Electrophysiological Analysis.*

Trends in Neurosciences. v. 1– . 1978– . New York: Elsevier. Monthly. $490/yr. ISSN 0166-2236.

Follows the established *Trends* format; refer to *Trends in Endocrinology . . .* (p. 345). A supplement to *Trends in Neurosciences, Neurotoxins,* was published in 1994, providing a compilation of toxins, listing their sources, general chemical nature, targets, modes of action, and key references.

Physiology

Advances in Comparative and Environmental Physiology. v. 1– , 1988– . New York: Springer-Verlag. Irregular. Volumes priced separately. ISSN 0938-2763.

Assists biologists, physiologists, and biochemists in keeping track of the extensive literature in the field by providing comprehensive, integrated reviews and summaries.

Annual Review of Physiology. v. 1– , 1939– . Palo Alto, CA: Annual Reviews. Annual. $49/yr. ISSN 0066-4278.

Follows the established format of other *Annual Review* series; refer to *Annual Review of Neuroscience* (p. 345).

Monographs of the Physiological Society. v. 1– , 1953– . New York: Oxford University Press.

Each volume deals with a particular topic in depth; a recent example is the 1993 publication *Intramembrane Charge Movements in Striated Muscle,* by Christopher L.-H. Huang. 292 p. $97.50. ISBN 0198577494.

News in Physiological Sciences. v. 1– , 1986– . Baltimore, MD: Williams & Wilkins for the American Physiological Society and the International Union of Physiological Sciences. Bimonthly. $80/yr. ISSN 0886-1714.

Brief review articles on major physiological developments.

Nobel Lectures in Physiology and Medicine, 1901–1970. New York:
Elsevier, 1964–1973. 4 v. Vol. 1: $87.25. ISBN 0444404198;
Vol. 2: $87.25. ISBN 0444404201; Vol. 4: 87.25. ISBN
04444409947. *Nobel Lectures in Physiology and Medicine 1971–
1980*; *Nobel Lectures in Physiology and Medicine 1981–1990.*
River Edge, NJ: World Scientific Publishing, 1992-1993. 2 v.
1971–1980: $96.00. ISBN 9810207905; 1981–1990: $94.00.
ISBN 9810207921.

The Nobel lectures provide fascinating accounts of research of Nobel laureates since the turn of the century.

Physiological Reviews. v. 1– , 1921– . Bethesda, MD: American
Physiological Society. Quarterly. $204/yr. ISSN 0031-9333.

State of the art coverage of issues in physiological and biomedical sciences, including about 25 review articles each year.

Reviews of Physiology, Biochemistry, and Pharmacology. v. 70– ,
1974– . New York: Springer-Verlag. Irregular. $98/v. 125/1994.
ISBN 0387579303. ISSN 0303-4240.

Continues *Ergebnisse der Physiologie Biologischen Chemie und Experimentellen Pharmakologie.* Review articles for researchers and scientists.

Societies

For more information about societies and associations, consult the *Encyclopedia of Associations* (Chapter 3, p. 39), in print or electronically, and on the Internet.

American Association of Anatomists (AAA)
c/o Dr. Robert D. Yates, Tulane Medical Center, 1430 Tulane Ave.,
New Orleans, LA 70112.

Founded: 1888, now with over 2,000 members. Professional society of anatomists and scientists in related fields. Presents awards and maintains placement services. Committees on anatomical nomenclature, educational affairs. Formerly: Association of American Anatomists. Publications: *American Journal Anatomy, Anatomical News, Anatomical Record, Directory* (Departments of Anatomy in the United States and Canada). Annual spring meeting.

American Physiological Society (APS)
9650 Rockville Pike, Bethesda, MD 20814.

Founded: 1887. 6,600 members. Professional society of physiologists devoted to fostering scientific research, to furthering education, and to the dissemination of scientific information. Maintains biographical archives and bestows awards. Various committees and affiliations with national and international related associations. Publications: *Advances in Physiology Education, American Journal of Physiology* and all of the sections, *FASEB Directory, Handbooks, Journal of Applied Physiology, Journal of Neurophysiology, News in Physiological Sciences, Physiological Reviews, The Physiologist.* Information about the Society including announcements, meeting schedules, employment opportunities, information for authors, reports from committees and award programs, directory of staff members, etc., is available on the Internet. Semiannual meetings are held in April and in the fall.

American Society for Neurochemistry (ASN)
Veterans Administration Medical Center, Dept. of Pathology
Research, 151B, 3801 Miranda Ave., Palo Alto, CA 94304.

Founded 1969. Over 1,000 members; investigators in the field of neurochemistry and scientists who are qualified specialists in other disciplines and are interested in the activities of the Society. Advances and promotes the science of neurochemistry and related neurosciences and to increase and enhance neurochemical knowledge; to facilitate the dissemination of information concerning neurochemical research through scientific meetings, seminars, publications, and related activities; to encourage the research of individual neurochemists. Conducts symposia, distributes research communications, maintains placement service. Publications: *American Society for Neurochemistry Membership Directory, Newsletter, Transactions.* Annual meeting.

Endocrine Society (ES)
9650 Rockville Pike, Bethesda, MD 20814-3998.

Founded: 1918. 7,000 members. Promotes excellence in research, education, and clinical practice in endocrinology and related disciplines. Maintains placement service and bestows awards for work of distinction in endocrinology. Formerly: Association for Study of Internal Secretions. Publications: *Endocrine News, Endocrine Reviews, Endocrine Society Membership Directory, Endocrinology, Journal of Clinical Endocrinology and Metabolism, Molecular Endocrinology.* Annual meeting, usually in October. See also, Women's Caucus of the Endocrine Society (p. 351).

European Neuroscience Association (ENA)
Postbus 238, NL-1400 AE Bussum, Netherlands.

Founded: 1976. 2,000 members. Neuroanatomists, neurochemists, neurocytologists, neuroendocrinologists, neuropharmacologists, neurophysiologists, neuropsychologists, and behavioral scientists in 41 countries. Objectives are to enhance comparative research, promote the exchange of technical information and theoretical knowledge, establish high professional standards in the neurosciences. Publications: *Abstracts, ENA Membership Directory, ENA Newsletter, European Journal of Neuroscience*. Annual meeting; periodic symposium.

International Brain Research Organization (IBRO)
    c/o Dr. David Ottoson, 51, blvd. de Montmorency, F-75016 Paris,
    France.

Founded: 1958. 27,000 members. Scientists working in neuroanatomy, neuroendocrinology, the behavioral sciences, neurocommunications and biophysics, brain pathology, and clinical and health-related sciences. Works to promote international cooperation in research on the nervous system. Sponsors fellowships, exchange of scientific workers, and traveling teams of instructors to supplement local teachings. Organizes international neuroscience symposia and workshops. Publications: *Directory of Members, Neuroscience*, symposia monograph series and handbook series. Quadrennial IBRO World Congress of Neuroscience.

International Society for Developmental Neuroscientists (ISDN)
    c/o Bernard Haber, Ph.D., University of Texas Medical Branch,
    200 University, No. 519, Galveston, TX 77550.

Founded: 1978. 850 members. Independent researchers who have produced meritorious work in the field of developmental neuroscience; individuals who have made outstanding contributions to developmental neuroscience; doctoral students. Aims to advance research and knowledge concerning the development of the nervous system and to support the effective application of this information for the improvement of human health. Publications: *International Journal of Developmental Neurosciences, ISDN Newsletter*. Biennial congress.

International Union of Physiological Sciences (IUPS)
    Lab de Physiologie Nerveuese, Groupe de Labs du CNRS, Gif sur
    Yvette, F-91198 France.

Objectives are to encourage the advancement of the physiological sciences; to facilitate the dissemination of knowledge in the field of physiological sciences; to foster and encourage research in the field of physiological sciences; to promote such other measures as will contribute to the development of physiological sciences in developing countries. Member of the International Council of Scientific Unions.

Physiological Society (PS)
    Société de Physiologie, 91, Blvd. de l'Hôpital, F-75634 Paris, cedex
    13, France.

Founded: 1926. Over 1,000 members. Physiologists from universities, hospitals, and research organizations. Purpose is to promote and facilitate scientific contacts among professionals in the field of physiology. Bestows awards. Publications: *Directory, Journal.* Annual conference, quarterly symposium.

Society for Neuroscience (SN)
    11 Dupont Circle, Ste. 500, Washington, DC 20036.
Founded: 1969. 18,000 members. Scientists who have done research relating to the nervous system; affiliate members are individuals and organizations interested in the society's objectives but not actively engaged in research or who live outside North America. Seeks to advance understanding of nervous systems, including their relation to behavior, by bringing together scientists of various backgrounds and by facilitating integration of research directed at all levels of biological organizations. Conducts workshops. Publications: *Abstracts, Journal of Neuroscience, Membership Directory, Neuroscience Newsletter, Neuroscience Training Programs in North America.* Annual meeting.

Society of General Physiologists (SGP)
    PO Box 257, Woods Hole, MA 02543.
Founded: 1946. 1,000 members. Biologists interested in fundamental physiological principles and phenomena. Publications: *Journal of General Physiology, Proceedings of Annual Symposium, SGP Constitution and Membership List.* Annual meeting with symposium in September.

Women's Caucus of the Endocrine Society (WE)
    University of Maryland School of Medicine, Dept. of Physiology, 655 W. Baltimore Street, Baltimore, MD 21201.
Founded: 1975. 850 members. Promotes professional advancement of women and younger members of the Endocrine Society. Maintains biographical archives, compiles statistics, conducts seminars and workshops. Publications: *Letter to Membership.* Annual meeting in conjunction with the Endocrine Society; also, holds annual meeting, with leadership conference.

Textbooks

Bagshaw, C. R. *Muscle Contraction*, 2nd ed. Chapman & Hall, 1993. 168 p. $27.50. ISBN 0412403706.

Basic Neurochemistry: Molecular, Cellular, and Medical Aspects, 5th ed. Editor-in-Chief, George J. Siegel. New York: Raven, 1994. 1,080 p. $67.00. ISBN 078170104X. Set of 495 slides, $350.00, ISBN 0781701341.

Bolander, Franklyn F. *Molecular Endocrinology*, 2nd ed. Sand Diego, CA: Academic Press, 1994. 569 p. $69.95. ISBN 0121112314.

Brown, M. C., W. G. Hopkins, and R. J. Keynes. *Essentials of Neural Development*, rev. ed. Cambridge, England: Cambridge University Press, 1991. 176 p. $29.95 (paper). ISBN 0521375568 (hardcover), ISBN 052137698X (paper).

Brown, Richard E. *An Introduction to Neuroendocrinology*. New York: Cambridge University Press, 1993. 408 p. $79.95. ISBN 0521416450.

Burton, Richard F. *Physiology by Numbers; An Encouragement to Quantitative Thinking*. New York: Cambridge University Press, 1994. 185 p. $59.95, $19.95 (paper). ISBN 0521420679, 0521421381 (paper).

Comparative Molecular Neurobiology. Edited by Y. Pichon. Basel, Switzerland: Birkhauser, 1993. (Experientia. Supplementum 63). 433 p. $80.00. ISBN 003764327855.

Endocrinology. Edited by Leslie J. DeGroot et al, 3rd ed. Philadelphia: Saunders, 1994. 3 v. $450.00 (set). ISBN 0721642624 (set).

Hall, Zach W. *An Introduction to Molecular Neurobiology*. Sunderland, MA: Sinauer, 1992. 555 p. $46.95. ISBN 0878933077.

Huguenard, John and David A. McCormick. *Electrophysiology of the Neuron; An Interactive Tutorial*. New York: Oxford University Press, 1994. 80 p; 1 disk MS-DOS or Macintosh, $18.95.
A companion to *Neurobiology* by Gordon Shepherd (p. 353).

Jacobson, Marcus. *Developmental Neurobiology*, 3rd ed. New York: Plenum, 1991. 786 p. $107.40. ISBN 03064397X.

Jacobson, Marcus. *Foundations of Neuroscience*. New York: Plenum, 1993. 387 p. $79.50. ISBN 0306445409.

Junge, Douglas. *Nerve and Muscle Excitation*, 3rd ed. Sunderland, MA: Sinauer, 1992. 263 p. $22.95 (paper). ISBN 0878934065.

Lecture Notes on Human Physiology, 3rd ed. Edited by John J. Bray et al. Oxford, England: Blackwell, 1994. 726 p. $29.95 (paper). ISBN 0632036443 (paper).

Levitan, Irwin B. and Leonard K. Kaczmarek. *The Neuron: Cell and*

Molecular Biology. New York: Oxford University Press, 1991. 450 p. $39.95 (paper). ISBN 0195070712.

Moffat, David Burns. *Lecture Notes on Anatomy*, 2nd ed. Boston: Blackwell, 1993. $80.00. ISBN 0632036966.

Moffett, David F., Stacia B. Moffett, and Charles L. Schauf. *Human Physiology: Foundations and Frontiers*, 2nd ed. St. Louis, MO: Mosby, 1993. 851 p. $56.95. ISBN 0801669030.

Neural Modeling and Neural Networks. Edited by F. Ventriglia. New York: Pergamon, 1994. (Pergamon Studies in Neuroscience, v. 11). 351 p. $125.00. ISBN 0080422772.

Nicholls, John G., A. Robert Martin, and Bruce G. Wallace. *From Neuron to Brain*, 3rd ed. Sunderland, MA: Sinauer, 1992. 700 p. $48.50. ISBN 0878935800.

Partridge, Lloyd D. and L. Donald Partridge. *The Nervous System: Its Function and Its Interaction with the World*. Cambridge, MA: MIT Press, 1993. 579 p. $65.00. ISBN 0262161346.

Principles of Neural Science, 3rd ed. Edited by Eric R. Kandel, James H. Schwartz, and Thomas M. Jessell. New York: Elsevier, 1991. 1,135 p. $65.00. ISBN 0444015620.

Shepherd, Gordon M. *Neurobiology*, 3rd ed. New York: Oxford University Press, 1994. 760 p. $65.00. ISBN 0195088425.
Also, see Huguenard and McCormick (p. 352) for a companion manual.

Thompson, Richard F. *The Brain; A Neuroscience Primer*. New York: Freeman, 1993. 475 p. $22.95. ISBN 0716723387.

Tortora, Gerard J. *Introduction to the Human Body; The Essentials of Anatomy and Physiology*, 3rd ed. New York: HarperCollins, 1994. 547 p. $54.50. ISBN 0065013638.

Vander, Arthur J. et al. *Human Physiology: The Mechanisms of Body Function*. Blue Ridge Summit, PA: McGraw-Hill, 1994. 754 p. $75.00. ISBN 0070669929.

Williams' Textbook of Endocrinology, 8th ed. Edited by Jean D. Wilson and Daniel W. Foster. Philadelphia: Saunders, 1992. 1,712 p. ISBN 0721695140.

12

Entomology

Entomology is "A branch of the biological sciences that deals with the study of insects," according to the *McGraw-Hill Dictionary of Scientific and Technical Terms*. Strictly speaking, the true insects are only those belonging to the class Insecta, which does not include the spiders and other animals often thought of as insects such as millipedes or ticks. This chapter includes material on both insects and their close relatives. However, applied entomology is largely excluded, although some basic tools are mentioned.

Abstracts and Indexes

Current Sources

Abstracts of Entomology. v. 1– , 1970– . Philadelphia: BIOSIS.
    Monthly. $230.00 ($115.00 cumulative index). ISSN 0001-3579.

Indexes all areas of entomology, with 19,000 references published in 1994. Has author, biosystematic, generic, and subject indexes.

Apicultural Abstracts. v. 1– , 1950– . Cardiff, Wales: International Bee
    Research Association. Quarterly. $240.00. ISSN 0003-648X.

Covers the world literature on bees, including their biology and pollination activities. Also available on DIALOG as part of CAB Abstracts (1972 to present).

Entomology Abstracts. v. 1– , 1969– . Bethesda, MD: Cambridge
    Scientific Abstracts. Monthly. $855.00. ISSN 0013-8924.

Covers all areas of entomology, including systematics, paleontology, behavior, genetics, and physiology. Author and subject indexes. Available online, on CD-ROM, and via the Internet as part of the Life Sciences Collection (see Chapter 4, p. 77).

Helminthological Abstracts, Series A (Animal Helminthology).
    v. 1– , 1932– . CAB International. Monthly. $496.00. ISSN
    0957-6789.

Includes journal articles, reports, conferences, and books on all aspects of parasitic helminths. Also available on DIALOG as part of CAB Abstracts (1972 to present). Includes monthly and annual author and subject indexes.

Helminthological Abstracts, Series B (Plant Nematology). v. 1– ,
    1932– . CAB International.

Index-Catalogue of Medical and Veterinary Zoology. v. 1– , 1932– .
    Animal Parasitology Institute, USDA.

Index to the literature of animal parasites of humans and animals. Issued as series of supplements.

Odonatological Abstracts. v. 1– , 1972– . Bilthoven, Netherlands: Societas Internationalis Odonatologica.

Indexes the literature of dragonflies.

Review of Agricultural Entomology. v. 1– , 1913– . Wallingford, UK: CAB International. Monthly. $658.00. ISSN 0957-6762.

Formerly *Review of Applied Entomology, Series A (Agricultural).* Covers arthropod pests of cultivated plants, trees, and stored products. Most citations also have abstracts. Monthly and annual author and subject indexes.

Review of Medical and Veterinary Entomology. v. 1– , 1913– . Wallingford, UK: CAB International. $342.00. ISSN 0957-6770.

Formerly *Review of Applied Entomology, Series B (Medical and Veterinary).* Covers disease-transmitting insects of importance to humans and animals. Includes journal articles, books, reports, and conferences. Monthly and annual author and subject indexes are included.

Termite Abstracts. v. 1– , 1980– . Taylor & Francis. Quarterly. ISSN 0144-5995.

Covers all aspects of termite biology and control.

Tropical Diseases Bulletin. v. 1– , 1912– . Bureau of Hygiene & Tropical Diseases. Monthly. $270.00. ISSN 0041-3240.

Includes section on medical entomology.

Zoological Record, Insecta. v. 1– , 1864– . Philadelphia: BIOSIS. Annual. $860 (6 sections).

Covers all insects in six sections (General Insecta and Smaller Orders, Coleoptera, Diptera, Lepidoptera, Hymenoptera, and Hemiptera). Has author, subject, geographical, paleontological, and systematic indexes. Also available on CD-ROM from SilverPlatter (1978 to present) and online from DIALOG (1978 to present).

See also *Bibliography of Agriculture* (p. 73), *Biological Abstracts* (p. 74), *Biological Abstracts/RRM* (p. 74), *Current Contents/Agriculture, Biology, and Environmental Sciences* (p. 76).

Sources for Retrospective Searches

Derksen, W. and U. Scheiding. 1963–75. *Index Litteraturae Entomo-logicae*. Ser. 2, *Die Welt-Literatur uber die gesamte Entomologie von 1864 bis 1900*. 5 v. Berlin: Akademie der Landwirtschafts-wissenschaften der Deutschen Demokratischen Republik.

Continues the entomological literature from Horn and Schenkling (below). v. 5 contains a list of journal titles.

Experiment Station Record. v. 1–95, 1889–1946. Washington, DC: Government Printing Office, Department of Agriculture Office of Experiment Stations.

Important source for retrospective applied entomological literature. Abstracts are arranged by topic with name and subject indexes.

Horn, W. and S. Schenkling. *Index Litteraturae Entomologicae*. Die Weltliteratur Uber die gesamte Entomologie bis inklusive 1863. 4 v. Berlin: Dahlem, 1928–29.

Revision of Hagen, *Bibliotheca Entomologica* (1862–63), containing 8,000 additional titles. Covers the literature of entomology from its beginning to 1862.

Index to the Literature of American Economic Entomology. v. 1–18, 1905/14–1959. College Park, MD: Entomological Society of America (Special Publication 1-18 of Entomological Society of America).

"Presents articles on economic entomology selected from pamphlets, periodicals, and books received in the National Agriculture Library." The index provides citations arranged by scientific and common names of insects of economic importance.

Royal Entomological Society of London. *Catalogue of the Library of the Royal Entomological Society of London*. Boston: G. K. Hall, 1980. 5 v.

Contains copies of the card catalog of the Royal Society, which holds many rare works.

U.S. Bureau of Entomology. *Bibliography of the More Important Contributions to American Economic Entomology*. Washington, DC: Government Printing Office, 1889–1905. Pt. 1–3: The more important writings of B. D. Walsh and C. V. Riley. Pt. 4–5: The more important writings of government and state entomologists and of other contributors to the literature of American economic

entomology. Pt. 6–8: The more important writing published
between June 30, 1888 and January 1, 1905.

Continued by *Index to the Literature of American Economic Entomology*.

Zoological Record.

See information (p. 356) in the Abstracts and Indexes section.

Dictionaries and Encyclopedias

Erickson, Ruth Olive. *A Glossary of Some Foreign-Language Terms in
Entomology*. 1961. (U.S. Department of Agriculture Agriculture
Handbook series, 218).

Languages include Czech, Danish, Dutch, French, German, Polish, Russian, and
Swedish.

Foote, Richard H. *Thesaurus of Entomology*. College Park, MD: Entomological Society of America, 1977. 188 p.

Rather than being a true dictionary, this is a thesaurus of indexing terms. About
9,000 terms are included, both by hierarchical classification and in an alphabetical list. While dated, it is still a good source for related terminology.

Greiff, Margaret, compiler. *Spanish-English-Spanish Lexicon of Entomological and Related Terms*. Slough, UK: Commonwealth Agricultural Bureaux, 1985. 255 p. $85.00. ISBN 0851985602.

Provides Spanish-to-English and English-to-Spanish translations of common
entomological terms, along with indexes of Latin names with Spanish and
English common names and Spanish common names with their Latin equivalent.

Harbach, Ralph E. and Kenneth L. Knight. *Taxonomists' Glossary of
Mosquito Anatomy*. Marlton, NJ: Plexus, 1980. 415 p. $24.95.
ISBN 0937548006.

This glossary includes many figures in addition to definitions of terms. It is
divided by life cycle stage (egg, larva, pupa, and adult), with a final section on
vesture (the surface of the mosquito and its structures).

O'Toole, Christopher, ed. *The Encyclopedia of Insects*. New York: Facts
on File Inc., 1987. 151 p. $24.95. ISBN 0816013586.

This one-volume encyclopedia provides an excellent introduction to the arthropods of the world, including myriapods (millipedes and centipedes), insects, and
arachnids. The entries are by taxonomic groups, and include information on sys-

tematics and behavior. There are numerous excellent photographs and line drawings.

Preston-Mafham, Rod and Ken Preston-Mafham. *Encyclopedia of Land Invertebrate Behavior*. Cambridge, MA: MIT Press, 1993. 320 p. ISBN 0262161370.

While this well-illustrated encyclopedia covers more than just insects and arachnids, the majority of the entries relate to those two groups. The authors provide an excellent, authoritative introduction to primarily sexual, egg-laying, parental care, feeding, and defensive behaviors.

Torre-Bueno, J. R. de la, et al. *The Torre-Bueno Glossary of Entomology*, rev. ed. New York: New York Entomological Society in cooperation with the American Museum of Natural History, 1989. 840 p. $60.00. ISBN 0913424137.

This is a revised and expanded edition of Torre-Bueno's 1937 *Glossary of Entomology*, and includes the 1960 *Supplement A* by George S. Tulloch. The terms covered include systematic, descriptive, and general terms. The editors have also included an extensive list of sources and other references.

Directories

Arnett, Ross H., Jr., G. Allan Samuelson, and Gordon M. Nishida. *The Insect and Spider Collections of the World*, 2nd ed. Gainesville, FL: Sandhill Crane Press, 1993. (Flora and Fauna Handbook, 11). 310 p. $30.00. ISBN 1877743151.

"The main purposes of this compilation are to inform those doing systematic research of the availability of stored data in the form of specimens and associated information." It is arranged by country, with general information on each country (population, size, biogeographical region) as well as detailed information on each major insect collection in the country.

International Pesticide Directory. v. 1– , 1981– . London: McDonald Publications. Annual. $50.00 (free with subscription to *International Pest Control*).

Consists of three parts: alphabetical list of companies and products, alphabetical list of pesticides with active ingredients, and alphabetical list of active ingredients with pesticides based on them.

Resources in Entomology. College Park, MD: Entomological Society of America, 1987. 269 p. $20.00. ISBN 0938522329.

A listing by state and country of academic, private, and governmental institutions dealing with entomology. Each institution is described, with information
on museum collections, interests, history, function, faculty or staff names, and
degrees or employment opportunities. There is no index, and foreign institutions
have addresses only.

Guides to Internet Resources

There are a number of Internet-accessible resources for the field of entomology,
particularly economic entomology. Many of the Gopher servers and other resources discussed in Chapters 9, "Ecology, Evolution, and Animal Behavior,"
and 13, "Zoology," also include information on insects. In addition, there are
a number of entomology Gopher servers, discussion groups, and other materials.
Information available through Gophers includes online insect databases, the
University of Minnesota bibliography on attractants of blood-sucking insects, and
insect fact sheets (available from many sites). Listservs and USENET groups
include Bugnet, a list for non-professionals (listserv@wsuvm1.csc.wsu.edu); ent-
list, a list for insect systematics and curation (send requests for information to
mfobrien@umich.edu); entomo-l, a general list (listserv@uoguelph@ca), and the
USENET groups bionet.drosophila and sci.bio.entomology.lepidoptera.
    Entomology is particularly well-supplied with useful World Wide Web
home pages. The home pages at the University of Illinois at Urbana-Champaign,
Colorado State University, and Iowa State University are good examples. All
offer information about the entomology program at that institution, as well as
an extensive list of other Internet accessible entomology resources. These
resources include images of insects, Gopher entomology servers, and information
on entomological discussion groups. Iowa State also has a list of entomology-
related CD-ROM titles.

Guides to the Literature

British Museum (Natural History). *List of Serial Publications in the
    Libraries of the Departments of Zoology and Entomology*. London:
    Trustees of the British Museum (Natural History), 1967.
Includes reports and publications of organizations.

Chamberlain, W. J. *Entomological Nomenclature and Literature*, 3rd ed.,
    rev. and enl. Westport, CT: Greenwood Press, 1970. 141 p. ISBN
    0837138108.

Excellent guide for its time; still useful for retrospective work.

Gilbert, Pamela and Chris J. Hamilton. *Entomology: A Guide to Information Sources*, 2nd ed. London: Mansell, 1990. 259 p. $90.00. ISBN 0720120527.

This guide has extensive coverage of the literature of entomology, including information on entomological collections, suppliers, and sources of illustrations, as well as the standard primary and secondary literature. Has a European slant.

Hammack, G. M. *The Serial Literature of Entomology: A Descriptive Guide*. College Park, MD: Entomological Society of America, 1970. 85 p.

A list of the journals publishing the majority of entomological studies, including complete bibliographic information. There are language and geographic indexes.

Hogue, Charles L. *Latin American Insects and Entomology*. Berkeley: University of California Press, 1993. 536 p. $85.00. ISBN 0520078497.

See p. 380 in the Treatises and General Works section for full annotation; has extensive information on locating information on Latin American entomology.

Liste de Périodiques d'Entomologie/Lijst van entomologische periodieken. Brussels: Institut Royal des Sciences Naturelles de Belgique, 1966. unpaged.

Introduction in French and Flemish. Provides bibliographic information on entomological journals.

Handbooks and Manuals

Advances and Challenges in Insect Rearing. E. G. King and N. C. Leppla, eds. New Orleans, LA: Agricultural Research Service, Southern Region, U.S. Dept. of Agriculture, 1984. 306 p.

Revised versions of papers presented at 1980 conference. Covers both basic and state-of-the-art topics in insect rearing.

Distribution Maps of Pests. v. 1– , 1951– . Wallingford, UK: CAB International. Biannual. $99.00. ISSN 0952-634X.

"A series of maps gives the world distribution, together with supporting references, of a particular arthropod pest. Those at present being issued deal with pests of importance in relation to agriculture, forestry, or their products. Eighteen maps are issued each year in two batches of nine each."

Handbook of Insect Rearing. Pritam Singh and R. F. Moore, eds. New York: Elsevier, 1985. 2 v. $115.50 (per volume). ISBN 0444424679 (set).

Handbook offering step-by-step instructions for rearing insects. Arranged by insect order.

Imes, Rick. *The Practical Entomologist*. New York: Simon and Schuster, 1992. 160 p. $15.00 (paper), $27.95. ISBN 06170746952 (paper), 0671746960.

For the amateur. Includes information on capturing and keeping live insects, and making an insect collection, and tips on insect photography as well as chapters on each major order of insects.

Kearns, Carol Ann and David William Inouye. *Techniques for Pollination Biologists*. Niwot, CO: University Press of Colorado, 1993. 583 p.

See main entry (p. 294) in Chapter 10, "Plant Biology." Includes chapter on collecting, preparing, and identifying pollinators, as well as methods of study.

Laboratory Training Manual on the Use of Nuclear Techniques in Insect Research and Control: A Joint Undertaking, 3rd ed. Food and Agriculture Organization of the United Nations and the International Atomic Energy Agency. Vienna: International Atomic Energy Agency, 1992. (Technical Reports Series, 336). 183 p.

"The purpose of this Manual is to help entomologists . . . in developing countries become familiar with the potential use of isotopes and radiation in solving some of their research and insect control problems." After opening sessions discussing safety and the use of radiation in research, the manual goes on to offer applications relating to entomological research, in particular the use of sterile insects in controlling populations.

Leppla, N. C. and Thomas E. Anderson. *Advances in Insect Rearing for Research and Pest Management*. Boulder, CO: Westview Press, 1992. (Westview's studies in insect biology). 519 p. $69.50. ISBN 0813378354.

Covers rearing insects for research in genetics, molecular biology, and nutrition, as well as pest management, and commercial purposes.

Mayer, Marion S. and John R. McLaughlin. *Handbook of Insect Phero-*

mones and Sex Attractants. Boca Raton, FL: CRC Press, 1990.
1083 p. $295.00. ISBN 0849329345.

Provides a "guide to the literature published before 1988 on chemicals that effect aggregation for mating and/or elicit sexual behavior in insects, mites, and ticks." The bulk of the handbook consists of references to the sex pheromones of insects, in order of genera. There are also indexes to footnoted species, common names, and chemicals. The second part consists of entries dealing with the chemistry of the pheromones, including synthesis and analytical methods.

Poinar, George O. and Gerard M. Thomas. *Laboratory Guide to Insect Pathogens and Parasites*. New York: Plenum, 1984. 392 p. $89.50. ISBN 0306416808.

According to the preface, this is "unique in covering all types of biotic agents which are found inside insects and cause them injury or disease." Has pictorial identification and information on the availability of insect pathogens as well as techniques for identification of pathogens, such as agars and staining.

Service, M. W. *Mosquito Ecology: Field Sampling Methods*, 2nd ed. New York: Elsevier Applied Science, 1993. 988 p. $140.00. ISBN 1851667989.

A manual for field workers, covering various methods for sampling eggs, larvae, and adults. Has author, mosquito species, and subject indexes.

Springer Series in Experimental Entomology. T. A. Miller, series editor. New York: Springer-Verlag, 1979– .

"It is the purpose of this series (1) to report new developments in methodology, (2) to reveal sources of groups who have dealt with and solved particular entomological problems, and (3) to describe experiments which may be applicable to use in biology laboratory courses." Sixteen volumes have been published by 1993; previous volumes include *Neuroanatomical Techniques, Techniques in Pheromone Research, Insect–Plant Interactions*, and *Rice Insects: Management Strategies*.

Walker, Annette K. and Trevor K. Crosby. *The Preparation and Curation of Insects*, Rev. ed. Wellington, New Zealand: Science Information Publishing Centre, DSIR, 1988. (DSIR Information Series, 163). 91 p.

Explains the methods used by workers at the New Zealand Arthropod Collection in preparing specimens and curating its collection. Some of the material covered is of only local use, but the techniques are described in detail and are useful for most collections. The handbook also includes a helpful checklist of entomological supplies to keep on hand.

Histories

Barnes, Jeffrey K. "Insects in the New Nation: A Cultural Context for the
   Emergence of American Entomology." *Bulletin of the Entomological
   Society of America* 31(1): 21–30. Spring 1985.

Discusses the history of entomology and natural history in America, with emphasis on America's early inferiority complex in matters zoological and on the
importance of the expansion of American agriculture.

Bulletin of the Entomological Society of America. 35(3). Fall 1989.

This special issue was published for the centennial of the Entomological Society
of America and has articles covering the history of the Society, biographies of
its past presidents, histories of each of the branches and sections of the Society,
as well as discussion of the services and publications of the Society.

Carpenter, M. M. Bibliography of Biographies of Entomologists.
   American Midland Naturalist, 33:1–116, 1945. Supplement
   50:257–348, 1953.

Bibliographic information includes obituaries, birthdays, portraits, anniversaries,
biographies, and disposition of collections.

Evans, Howard Ensign. *The Pleasures of Entomology: Portraits of Insects
   and the People Who Study Them*. Washington, DC: Smithsonian
   Institution Press, 1985. 238 p. ISBN 0874744210.

This work could just as well be treated as a general work, since the author has
written on interesting insects and the people who study them; also has a brief
discussion of the history of entomology and the lives of American entomologists.
Well written for the general public.

Gilbert, P. *A Compendium of the Biographical Literature on Deceased
   Entomologists*. London: British Museum (Natural History), 1977.

Includes and enhances Carpenter's *Bibliography of Biographies of Entomologists*
(above). The bibliography attempts to be complete through 1975, listing the
names of 7,500 entomologists.

History of Entomology. Ray F. Smith, Thomas E. Mittler, and Carroll N.
   Smith, eds. Palo Alto, CA: Annual Reviews, 1973. 517 p. ISBN
   0824321017.

A supplement volume to *Annual Review of Entomology*, this volume covers the
world-wide history of entomology, with emphasis on "the personalities of past
scientists who contributed to the development of entomological ideas and principles." Various volumes of the *Annual Review of Entomology* include chapters
on the history of entomology.

Identification Manuals

There are innumerable excellent field guides for identifying insects. Only a very few are listed below, along with more technical manuals and other identification aids.

ANI-CD: Arthropod Name Index on CD-ROM. Wallingford, England: CAB International, 1995. CD-ROM disc. $900.00 (new order), $215 (annual update).

An authority file for arthropod names, including preferred terms, synonyms, common names, taxonomy, and major bibliographic references for arthropods of economic importance. Originally created as the authority file for arthropod names for the **CAB Abstracts** database (see Chapter 4, p. 75).

Arnett, Ross H., Jr. *American Insects: A Handbook of the Insects of America North of Mexico*. Gainesville, FL: Sandhill Crane Press, 1993. 850 p. $59.50. ISBN 1877743194. Reprint of the 1985 edition.

A synopsis of the insects of North America, with keys to the generic level, and descriptions of orders, families, and some subfamilies, as well as a number of representative species. Also includes introductory material on insect biology, systematics, and preparation of specimens. Over 22,000 species are described. An authoritative work; there is nothing else quite as comprehensive for North American insects.

Borror, Donald J. and Richard E. White. *A Field Guide to Insects of America North of Mexico*. Boston: Houghton Mifflin, 1970. (Peterson Field Guide Series, 19). 404 p. $17.95, $13.95 (paper). ISBN 0395074363, 0395185238 (paper).

Burgess, N. R. H. and G. O. Cowan. *A Colour Atlas of Medical Entomology*. New York: Chapman and Hall, 1993. 152 p. $109.95. ISBN 0412323400.

An identification guide to insects of medical importance, mainly of the tropics. The atlas includes many photographs of the insects, their habitats, and the diseases caused by them. Also includes information on the life cycle, habits, and medical problems caused by each species.

CABIKEY. Wallingford, England: CAB International, 1995– . IBM-compatible diskettes.

This series of programs provides taxonomic keys to insects and other arthropods, most of economic or medical importance. Presently availabe: **CABIKEY to the Major Beetle Families**. 1995. $189.00 (single user). On 3½" high-density disks. Upcoming keys include European thrips, termite genera, and mosquito genera.

Chinery, Michael. *Insects of Britain and Northern Europe*, 3rd ed. New York: HarperCollins, 1993. (Collins Field Guide series). 320 p. ISBN 0002199181.

Covell, Charles V., Jr. *A Field Guide to the Moths of Eastern North America*. Boston: Houghton Mifflin, 1984. (Peterson Field Guide Series, 30). 496 p. $18.95. ISBN 0395361001.

Goddard, Jerome. *Physician's Guide to Arthropods of Medical Importance*. Boca Raton, FL: CRC Press, 1993. 332 p. $110.00. ISBN 084935160X.

Covers the identification of insects, mites, scorpions, and spiders of public health importance. Intended to assist doctors and other medical entomologists in identifying and diagnosing arthropods and the injuries they cause.

Hogue, Charles L. *Latin American Insects and Entomology.*

See entry (p. 380) in the Treatises and General Works section.

Line, Richard P. and Roger W. Crosskey, eds. *Medical Insects and Arachnids*. New York: Chapman and Hall, 1993. 744 p. $170.00. ISBN 0412400006.

An identification guide for medically important insects and arachnids. There are two extensive introductions with extensive references. Each chapter covers a major group and includes information not only on identification, but also control, biology, medical importance, and collecting of specimens. Designed for students and field workers as well as researchers.

Michener, Charles D., Ronald J. McGinley, and Bryan N. Danforth. *The Bee Genera of North and Central America*. Washington, DC: Smithsonian Institution Press, 1994. 209 p. $53.95. ISBN 156098256X.

An excellent key to the 169 species of bees found north of the Colombia–Panama border. The key has parallel text in English and Spanish. There are numerous line drawings and photographs, and the introduction includes information on collecting and preserving specimens as well as terminology used in the keys.

Milne, Lorus and Margery Milne. *The Audubon Society Field Guide to North American Insects and Spiders*. New York: Knopf, 1980. (The Audubon Society Field Guide Series). 959 p. $18.00. ISBN 0394507630.

Opler, Paul A. *A Field Guide to Eastern Butterflies*, 2nd ed. Boston: Houghton Mifflin, 1992. (Peterson Field Guide Series, 4). 392 p. $16.95 (paper). ISBN 0395364523 (paper).

First edition by Alexander Klots, *A Field Guide to Butterflies East of the Great Plains*, 1951.

Pyle, Robert Michael. *The Audubon Society Field Guide to North American Butterflies*. New York: Knopf, 1981. (The Audubon Society Field Guide Series). 916 p. $18.00. ISBN 0394519140.

Stehr, Frederick W., ed. *Immature Insects*. Dubuque, IA: Kendall/Hunt, 1987–1991. 2 v: $79.95 (v. 1), $215.00 (v. 2). ISBN 0840337027 (v. 1).

Designed to serve as identification guide and textbook, this set includes keys, tables of features, and extended literature references. Covers mainly North American insects.

Tilden, James W. and Arthur Clayton Smith. *A Field Guide to Western Butterflies*. Boston: Houghton Mifflin, 1984. (Peterson Field Guide Series, 33). 370 p. $19.95, $14.95 (paper). ISBN 0395354072, 039541654X (paper).

White, Richard E. *A Field Guide to the Beetles of North America*. Boston: Houghton Mifflin, 1983. (Peterson Field Guide Series, 29). 368 p. $17.95, $12.95 (paper). ISBN 0395318084, 0395339537 (paper).

Nomenclature

Benoit, Paul. *Noms français d'insectes au Canada. French Names of Insects in Canada, with Corresponding Latin and English Names*, 4th ed. Agriculture Quebec, 1975. 214 p.

The names of insects are listed in a single alphabetical list, with French, Latin, and English names intermingled; three columns provide the corresponding names in the other languages.

Common Names of Insects and Related Organisms. Committee on Common Names of Insects, 3rd ed. College Park, MD: Entomological Society of America, 1989. 199 p. $35.00. ISBN 0938522345.

This list of standardized common names of insects provides names for 2,018 insects from the United States. There are three major sections: insects listed by common name, scientific name, and higher taxonomic group, and a fourth section listing vernacular equivalents for higher taxonomic groups (Aleyrodidae = whiteflies, for instance).

Grzimek's Animal Life Encyclopedia. Volume 2, *Insects*. (Also see main listing (p. 396) in Chapter 13, "Zoology.")

Has common names in English, German, French and Russian (Cyrillic alphabet) in the "Animal Dictionary" section (pp. 565–618).

Miller, Jacqueline Y., ed. *Common Names of North American Butterflies*. Washington: Smithsonian Institution Press, 1992. 177 p. ISBN 1560981229.

Provides preferred and alternate common names for butterflies north of Mexico, including Hawaii.

Miller, Lee D. and F. Martin Brown. *A Catalogue/Checklist of the Butterflies of America, North of Mexico*. Los Angeles: Lepidopterists' Society, 1981. 280 p. ISBN 0930282035.

Supplement published 1993.

Periodicals

Journals

Acarologia. v. 1– , 19– . West Bloomfield, MI: Indira Publishing House. Quarterly. $275.00. ISSN 0164-7954.

"An International Journal of Acarology." Covers agricultural, aquatic, general, medical, and veterinary aspects of Acarina. Primarily taxonomic studies.

Acta Entomologica Sinica. v. 1– , 1954– . Quarterly. Beijing: Science Press (Beijing). $10.00 per number. ISSN 0454-6296.

Also known as *K'un Ch'ung Hsueh Pao* or *Kunchong Xuebao*. The leading Chinese entomological journal, covering all aspects of basic and applied entomology.

American Entomologist. v. 36– , 1990– . Lanham, MD: Entomological Society of America. Quarterly. $55.00. ISSN 1046-2821.

Formerly *Bulletin of the Entomological Society of America*. "Contributions consist of articles and information of general scientific interest, particularly to entomologists. Articles need not be based on primary research; they can include historical, synthetic, or speculative material."

Annals of the Entomological Society of America. v. 1– , 1908– Lanham, MD: Entomological Society of America. Bimonthly. $150.00. ISSN 0013-8746.

Published by the Entomological Society of America. "Contributions report on the basic aspects of the biology of arthropods."

Applied Entomology and Zoology. v. 1– , 1966– . Tokyo: Japanese Society of Applied Entomology. Quarterly. $84.00. ISSN 0003-6892.

"The journal publishes articles and short communications concerned with applied zoology, applied entomology, applied and experimental agricultural chemistry and pest control equipment in English."

Aquatic Insects: An International Journal of Freshwater Entomology. v. 1– , 1979– . Lisse, Netherlands: Swets and Zeitlinger. Quarterly. $206. ISSN 0165-0424.

"The journal publishes original research on the taxonomy and ecology of aquatic insects. Purely faunistic studies and other papers of only regional interest are not considered."

Archives of Insect Biochemistry and Physiology. v. 1– , 1983– . New York: Wiley-Liss. Monthly. $686.00. ISSN 0739-4462.

Published in collaboration with the Entomological Society of America. "An international journal that publishes articles in English that are of interest to insect biochemists and physiologists."

Bee World. v. 1– , 1919– . Cardiff, UK: International Bee Research Association. Quarterly. $60.00. ISSN 0005-772X.

"Containing original articles and features which are peer-reviewed & comprises the official organ of the International Commission for Plant-Bee Relationships."

Bulletin of the Entomological Research. v. 1– , 1910– . Wallingford, UK: CAB International. Quarterly. $305.00. ISSN 007-4853.

Edited by the International Institute of Entomology. "Publishes original research papers concerning insects, mites, ticks or other arthropods of economic importance. . . . The geographical scope of the *Bulletin* is worldwide but with emphasis on the tropics."

Bulletin of the Natural History Museum, Entomology Series. v. 1– , 1949– . London: Natural History Museum. Semiannual. ISSN 0968-0454.

"Papers in the *Bulletin* are primarily the results of research carried out on the unique and ever-growing collections of the Museum, both by the scientific staff and by specialists from elsewhere who make use of the Museum's resources."

Canadian Entomologist. v. 1– , 1868– . Ottawa: Entomological Society of Canada. Bimonthly. $190.00. ISSN 0008-347X.

Published by the Entomological Society of Canada (which also publishes *Memoirs* and *Bulletin*). "Will publish results of original observations and research on all aspects of entomology."

Ecological Entomology. v. 1– , 1976– . Osney Mead, UK: Black-
     well Scientific Publications. Quarterly. $239.00. ISSN 0307-
     6946.

Published for the Royal Entomological Society. Publishes "original research papers on insect ecology . . . Reviews, descriptive papers and short communications may on occasion be accepted."

Entomography: An Annual Review for Biosystematics. v. 1– , 1982– .
     Sacramento: Entomography Publications. Annual. $50.00. ISSN
     0734-9874.

Entomologia Experimentalis et Applicata. v. 1– , 1958– . Dordrecht:
     Kluwer Academic Publishers. Monthly. $620.83. ISSN 0013-
     8703.

Published for the Nederlandse Entomologische Vereniging. "Covers the field of experimental biology and ecology, both pure and applied, of insects and other land arthropods."

Entomologia Generalis. v. 1– , 1974– . Stuttgart: E. Schweizer-
     bartsche Verlagsbuchhandlung. Quarterly. ISSN 0171-8177.

"An International Journal of General and Applied Entomology, concerned with comparative and descriptive problems in all fields of research in insects and other terrestrial arthropods."

Entomological News. v. 1– , 1889– . Philadelphia: American
     Entomological Society. Bimonthly. $18.00. ISSN 0013-872X.

"Manuscripts on taxonomy, systematics, morphology, physiology, ecology, behavior and similar aspects of insect life and related terrestrial arthropods are appropriate for submission to *Entomological News*."

Entomological Review. v. 1– , 1957– . New York: Scripta Technica.
     Quarterly. $1,196.00. ISSN 0013-8738.

Consists of translations of papers from the Russian *Entomologicheskoye Obozreniye*. "*Entomological Review* deals with all aspects of entomology, covering systematics, faunistics, ecology, morphology, physiology and biochemistry of insects, as well as biological and chemical control of pests. Each issue of *Entomological Review* consists of papers selected from one issue of *Entomologicheskoye Obozreniye*."

Entomologist's Gazette: A Journal of Palearctic Entomology. v. 1– ,
    1950– . Wallinford, UK: Gem Publishing Co. $50.00. ISSN 0013-
    8894.

"The publication of notes and papers in the *Entomologist's Gazette* is restricted
to those dealing with Palearctic entomology."

Entomon. v. 1– , 1976– . Kariavattom, Trivandrum, India: Association
    for Advancement of Entomology. Quarterly. ISSN 0377-9335.

"Devoted to publication of research work on various aspects of insects and other
land arthropods."

Entomophaga: A Journal of Biological and Insect Control. v. 1– ,
    1956– . Paris: Lavoisier Abonnements. Quarterly. $142.00.
    ISSN 0013-8959.

The official journal of the International Organization for the Biological Control
of Noxious Animals and Plants. "Publishes original papers dealing with basic
and/or applied research in the various aspects of the biological control of pest
organisms."

Environmental Entomology. v. 1– , 1972– . Lawrence, KS: Allen Press.
    Bimonthly. $150.00. ISSN 0046-225X.

Published by the Entomological Society of America. "Contributions report on
the interaction of insects with the biological, chemical, and physical aspects of
their environment."

Experimental and Applied Acarology. v. 1– , 19– . Northwood, UK:
    Scientific and Technical Letters. Monthly. $512.00. ISSN 0168-
    8162.

"The journal is concerned with the publication of original scientific papers in the
field of experimental and applied acarology. . . . The scope encompasses
different aspects of working on agricultural mites, stored-products mites,
parasitic mites (ticks, Varroa, etc.), and mites of environmental significance."

Florida Entomologist. v. 1– , 1920– . Winter Haven, FL: Florida Ento-
    mological Society. Quarterly. $30.00. ISSN 0015-4040.

Published by the Florida Entomological Society. "Manuscripts from *all* areas
of the discipline of entomology are accepted for consideration. At least one
author must be a member of the Florida Entomological Society."

General and Applied Entomology. v. 1– , 1964– . Sydney: Entomo-
    logical Society of New South Wales. Annual. $20.00. ISSN 0158-
    0760.

Formerly *Journal of the Entomological Society of Australia (NSW)*. "Entomological contributions of a high standard will be eligible for publication if submitted by a member of the Entomological Society of New South Wales."

Great Lakes Entomologist. v. 1– , 1966– . East Lansing, MI: Michigan Entomological Society. Quarterly. $20.00. ISSN 0090-0222.

Published by the Michigan Entomological Society. "Papers dealing with any aspect of entomology will be considered for publication in *The Great Lakes Entomologist.* Appropriate subjects are those of interest to professional and amateur entomologists in the North Central States and Canada, as well as general papers and revisions directed to a larger audience while retaining an interest to readers in our geographical area."

Insect Biochemistry and Molecular Biology. v. 1– , 1971– . Exeter, UK: Pergamon. 8/yr. $780.00. ISSN 0965-1748.

"Publishes original contributions and mini-reviews in the fields of insect biochemistry and insect molecular biology." Formerly *Insect Biochemistry.*

Insect Molecular Biology. v. 1– , 1992– . Osney Mead, UK: Blackwell Scientific Publications. Quarterly. $219.00. ISSN 0962-1075.

Published for the Royal Entomological Society. "An international forum for all those applying molecular genetic techniques to the study of insects."

Insect Science and Its Application: The International Journal of Tropical Insect Science. v. 1– , 1980– . Nairobi: ICIPE Science Press. Bimonthly. $410.00. ISSN 0191-9040.

Sponsored by the International Centre of Insect Physiology and Ecology (ICIPE) and the African Association of Insect Scientists (AAIS). "Deals comprehensively with all aspects of scientific research targeted on tropical insects (and related arthropods), and the application of new discoveries and innovations to such diverse fields as pest and vector management and use of insects for human welfare."

Insectes Sociaux: International Journal for the Study of Social Arthropods. v. 1– , 1954– . Basel: Birkhauser. Quarterly. $251.88. ISSN 0020-1812.

Official journal for the International Union for the Study of Social Insects. Publishes "original research papers and reviews on all aspects related to the biology and evolution of social insects and other presocial arthropods."

International Journal of Insect Morphology and Embryology. v. 1– , 1971– . Oxford, UK: Pergamon. Quarterly. $470.00. ISSN 0020-7322.

"The Journal will publish original contributions on all aspects of gross morphology, paleomorphology, macro- and microanatomy, ultrastructure . . . , molecular . . . , functional and experimental morphology, and development. . . ."

Invertebrate Reproduction and Development. v. 1– , 1979– . Rehovot, Israel: Balaban. Bimonthly. $352.00.ISSN 0168-8170.

Published in collaboration with the International Society of Invertebrate Reproduction. "The journal publishes original papers and reviews with a wide approach to the sexual, reproductive and developmental (embryonic and post-embryonic) biology of the Invertebrata." Formerly *International Journal of Invertebrate Reproduction and Development*.

Journal of Apicultural Research. v. 1– , 1962– . Cardiff, UK: IBRA. $125.00. ISSN 0021-8839.

"Publishes descriptions of new findings on the scientific aspects of behavior, ecology, natural history and culture of Apoidea in general and Apis species in particular. The Journal also publishes theoretical papers where these relate to Apis, and letters on Apoidea-related subjects which have recently appeared in the journal or elsewhere. Papers dealing with economics, techniques, or society of beekeeping are more usually suited to IBRA's journal *Bee World* or other technical periodicals."

Journal of Applied Entomology. v. 1– , 1914– . Hamburg: Paul Parey. Semi-quarterly. $784.00. ISSN 0931-2048.

Formerly *Zeitschrift für Angewandte Entomologie*.

Journal of Arachnology. v. 1– , 19– . Lawrence, KS: Allen Press. 3 times/year. $80.00. ISSN 0160-8202.

"Official organ of the American Arachnological Society." Publishes in English, Spanish, French, and Portuguese. Mostly taxonomic.

Journal of Economic Entomology. v. 1– , 1908– . Lanham, MD: Entomological Society of America. Bimonthly. $210.00. ISSN 0022-0493.

Published by the Entomological Society of America. "Contributions report on the economic significance of insects."

Journal of Entomological Research. v. 1– , 1977– . New Delhi: Malhotra Publishing House. Quarterly. ISSN 0378-9519.

"Original contributions covering every aspect of fundamental and applied entomological research will be considered for publication."

Journal of Entomological Science. v. 1– , 1985– . Griffin, GA: Georgia
    Entomological Society. Quarterly. $30.00. ISSN 0749-8004.

Formerly *Journal of the Georgia Entomological Society.*

Journal of Insect Behavior. v. 1– , 1988– . New York: Plenum.
    Bimonthly. $225.00. ISSN 0892-7553.

"An international forum for the publication of peer-reviewed original research
papers and critical reviews on all aspects of the behavior of insects and other
terrestrial arthropods."

Journal of Insect Physiology. v. 1– , 1957– . Exeter, UK: Pergamon.
    $1,005.00. ISSN 0022-1910.

"All aspects of insect physiology are published in this journal. The journal will
also accept papers on the physiology of other arthropods, if the referees consider
the work to be of general interest."

Journal of Medical Entomology. v. 1– , 1964– . Lanham, MD: Entomo-
    logical Society of America. Bimonthly. $210.00. ISSN 0022-2585.

Published by the Entomological Society of America. "Contributions report on
all phases of medical entomology and medical acarology, including the systematics and biology of insects, acarines, and other arthropods of public health and
veterinary significance."

Journal of the American Mosquito Control Society. v. 1– , 1985– .
    Lake Charles, LA: American Mosquito Control Society. Quarterly.
    $90.00. ISSN 8756-971X.

Published for the American Mosquito Control Society. Publishes "previously
unpublished papers which contribute to the advancement of knowledge of mos-
quitoes and other vectors, and their control."

Journal of the Kansas Entomological Society. v. 1– , 1928– .
    Lawrence, KS: Allen Press. Quarterly. $75.00. ISSN 0022-8567.

"Membership in the Society is required of authors who submit manuscripts to
the *Journal.*"

Journal of the Lepidopterists' Society. v. 1– , 1947– . Los Angeles:
    Lepidopterists' Society. Quarterly. $40.00. ISSN 0024-0966.

"Contributions to the *Journal* may deal with any aspect of Lepidoptera study."

Medical and Veterinary Entomology. v. 1– , 1987– . Osney Mead, UK:
    Blackwell Scientific Publications. Quarterly. $230.00. ISSN 0269-
    283X.

Publishes "original research papers on the biology and control of arthropods of medical or veterinary importance such as ectoparasites, endoparasites, vectors of pathogens affecting man and other animals and arthropods of forensic importance."

Memoirs of the American Entomological Society. v. 1– , 1916– .
Gainesville, FL: Association Publishers. Irregular. ISSN 0065-8162.
"Monographic works on insects are published as Memoirs."

Oriental Insects. v. 1– , 1968– . Gainesville, FL: Association
Publishers. Annual. $55.00. ISSN 0030-5316.
"An international journal of taxonomy and zoogeography of insects and other land arthropods of the Old World Tropics. . . . Devoted to publication of original papers and reviews on the taxonomy, ecology, distribution, and evolution of insects and other land arthropods of the Old World Tropics."

Pan-Pacific Entomologist. v. 1– , 1924– . Lawrence, KS: Allen Press.
Quarterly. $40.00. ISSN 0031-0603.
Published by the Pacific Coast Entomological Society in cooperation with the California Academy of Science. Mostly North American systematics.

Physiological Entomology. v. 1– , 1976– . Osney Mead, UK: Blackwell
Scientific Publishers. Quarterly. $259.00. ISSN 0307-6962.
Published for the Royal Entomological Society. "*Physiological Entomology* is a journal designed primarily to serve the interests of experimentalists who work on the behaviour of insects and other arthropods."

Psyche: A Journal of Entomology. v. 1– , 1874– . Lexington, MA:
Lexington Press. Quarterly. $30.00. ISSN 0033-2615.
Founded by the Cambridge Entomological Society. Emphasis on neotropical insects, including mites and spiders.

Series Entomologica. v. 1– , 1966– . Dordrecht, Netherlands: Kluwer
Academic. Irregular. Price varies. ISSN 0080-8954.
A series of monograph-length works, mostly systematics but also including ecology, development, biology, etc. Volume 51 was published in 1994.

Southwestern Entomologist. v. 1– , 1976– . Dallas, TX: Southwestern
Entomological Society. Quarterly. $20.00. ISSN 0147-1724.
Covers original research in the southwestern United States and Mexico.

Systematic Entomology. v. 1– , 1976– . Osney Mead, UK: Blackwell
Scientific Publishers. Quarterly. $239.00. ISSN 0307-6970.

Published for the Royal Entomological Society. Publishes "original contributions to insect taxonomy and systematics, although descriptive morphology and other subjects bearing on taxonomy may be considered."

Thomas Say Publications in Entomology. 1992– . Lanham, MD: Entomological Society of America. Irregular.

Formed by the merger of the Entomological Society of America's (ESA's) *Miscellaneous Publications* and the Thomas Say Foundation Monograph Series. Consists of the *Systematic Monographs of the ESA*, the *Memoirs of the ESA*, and *Proceedings of the ESA*.

Reviews

Advances in Insect Physiology. v. 1– , 1963– . New York: Academic Press. Annual. $85.00. ISSN 0065-2806.

Review articles on any aspect of insect physiology.

Annual Review of Entomology. v. 1– , 1956– . Palo Alto, CA: Annual Reviews Inc. Annual. $47.00. ISSN 0066-4170.

Authoritative reviews on various topics in entomology.

Societies

American Entomological Society
    Academy of Natural Sciences of Philadelphia, 1900 Race St., Philadelphia, PA 19103.

Founded: 1859. 350 members. Publications: *Entomological News*, *Memoirs of the American Entomological Society*, and *Transactions of the American Entomological Society*.

Coleopterist's Society
    c/o Ed Zuccaro, P.O. Box 767, Natchez, MS 39121.

Founded: 1969. 600 members. Publications: *Coleopterists Bulletin: An International Study of Beetles*.

Dragonfly Society of America
    469 Crailhope Rd., Center, KY 42214.

Founded: 1989. 200 members. Publications: *Argaia* and *Bulletin of American Odonatology*.

Entomological Society of America
    9301 Annapolis Rd., Lanham, MD 20706-3115.

Founded: 1953. 9,200 members. The largest U.S. entomological society; publishes *American Entomologist, Annals of the Entomological Society of America, Entomological Society of America-Newsletter, ESA Newsletter, Insecticide and Acaricide Tests, Journal of Economic Entomology, Journal of Medical Entomology,* and *Discover Entomology (Careers).* Also publishes many monographic works. Formed by the merger of the American Association of Economic Entomologists and the former Entomological Society of America. Absorbed the American Registry of Professional Entomologists.

International Centre for Insect Physiology and Ecology (ICIPE)
P.O. Box 30772, Nairobi, Kenya.

Founded: 1970. Publications: *Annual Report, Directory of Insect Scientists in Africa, Dudu,* and *ICIPE Profile.*

International Union for the Study of Social Insects
c/o Dr. H. H. W. Velthuis, Laboratory of Comparative Physiology, University of Utrecht, Postbus 80086, NL-3508 TB, Utrecht, The Netherlands.

Founded: 1951. 800 members. Publications: *Insectes Sociaux/Social Insects.*

Lepidopterists' Society
c/o Dr. William D. Winter, Jr., 257 Common Street, Dedham, MA 02026-4020.

Founded: 1947. 1,600 members. "Open to all persons interested in any aspect of Lepidopterology." Publications: *Journal of the Lepidopterists' Society, News of the Lepidopterists' Society,* a biennial membership directory and *Memoirs of the Lepidopterists' Society.* Also makes available *Catalogue/Checklist of the Butterflies of America North of Mexico.*

Young Entomologists' Society
1915 Peggy Pl., Lansing, MI 48910-2553.

Founded: 1965. 750 members. Publications: *Flea Market, Insect World, YES International Entomological Resource Guide, YES Membership Directory, YES Quarterly,* and *Buggy Books: A Guide to Juvenile and Popular Books on Insects and Their Relatives.*

Textbooks

Borror, Donald J., Charles A. Triplehorn, and Norman F. Johnson.
An Introduction to the Study of Insects, 6th ed. Philadelphia: Saunders College Pub., 1989. 875 p. $40.00. ISBN 0030253977.

Includes keys for all families, and some subfamilies, of insects in North America. Also has information on collecting and preserving insects. Includes chapter on non-insect arthropods (crustaceans, arachnids, etc.).

Davies, R. G. *Outlines of Entomology*, 7th ed. New York: Chapman and Hall, 1988. 408 p. $82.00, $41.50 (paper). ISBN 0412266709, 0412266806 (paper).

An enlarged edition of Imms' *Outlines of Entomology*. Covers insect biology, ecology, and classification as well as a section on injurious insects.

Evans, Howard Ensign, et al. *Insect Biology: A Textbook of Entomology*. Reading, MA: Addison-Wesley, 1984. 448 p. $52.75. ISBN 0201119811.

For undergraduates.

Gullan, Penny J. and Peter S. Cranston. *The Insects: An Outline of Entomology*. New York: Chapman and Hall, 1994. 491 p. $43.00 (paper). ISBN 0412493608 (paper).

Metcalf, Robert L. and Robert A. Metcalf. *Destructive and Useful Insects: Their Habits and Control*. New York: McGraw-Hill, 1993. var. pagings. ISBN 0070416923.

Designed for undergraduates and non-scientists. Provides keys, life histories, and further references to over 600 species of North American pests. Arranged by type of problem (e.g., pests of cotton, stored grains). Also useful as a general reference.

Pedigo, Larry P. *Entomology and Pest Management*. New York: Macmillan, 1989. 646 p. ISBN 0023933100.

An introduction to applied entomology for undergraduates and beginning graduate students. Also includes list of common insecticides and insect common names.

Richards, O. W. and R. G. Davies. *Imms' General Textbook of Entomology*, 10th ed. New York: Chapman and Hall, 1978. 2 v. $47.50 (v. 1), $97.50 (v. 2). ISBN 041215210X (v. 1), 0412152304 (v. 2).

Volume 1 covers anatomy and physiology and development and metamorphosis, while volume 2 covers classification and biology. Only insects (not arachnids) are covered.

Ross, Herbert H., Charles H. Ross, and June R. P. Ross. *A Textbook of Entomology*. Reprint edition. Malabar, FL: Krieger, 1991. 666 p. $69.50. ISBN 0894644971. Reprint of 4th edition, 1982.

Treatises and General Works

Ashburner, M. and E. Novitski, eds. *The Genetics and Biology of* Drosophila. New York: Academic Press, 1976– . v. 1A– .

This series contains review articles from a variety of experts covering numerous
aspects of *Drosophila* genetics and biology, including taxonomy, ecology, parasites, population genetics, molecular genetics, and behavior. Currently up to
volume 3E.

Beckage, N.E., S. N. Thompson, and B. A. Federici, eds. *Parasites and
Pathogens of Insects*. San Diego, CA: Academic Press, 1993– . 2
v. ISBN 0120844419 (v. 1), 0120844427 (v. 2).

"The focus of this two-volume set is the interface between insects and their
associated parasites and pathogens, with particular emphasis placed on the basic
biology, biochemistry, and molecular biology of these intimate and intriguing
relationships." Volume 1 deals with parasites, and volume 2 with pathogens.

Berenbaum, May R. *Ninety-nine Gnats, Nits, and Nibblers*. Urbana, IL:
University of Illinois Press, 1989. 254 p. $10.95 (paper). ISBN
025206027X (paper), 0252015711.

Based on short radio spots, these humorous essays on common insects are informative and easily understood. Also by the same author is *Ninety-nine More
Maggots, Mites, and Munchers* (1993), covering more unusual insects and arachnids.

Heinrich, Bernd. *The Hot-Blooded Insects: Strategies and Mechanisms of
Thermoregulation*. Cambridge, MA: Harvard University Press, 1993.
601 p. $75.00. ISBN 0674408381.

Provides a review and critique of the literature on insect thermoregulation,
designed for researchers and advanced students.

Hermann, Henry R., ed. *Social Insects*. New York: Academic Press,
1978–82. 4 v.

Extensive coverage of the sociobiology of social insects, including bees, wasps,
ants, termites, and arachnids.

Hinton, H. E. *Biology of Insect Eggs*. New York: Pergamon Press, 1981.
3 v. $670.00 (set). ISBN 0080215394 (set).

Covers all areas relating to insect eggs. Volume 1 offers general topics (parental
care, oviposition, etc.), while volume 2 covers each insect order. The final
volume provides references; species, author, and subject indexes; and a bibliography of the author's works.

Hogue, Charles L. *Latin American Insects and Entomology.* Berkeley: University of California Press, 1993. 536 p. $85.00. ISBN 0520078497.

This work belongs among the identification materials as well; it consists of an extensive introductory section covering the history and present state of entomology in Central and South America, and a discussion of general entomology followed by accounts of selected families and orders. There is an extensive section listing sources for further information such as Latin American journals, institutions, insect collections, and other resources. The intended audience is anyone from a tourist interested in insects to a professional entomologist. In English.

Insect Learning: Ecological and Evolutionary Perspectives. Daniel R. Papaj and Alcinda C. Lewis, eds. New York: Chapman and Hall, 1993. 412 p. $60.50. ISBN 0412025612.

A multi-authored review of learning in both social and non-social insects. Includes applications for pest control.

Insect Ultrastructure. Robert C. King and Hiromu Akai, eds. New York: Plenum Press, 1982– . v. $110.00 (v. 1), $130.00 (v. 2). ISBN 0306409232 (v. 1), 0306415453 (v. 2).

"The purpose . . . is to provide the interested reader with a series of up-to-date, well-illustrated reviews of selected aspects of Insect Ultrastructure by authorities in the field."

The Insects and Arachnids of Canada. Pt. 1– , 1977– . Ottawa: Agriculture Canada. Irregular. ISSN 0706-7313.

A complete listing of all insects and arachnids of Canada and the adjacent United States. Part 1 consists of a guide to collecting and preserving insects. The remaining volumes feature keys to species. Part 22, *Aphids*, was published in 1993.

Kerkut, Gerald A. and Lawrence I. Gilbert, eds. *Comprehensive Insect Physiology and Pharmacology.* New York: Pergamon Press, 1985. 13 v. $4,000.00. ISBN 0080268501 (set).

This massive set contains 200 articles written by 220 researchers, and refers to 5,000 species of insects. It is designed for both practitioners and students. Topics covered include all areas of insect physiology, behavior, biochemistry, pharmacology, and control. Each article contains extensive references and numerous illustrations. The final volume contains species, author, and subject indexes.

13
Zoology

Zoology is "the science that deals with knowledge of animal life," according to the *McGraw-Hill Dictionary of Scientific and Technical Terms*, 5th ed. Entomology is treated separately in its own chapter (Chapter 12). The other branches of zoology, such as ornithology or nematology, are not separated separately in this chapter; rather, the arrangement is by type of material following the pattern of the other chapters.

Abstracts and Indexes

Current Sources

Aquatic Sciences and Fisheries Abstracts. Part 1: Biological Sciences and Living Resources. v. 8– , 1978– . Bethesda, MD: Cambridge Scientific Abstracts. Monthly. $955. ISSN 0140-5373. Also available on CD-ROM and via the Internet.

Covers all aspects of marine and freshwater organisms, including biology and exploitation. Contains author, subject, taxonomic, and geographic indexes. Part 2: *Ocean Technology, Policy, and Non-Living Resources* and Part 3: *Aquatic Pollution and Environmental Quality* are also available.

Bibliography of Reproduction: A Classified Monthly List of References Compiled from the Research Literature. v. 1– , 1963– . Cambridge, UK: Reproduction Research Information Service. Monthly. $450.00. ISSN 0006-1565.

A classified monthly list of references compiled from research literature.

Current References in Fish Research. v. 1– , 1976– . Eau Claire, WI: V. A. Cvancara. Annual. $18.00. ISSN 0739-540X.

Titles of fish papers published annually. Scans over 130 journals. Includes list of journals scanned, author index, key word index, scientific name, title listing. Indexes most popular fish journals.

Fisheries Review. v. 31– , 1986– . Fort Collins, CO: Fish and Wildlife Service. Quarterly. $17.50. ISSN 1042-6299.

Continues *Sport Fishery Abstracts* and absorbed *Fish Health News*. Abstracts arranged by broad topic. Author, geographic, and systematic indexes.

Helminthological Abstracts. v. 59– , 1990– . Wallingford, UK: CAB International Bureau of Animal Health and CAB International Institute of Parasitology. Monthly. $496.00. ISSN 0957-6789.

Covers journal articles, books, reports, and conferences on all aspects of parasitic helminths. Has author and subject indexes. Continues *Helminthological Abstracts Series A: Animal and Human Helminthology.*

Nematological Abstracts. v. 59– , 1990– . Wallingford, UK: CAB International Bureau of Crop Protection. Quarterly. $196.00. ISSN 0957-6797.

Covers journal articles, books, reports, and conferences on all aspects of nematodes. Has author and subject indexes and occasional review articles. Continues *Helminthological Abstracts Series B: Plant Nematology.*

Protozoological Abstracts. v. 1– , 1977– . Wallingford, UK: CAB International. Monthly. $466.00. ISSN 0309-1287.

Covers journal articles, books, reports, and conferences on all aspects of protozoa and protozoan diseases. Has author and subject indexes and occasional review articles.

Recent Ornithological Literature. 1986– . Washington, DC: American Ornithologists' Union, British Ornithologists' Union, and Royal Australasian Ornithologists' Union. Quarterly.

This publication is a joint supplement to *The Auk*, *The Emu*, and *Ibis*, and attempts to provide comprehensive coverage of the world literature in ornithology, scanning about 900 titles. A "List of Journals Scanned" is published each year in the fourth supplement. The bibliography is divided by broad subject category and includes information on new and renamed journals.

Wildlife Review. v. 1– , 1935– . Washington, DC: Fish and Wildlife Service. Quarterly. $36.25. ISSN 0043-5511.

Classified by subject. Has author and geographic indexes.

Sources for Retrospective Searches

British Museum (Natural History). *List of Serial Publications in the Libraries of the Department of Zoology and Entomology* London: British Museum (Natural History), 1967. 281 p. (Publication 664).

Bronn, H. G. *Dr. H. G. Bronn's Klassen und Ordnungen des Thierreichs, Wissenschaftlich Dargestellt in Wort und Bild. . . .* Leipzig: Winter, 1866–1964.

Bibliography of zoology lists titles in morphology, histology, ontogeny, physiology, ecology, phylogeny, and classification.

Coues, Elliott. "List of Faunal Publications Relating to North
American Ornithology." pp. 567–784 in *Birds of the Colorado
Valley*. New York: Arno Press, 1974. $54.00. ISBN
040505730X.

A reprint of the first installment of Coues' bibliography, covering the period
1600–1878. Originally published in 1878.

Coues, Elliott. *American Ornithological Bibliography*. New York: Arno
Press, 1974. 650 p. $43.00. ISBN 0405057040.

A reprint of the second (faunal publications for South and Central America) and
third (systematic publications) installments of Coues' bibliography, originally
published in *Bulletin of the U.S. Geological and Geographical Survey of the
Territories*. v. 5 (1879). Covers the late 16th century to the late 1870s.

Dean, B. *A Bibliography of Fishes*. Enlarged and edited by Charles R.
Eastman. v. 1–3. New York: American Museum of Natural History,
1916–1923.

Index to the literature of fishes by author, title, pre-Linnean publications,
voyages and expeditions, periodicals, and subject from 1758. Continued by the
American Museum of Natural History under this title until it was incorporated
into *Zoological Record* as part of the Pisces section.

Englemann, W. *Bibliotheca Historieco-Naturalis*. Leipzig: Englemann,
1846. *Bibliotheca Zoologica I*. Leipzig: Englemann, 1861. v.
1–2. *Bibliotheca Zoologica II*. Leipzig: Englemann, 1887–1923.
v. 1–8.

Covers the literature of zoology from 1700 to 1800.

Harvard University, Museum of Comparative Zoology. *Catalogue of the
Library of the Museum of Comparative Zoology, Harvard University*.
v. 1–8. Boston: Hall, 1968. $835.00. ISBN 081610767X (set).
Supplement. v. 1– , 1976– . $140.00. ISBN 0816108110 (sup.
1).

Catalog of the volumes, manuscripts, photographs, and maps in Harvard's
Library of the Museum of Comparative Anatomy. Arrangement is alphabetical;
this is an important source for verification of particular authors' works in
monographs as well as serials.

Index-Catalogue of Medical and Veterinary Zoology. v. 1–18, 1932–52.
Supplement. v. 1– , 1953– . Washington, DC: Government
Printing Office.

Indispensable source for parasitologists.

International Catalogue of Scientific Literature: N, Zoology. 1st–14th annual issues. London: Harrison, 1902–1916.

See annotation (p. 81) in Chapter 4, "Abstracts and Indexes."

Irwin, Raymond. *British Bird Books: An Index to British Ornithology, A.D. 1481 to A.D. 1948.* London: Grafton & Co., 1951. 398 p.

Divided into several sections, including subject lists, regional lists, systematic list, and indexes to authors, subjects, species, and places. Also includes a supplement bringing coverage to 1950.

Reuss, J. D. *Repertorium Commentationum a Societatibus Litterariis Editarum. . . . T. 1, Historia Naturalis, Generalis et Zoologia.* Gottingae: Dieterich, 1801. (Reprint: New York: Burt Franklin, 1961).

An index to society publications to 1800.

Ronsil, R., ed. *Bibliographie Ornithologique Française.* v. 1–2. Travaux publiés en langue française et en latin en France et dans les Colonies françaises de 1473 à 1944. (Encyclopedie ornithologique, VII and IX). Paris: Lechevalier, 1948–49. v. 1: Bibliography of ornithological literature published in France or one of the French colonies between 1473–1944. v. 2: Indexes of abbreviations of periodical publications; ornithological, geographic terms; history of ornithology.

Ruch, T. C. *Bibliographia Primatologica.* A classified bibliography of primates other than man. Springfield, IL: Thomas, 1941. (Yale Medical Library, Historical Library, publication no. 4).

Excellent bibliography, covers to 1939. Over 4,000 entries covering anatomy, embryology, quantitative morphology, physiology, pharmacology, psychobiology, phylogeny, etc.

Sherborn, Charles Davies. *Index Animalium: Sive, Index Nominum quae ab A.D. MDCCLVIII Generibus et Speciebus Animalium Imposita Sunt, Societabus Eruditorum Adiuvantibus.* Bath, England: Chivers, 1969. 1195 p. Reprint.

First published 1902–1933, this index provides a list of animals named from 1758 to 1800, with citations to the first description.

Strong, R. M. *A Bibliography of Birds.* With special reference to anatomy, behavior, biochemistry, embryology, pathology, physiology, genetics, ecology, aviculture, economic ornithology, poultry culture, evolution, and related subjects. Chicago: Field Museum of Natural History, 1939–59. (Publications of the Field Museum of Natural

History, v. 25). Pt. 1: Author catalog, A–J. Pt. 2: Author catalog, K–Z. Pt. 3: Subject index. Pt 4: Finding index.

Comprehensive coverage of world literature until 1926, although there are some references later than that year.

Walker, Ernest P. *Mammals of the World*, 1st ed. Baltimore: Johns Hopkins Press, 1964. 3 v.

Volume 3 of this set consists of "A Classified Bibliography of Literature Regarding Mammals." It contains about 70,000 citations arranged by systematic, geographic, or subject categories. Subsequent editions of this work dropped the bibliography volume.

Wood, Casey A. *An Introduction to the Literature of Vertebrate Zoology*. New York, Arno Press, 1974. (Natural sciences in America). 643 p. $95.00. ISBN 0405057725. Reprint. Originally published by Oxford University Press in 1931.

Invaluable retrospective bibliographic reference. Some chapter headings: Beginnings of zoological records; Medieval writers on zoology; Travelogues of explorers; Oriental literature; Literature of zoogeography, etc.

Zoological Record. v. 1– , 1864– . London: The Zoological Society of London.

The world's most comprehensive index to systematic zoology. See annotation (p. 81) in Chapter 4, "Abstracts and Indexes."

Important Bibliographic Journals

Bibliographia Zoologica. v. 1–43, 1896–1934. Zurich Sumptibus Concilii Bibliographici, 1896–1934.

Zentralblatt für Zoologie, Allgemaine und Experimentelle Biologie. Bd. 1–6. Leipzig: Teubner, 1912–18. 6. v.

Zoologischer Anzeiger. v. 1–18, 1878– . Leipzig: Geest and Portig. Records of current literature which were issued separately 1896– , as *Bibliographica Zoologica*, supplement to the journal.

Zoologischer Bericht. Im Auftrage der Deutschen Zoologischen Gesellschaft. . . . Bd. 1–55. Jena: Fischer, 1922–43/44. 55 v.

Zoologischer Jahresbericht. Hrsg. von der zoologischen Station zu Neapel. 1879–1913. Leipzig, 1880–1924. v. 1–35.

Atlases, Checklists, and Identification Manuals

There are a large number of identification tools, including atlases, manuals, faunas, and field guides. This section annotates only a select portion of them. See p. 41 in Chapter 3, "General Sources," for descriptions of the major field guide series, for instance.

1994 IUCN Red List of Threatened Animals. Ed. by Brian Groombridge. Cambridge, England: IUCN, 1993. 286 p. (paper). ISBN 2831707945.

Lists over 5,000 threatened taxa, including scientific and English common names, IUCN category and CITES listing, and range. Also has index to order, family, generic, and common names.

Animal Identification: A Reference Guide. Ed. by R. W. Sims. London: British Museum (Natural History); New York: Wiley, 1980. 3 vol. v. 1: Marine and brackish water animals. v. 2: Land and freshwater animals (not insects). v. 3: Insects.

Principle references useful in identifying animals are chosen for the non-specialist scientist, student, or research worker. Arrangement for each volume is systematic by animal groups. Within each group there is further subdivision for general, systematic, and geographic sections. A very useful set.

Arnold, Edwin Nicholas and J. A. Burton. *A Field Guide to the Reptiles and Amphibians of Britain and Europe.* London: Collins, 1978. 272 p. ISBN 0002193183.

Bartholomew, J. G., W. Eagle Clarke, and Percy H. Grimshaw. *Atlas of Zoogeography: A Series of Maps Illustrating the Distribution of over Seven Hundred Families, Genera, and Species of Existing Animals.* Edinburgh: Published at the Royal Geographical Society by J. Bartholomew, 1911. (Physical atlas, v. 5).

Behler, John. *The Audubon Society Field Guide to North American Reptiles and Amphibians.* New York: Knopf, 1979. (Audubon Society Field Guide series). 719 p. $18.00. ISBN 0394508246.

Boschung, Herbert T., David K. Caldwell, Melba C. Caldwell, Daniel W. Gotshall, and James D. Williams. *The Audubon Society Field Guide to North American Fishes, Whales, and Dolphins.* New York: Knopf, 1983. (Audubon Society Field Guide series). 848 p. $18.00. ISBN 0394534050.

Bull, John L. and John Farrand, Jr. *The National Audubon Society Field Guide to North American Birds, Eastern Region*, Rev. ed. New York: Knopf, 1994. 797 p. $19.00. ISBN 0679428526.

Burt, William Henry. *A Field Guide to the Mammals: Field Marks of all North American Species Found North of Mexico*, 3d ed. Boston: Houghton Mifflin, 1976. (Peterson Field Guide series, 5). 289 p. $17.95, $13.95 (paper). ISBN 0395240824, 0395240840 (paper).

Catalogue of American Amphibians and Reptiles. New York: American Museum of Natural History for the Society for the Study of Amphibians and Reptiles, 1971– . $20.00. Published by the American Society of Ichthyologists and Herpetologists, 1963–1970.

"Series of individual accounts, each prepared by a separate author." Each entry lists previous references, content, definition, description, illustrations, distribution, fossil record, pertinent literature, remarks, etymology, comments, literature cited.

Check-list of Birds of the World, 2nd ed. Ernst Mayr and G. William Cottrell, eds. Cambridge, MA: Museum of Comparative Zoology, 1979–1986. 16 v.

"Revision of the work of James L. Peters." This massive work includes the Latin name, first description, and distribution of the birds of the world.

Check-List of North American Birds: Species of Birds of North America from the Arctic through Panama, Including the West Indies and Hawaiian Islands, 6th ed. Prepared by the Committee on Classification and Nomenclature of the American Ornithologists' Union. Lawrence, KS: American Ornithologists' Union, 1983. 877 p. ISBN 094361032X.

The checklist is updated irregularly (the previous one in 1957). Each species is listed with scientific and English name, original citation, habitat, distribution (summer and winter), and notes.

Conant, Roger and Joseph T. Collins. *A Field Guide to Reptiles and Amphibians: Eastern and Central North America*, 3rd ed. Boston: Houghton Mifflin, 1991. (Peterson Field Guide series, 12). 450 p. $24.95, $16.95 (paper). ISBN 0395370221, 039558396 (paper).

Corbet, G. B. and J. E. Hill. *A World List of Mammalian Species*, 3rd ed. New York: Oxford University Press, 1991. 243 p. $72.00. ISBN 0198540175.

This list provides Latin name, English name, and geographical range, with selected line drawings.

Ernst, Carl H. *Venomous Reptiles of North America*. Washington, DC: Smithsonian Institution Press, 1992. 236 p. $35.00. ISBN 1560961148.

Primarily natural history of the 20 or so venomous reptiles of North America; also includes information on venom and bites. Has key for identifying venomous reptiles.

Eschmeyer, William N. and Earl S. Herald. *A Field Guide to Pacific Coast Fishes of North America: From the Gulf of Alaska to Baja, California*. Boston: Houghton Mifflin, 1983. (Peterson Field Guide series, 28). 336 p. $19.95, $15.95 (paper). ISBN 0395268737, 0395331889 (paper).

Goddard, Jerome. *Physician's Guide to Arthropods of Medical Importance*. Boca Raton, FL: CRC Press, 1993. 332 p. $110.00. ISBN 084935160X.

Hall, E. Raymond. *Mammals of North America*. New York: John Wiley & Sons, 1981. 2 v. $112.50 (set). ISBN 0471054437 (v. 1), 0471054445 (v. 2), 0471055956 (set).

Summarizes taxonomic studies on recent mammals of North America. Includes keys, skulls, and distribution maps for most species and line drawings for some.

Halstead, Bruce W., in collaboration with Paul S. Auerbach. *Dangerous Aquatic Animals of the World: A Color Atlas: With Prevention, First Aid, and Emergency Treatment Procedures*. Princeton, NJ: Darwin Press, 1992. 264 p. ISBN 0878500456.

An oversized, heavily illustrated guide to dangerous marine animals, from molluscs to mammals. Divided by type of injury caused by the organism, such as bites, stings, poisoning, and electric shock. Includes photographs (many in color) of both the dangerous animal and the injuries it causes, extensive information on the mechanism of injury (teeth, stinging tentacles, etc.), and a chapter on treatment.

Haltenorth, Theodor and Helmut Diller. *The Collins Field Guide to the Mammals of Africa Including Madagascar*. Lexington, MA: S. Greene Press, 1988. 400 p. $32.50. ISBN 0828906998.

Handbook of the Birds of the World. Josep del Hoyo, et al., eds. Barcelona: Lynx Edicions; New York: ICBP, 1992– . v. $165.00 (v. 1). ISBN 8487334105 (v. 1). v. 1: Ostrich to Ducks.

This set is intended to consist of 10 volumes. The first volume begins with an overview of avian biology, with the remainder consisting of species accounts for 27 families. There are numerous illustrations.

Hickman, Graham C. National Mammal Guides: A Review of References to Recent Faunas. *Mammal Review* 11(2): 53-85. 1981.

This article cites over 400 reference guides to mammalian faunas of over 150 countries. The items referred to include field guides, checklists, atlases, and faunas.

Howard, Richard and Alick Moore. *A Complete Checklist of the Birds of the World*, 2nd ed. San Diego: Academic Press, 1991. 622 p. $49.50. ISBN 0123569109.

This checklist follows the taxonomic sequence set forth in Peters' classic *Check-List of Birds of the World* (second ed., p. 388). Fossil and extinct birds are not included. The entries include the Latin name, the English name and geographic distribution. The authors also provide an index of Latin names (arranged by species, not generic, name) and an index of English species names.

Iverson, John B. *A Checklist with Distribution Maps of the Turtles of the World*. Richmond, IN: Paust Printing, 1986. 282 p. $20.00. ISBN 0961743107.

Includes distribution maps as well as the usual taxonomic and nomenclatural information.

Jones, J. Knox and Richard W. Manning. *Illustrated Key to Skulls of North American Land Mammals*. Lubbock: Texas Tech University Press, 1992. 75 p. ISBN 0896722899 (paper).

Dichotomous key to mammalian skulls, to genus. Includes glossary.

Macdonald, David W. and Priscilla Barrett. *Mammals of Britain and Europe*. New York: HarperCollins, 1993. (Collins Field Guide). 312 p. ISBN 0002197790.

Mammal Species of the World: A Taxonomic and Geographic Reference, 2nd ed. Don E. Wilson and DeeAnn M. Reeder, eds. Washington, DC: Smithsonian Institution Press, 1993. 1206 p. $75.00. ISBN 1560982179.

A checklist, providing original citation, type locality, distribution, status, synonyms, and comments. Also available on the Internet at the Harvard Biodiversity Gopher.

Mammalian Species. v. 1– , 1969– . American Society of Mammalogists. Irregular. $15.00. ISBN 0076-3519.

Each number covers one species, including species name, context and content, diagnosis, general characters, distribution, ecology, etymology, function, reproduction, behavior, genetics, literature cited. Authoritative source.

Meinkoth, Norman August. *The Audubon Society Field Guide to North American Seashore Creatures.* New York: Knopf, 1981. (Audubon Society Field Guide series). 799 p. $18.00. ISBN 0394519930.

Monroe, Burt L. and Charles G. Sibley. *A World Checklist of Birds.* New Haven: Yale University Press, 1993. 393 p. ISBN 0300055471.

This checklist is based on the taxonomic system of Sibley and Monroe (see p. 393). As well as the usual Latin and English names and distribution of each species, the checklist provides a column for the dedicated birder to check off which of the 9,702 species he or she has seen.

Morris, Percy A. *A Field Guide to Shells of the Atlantic and Gulf Coasts and the West Indies*, 3rd ed. Boston: Houghton Mifflin, 1973. (Peterson Field Guide series, 3). 330 p. $17.95 (paper). ISBN 0395168090, 0395171709 (paper).

Morris, Percy A. *A Field Guide to Pacific Coast Shells, Including Shells of Hawaii and the Gulf of California.* 2d ed., rev. and enl. Boston: Houghton Mifflin, 1974. (Peterson Field Guide series, 6). 297 p. $17.95. ISBN 0395080290, 0395183227 (paper).

Nowak, Ronald M. *Walker's Mammals of the World*, 5th ed. Baltimore: Johns Hopkins University Press, 1991. 2 v. $89.95. ISBN 080183970X.

Revision of Ernest P. Walker's work. Includes photographs of representatives of each genera, taxonomic and biological information. Major resource.

Palmer, Ralph S. *Handbook of North American Birds.* New Haven: Yale University Press, 1962– . v. $45.00 (v. 5, pt. 2). ISBN 0300040601 (v. 5, pt. 2).

Currently up to v. 5, covering the accipiters.

Page, Lawrence M. and Brooks M. Burr. *A Field Guide to Freshwater Fishes: North America North of Mexico.* Boston: Houghton Mifflin, 1991. (Peterson Field Guide series, 42). 432 p. $24.95, $16.95 (paper). ISBN 0395353076, 0395539331 (paper).

Pennak, Robert W. *Fresh-Water Invertebrates of the United States: Protozoa to Mollusca*, 3rd ed. New York: John Wiley and Sons, 1989. 628 p. $69.95. ISBN 0471631183.

Covers the identification and biology of free-living freshwater invertebrates. Some groups are keyed to species, others to genera. Aquatic insects and parasitic invertebrates are excluded.

Peterson, Roger Tory. *A Field Guide to the Birds: A Completely New Guide to All the Birds of Eastern and Central North America*, 4th ed., completely rev. and enl. Boston: Houghton Mifflin, 1980. (Peterson Field Guide series, 1). 384 p. $17.99, $15.95 (paper), $15.95 (flexibook). ISBN 0395266211, 039526619X (paper), 0395361648 (flexibook).

Peterson, Roger Tory. *A Field Guide to Western Birds: A Completely New Guide to Field Marks of All Species Found in North America West of the 100th Meridian and North of Mexico*, 3rd ed., completely rev. and enl. Boston: Houghton Mifflin, 1990. (Peterson Field Guide series, 2). 432 p. $22.95, $16.95 (paper). ISBN 0395517494, 03951424X (paper).

Peterson, Roger Tory and Edward L. Chalif. *A Field Guide to Mexican Birds: Field Marks of All Species Found in Mexico, Guatemala, Belize (British Honduras), El Salvador*. Boston: Houghton Mifflin, 1973. 298 p. (Peterson Field Guide series, 20). $21.95, $16.95 (paper). ISBN 0395171296, 0395483549 (paper).

Peterson, Roger Tory, Guy Mountfort, and P. A. D. Hollom. *A Field Guide to Birds of Britain and Europe*, 5th ed. Boston: Houghton Mifflin, 1993. (Peterson Field Guide series, 8). 261 p. $24.95, $19.95 (paper). ISBN 0395669227 (paper), 0395669316.

Robbins, Chandler S., Bertel Bruun, and Herbert S. Zim. *Birds of North America: A Guide to Field Identification*, expanded, rev. ed. New York: Golden Press, 1983. (Golden Field Guide series). 360 p. $11.50 (paper). ISBN 030737002X, 0307336565 (paper).

Robins, C. Richard and G. Carleton Ray. *A Field Guide to Atlantic Coast Fishes of North America*. Boston: Houghton Mifflin, 1986. (Peterson Field Guide series, 32). 354 p. $20.95, $14.95 (paper). ISBN 0395318521, 0395391989 (paper).

Rodger, Robin W. A. *Fish Facts: An Illustrated Guide to Commercial Species*. New York: Van Nostrand Reinhold, 1990. 175 p. $42.95. ISBN 0442005431.

"This compact, basic reference offers pertinent data on the characteristics and biology of eighty of the most commercially important North American fish species." Information includes physical description, landings and values, commercial uses, and life cycle. The most important shellfish are also included.

Schmidt, Gerald D. *CRC Handbook of Tapeworm Identification*. Boca Raton, FL: CRC Press, 1986. 675 p. $265.00. ISBN 084933280X.

Keys to the identification of nearly 4,000 species of tapeworms with numerous illustrations of adult tapeworm morphology. The only worldwide key to tapeworm identification in print at the time of writing.

Sibley, Charles G. and Jon E. Ahlquist. *Phylogeny and Classification of Birds: A Study in Molecular Evolution*. New Haven: Yale University Press, 1990. 976 p. $100.00. ISBN 0300040857.

"The goals of our study were the reconstruction of the phylogeny of the groups of living birds and the derivation of a new classification scheme based on the phylogeny." For each family of birds, the authors describe past efforts at classifying the species, the present the evidence from their own DNA hybridization tests.

Sibley, Charles G. and Burt L. Monroe. *Distribution and Taxonomy of Birds of the World*. New Haven: Yale University Press, 1990. 1111 p. $125.00. ISBN 0300049692.

The goals of this work include delineating the present distribution of the birds of the world, list the species in a classification based on DNA (see Sibley and Ahlquist, above), and offer a gazetteer and maps for locating regions mentioned in the atlas. The indexes are by scientific and English name. A Supplement was published in 1993 with updates and corrections.

Smith, Hobart Muir and Edmund D. Brodie, Jr. *Reptiles of North America: A Guide to Field Identification*. New York: Golden Press, 1982. (Golden Field Guide series). 240 p. $11.95. ISBN 0307136663 (paper).

Threatened Birds of the Americas: The ICBP/IUCN Red Data Book, 3rd ed. Washington, DC: Smithsonian Institute Press in cooperation with International Council for Bird Preservation, 1992. 1150 p. $75.00. ISBN 150982675.

Provides in-depth analysis of problems affecting threatened birds of the Americas, and proposes measures to be taken. Extensive discussion of distribution, population, and ecology of each species.

Ubelaker, John E. *Stedman's ASP Parasite Names*. Baltimore: Williams and Wilkins, 1993. 149 p. $28.00. ISBN 0683079581.

Contains over 3,000 names of pathogenic parasites, including valid name, author citation, and year published.

Udvardy, Miklos D. F. *National Audubon Society Field Guide to North American Birds, Western Region*, rev. ed. New York: Knopf, 1994. 822 p. $19.00. ISBN 0679428518.

Walters, Michael. *The Complete Birds of the World*. North Pomfret, VT: David & Charles, 1980. 340 p. ISBN 0715376667.

This work "attempts to list every bird species known to exist or to have existed in recent (i.e., post-Pleistocene) times." More than just a checklist, it provides in very telegraphic form not only information on Latin name, authority, and English name, but also distribution, habitat, food preferences, nest site, clutch size, incubation share of the sexes, and fledging. The indexes, unfortunately, are only to the family name.

Whitaker, John O. *The Audubon Society Field Guide to North American Mammals*. New York: Knopf, 1980. (Audubon Society Field Guide series). 745 p. $18.00. ISBN 0394507622.

World Biodiversity Database. New york: Springer Verlag, 1994– .

This CD-ROM series (available in both Macintosh and IBM-compatible formats) offers a variety of resources dealing with biodiversity and taxonomy for both specialists and the general public. Among the discs currently available (all 1994):

> **Birds of Europe**. $99.00. ISBN 3540141898 (Macintosh), 3540141901 (IBM).
>
> Field guide, illustrations, quiz for testing the users knowledge of European birds, and glossary.
>
> **Five Kingdoms: Life on Earth**. $79.00. ISBN 3540145001 (Macintosh), 345014501X (IBM).
>
> A multimedia version of Margulis et al's *The Illustrated Five Kingdoms* (see p. 31). Has drawings, photographs, video segments, hyperlinked text and habitat key, plus extensive glossary.
>
> **North Australian Sea Cucumbers**. $79.00. ISBN 3540145109 (Macintosh), 3450141995 (IBM).
>
> Key and glossary for over 90 species of sea cucumbers.
>
> **Protoctist Glossary**. $79.00. ISBN 3540145109 (Macintosh), 3450141995 (IBM).

Includes two glossaries, one with 1,900 scientific terms and the other with 1,300 taxonomic names. Includes quiz.

Dictionaries and Encyclopedias

Allaby, Michael. *Concise Oxford Dictionary of Zoology*. New York: Oxford University Press, 1992. 508 p. $10.95 (paper). ISBN 0192860933 (paper). Hardcover edition 1991, ISBN 0198661622.

Based on the 1985 *Oxford Dictionary of Natural History*. For students and the general public.

Banister, Keith and Andrew Campbell. *The Encyclopedia of Aquatic Life*. New York: Facts on File, 1985. 349 p. $45.00. ISBN 0816012571.

Covers fishes, aquatic invertebrates, and aquatic mammals (whales, dolphins, and manatees only). The editors attempted to "distill their knowledge to give the reader a flavor of the essence of" aquatic animals.

Birds of North America: Life Histories for the 21st Century. Alan F. Poole, Peter Stettenheim, and Frank B. Gill, eds. Washington, DC: American Ornithologists' Union; Philadelphia: Academy of Natural Sciences, 1992– . v. $175 per volume.

Life history accounts of over 700 species of birds. Issued as separate self-contained profiles; expected to consist of 18 volumes of 40 accounts each. Each account includes extensive bibliography. Updates A. C. Bent's *Life Histories of North American Birds*.

Cambridge Encyclopedia of Ornithology. Edited by Michael Brooke and Tim Birkhead. New York: Cambridge University Press, 1991. 362 p. $49.50. ISBN 0521362059.

Contains extensive information on all aspects of ornithology, including anatomy, behavior, distribution, ecology, and the relationship of humans with birds. Also includes a survey of bird orders. Very well illustrated.

Choate, Ernest. *The Dictionary of American Bird Names*. Boston: Gambit, 1973. 261 p. $9.95. ISBN 0876450656.

Discusses common and scientific names for birds and their origins. Includes obscure and archaic terms, and has a biographical appendix listing American ornithologists and people who have had birds named after themselves.

A Dictionary of Birds. Bruce Campbell and Elizabeth Lack, eds. Vermillion, SD: Buteo Books, 1985. 670 p. $75.00. ISBN 0931130123.

Contains definitions and extended essays on ornithological terms, including systematics, behavior, and biology.

Encyclopedia of Mammals. David Macdonald, ed. New York: Facts on
    File, 1989. 960 p. $65.00. ISBN 0871968711.
Well-illustrated encyclopedia of mammals.

Grzimek's Animal Life Encyclopedia. Bernhard Grzimek, editor-in-chief.
    New York, Van Nostrand Reinhold Co., 1972–75. 13 v. Transla-
    tion of *Tierleben.* Also published in paperback, 1984.
The grand old classic, covering all animals. While rather old, there is still nothing to compare with *Grzimek's* for coverage of the animal world.

Grzimek's Encyclopedia of Mammals. New York: McGraw-Hill, 1990. 5
    v. $500.00. ISBN 0079095089 (set). Translation of *Grzimek's
    Enzyklopadie Saugetiere.*
A worthy successor to *Grzimek's Animal Life Encyclopedia*, above. Has beautiful color photographs and extensive discussion of mammalian species.

Hayssen, Virginia, Ari van Tienhoven, and Ans van Tienhoven. *Asdell's
    Patterns of Mammalian Reproduction: A Compendium of Species-
    Specific Data.* Ithaca, NY: Comstock Publishing Associates, 1993.
    1023 p. $75.00. ISBN 0801417538. Revised edition of the 2nd
    ed. of *Patterns of Mammalian Reproduction*, S. A. Asdell, 1964.
This compendium lists information on mammalian reproduction, with the exception of well-known domesticated and laboratory species such as cows and rats. It is organized taxonomically, with a summary of the reproduction of each family and order. This summary is followed by a table listing species-specific data, with citations to over 12,000 original articles. The authors also provide a list of the Mammalian Species Accounts published by the American Society of Mammalogists and a list of core journals. There are indexes to the common names of the most well-known mammals and to scientific names.

Jacobs, George J. *Dictionary of Vertebrate Zoology, English-Russian/
    Russian-English: Emphasizing Anatomy, Amphibians, and Reptiles.*
    Washington, DC: Smithsonian Institution Press, 1978. 48 p. $8.50
    (paper). ISBN 0874745519 (paper).
As the subtitle suggests, this dictionary is intended primarily for the herpetologist. Terms are not defined, but rather are translated between Russian and English. Russian common names for species are translated to the Latin name, not the English common name, for greater clarity.

Jobling, James A. *A Dictionary of Scientific Bird Names*. New York: Oxford University Press, 1991. 272 p. $29.95. ISBN 0198546343.

Gives derivation and meaning of all currently accepted scientific bird names.

Leahy, Christopher W. *The Birdwatcher's Companion: An Encyclopedic Handbook of North American Birdlife*. New York: Hill and Wang, 1982. 917 p. ISBN 089030365.

A dictionary of birding and ornithological terms, with accounts of the families of birds of North America, birding terms, birding "hot spots," and biographies, as well as a phylogenetic list, list of vagrant bird species, and a birder's calendar. One-stop shopping for the birder, but also useful for libraries.

Lodge, Walter. *Birds: Alternative Names: A World Checklist*. New York: Sterling Pub., 1991. 208 p. $19.95. ISBN 0713722673.

Multimedia Encyclopedia of Mammalian Biology. New York: McGraw-Hill, 1994. 1 CD-ROM disc. $995.00 (single user), $1250.00 (networked). ISBN 0077077008 (single user), 0077077016 (up to 16 users).

Contains all 5 volumes of the *Grzimek's Encyclopedia of Mammals* (p. 395), as well as additional articles, a hypertext glossary, and videos of mammals in their natural habitats. For IBM-compatible computers.

Nelson, Joseph S. *Fishes of the World*, 3rd ed. New York: John Wiley, 1994. 600 p. $72.50. ISBN 0471547131.

The purpose of this book is "to present a modern introductory systematic treatment of all major fish groups." To the family level. Includes number of species or genera per family, range, and description.

Preston-Mafham, Rod and Ken Preston-Mafham. *Primates of the World*. New York: Facts on File, 1992. 192 p. $24.95. ISBN 0816027455.

Provides an overview of primates. For students and the general public. One of the . . . *of the World* series published by Facts on File. Other titles include *Grasshoppers and Mantids of the World*, *Frogs and Toads of the World*, and *Whales of the World*.

Rojo, Alfonso L. *Dictionary of Evolutionary Fish Osteology*. Boca Raton: CRC Press, 1991. 273 p. $79.95. ISBN 0849342147.

In English, French, German, Latin, Russian, and Spanish.

Stachowitsch, Michael. *The Invertebrates: An Illustrated Glossary*. New
    York: Wiley-Liss, 1992. 676 p. $198.00. ISBN 0471832944,
    0471561924 (paper).

Contains over 10,000 entries and 1,100 figures in two sections, the first defining
anatomical features in a taxonomical arrangement, and the second with adjec-
tives describing modifications of the features. The German equivalent of each
term is also given.

Wheeler, Alwyne C. *The World Encyclopedia of Fishes*. London: Mac-
    donald and Queen Anne Press, 1985. 368 p. ISBN 0356107159.

Numerous color plates in addition to line drawings. The dictionary section lists
fish species by scientific name and includes information on distribution, size,
habits, etc. Common names are cross-referenced to the scientific name. Nelson
(p. 397) lists only families and subfamilies.

Guides to Internet Resources

There are a number of resources available through the Internet for zoologists.
One general guide to zoological sources is "The Electronic Zoo," prepared by
Ken Boschert and available at the University of Michigan library school Gopher.
Many of the resources listed in the ecology chapter are also applicable here. For
instance, the Harvard Biodiversity Gopher also includes a number of taxonomic
resources, such as electronic versions of Eschmeyer's *Catalog of the Genera of
Recent Fishes* (p. 403) and others.

Guides to the Literature

Bell, George H. and Diane B. Rhoades. *A Guide to the Zoological Litera-
    ture*. Englewood, CO: Libraries Unlimited, 1994. 504 p. ISBN
    1563080826.

An annotated guide to zoology, with general reference sources and sources for
major groups of animals, such as fishes or arthropods. A detailed, up-to-date
source, especially useful for the extensive lists of field guides, checklists, and
other identification sources.

Miller, Melanie Ann. *Birds: a Guide to the Literature*. New York: Garland
    Pub., 1986. (Garland reference library of the humanities). 887 p.
    $86.00. ISBN 0824087100.

An annotated list of books written about birds, both popular and technical.

Handbooks

Bibby, Colin J., Neil D. Burgess and David A. Hill. *Bird Census Techniques*. San Diego: Academic Press, 1992. 257 p. $42.00. ISBN 0120958309.

"This guide is offered largely to people lacking the time or facilities to read and assess the extensive and often conflicting bird census literature." The authors provide general information about the design of bird censuses and describe many methods for performing them, including line transects, and marking.

Cailliet, Gregor M., Milton S. Love, and Alfred W. Ebeling. *Fishes: A Field and Laboratory Manual on their Structure, Identification, and Natural History*. Belmont, CA: Wadsworth, 1986. 194 p. ISBN 0534055567 (paper).

Includes laboratory exercises and methods for studying the various aspects of ichthyology such as dissecting and identifying fish in the lab, and capture methods in the field.

CRC Handbook of Census Methods for Terrestrial Vertebrates. Edited by David E. Davis. Boca Raton, FL: CRC Press, 1982. 397 p. $209.95. ISBN 0849329701.

Brief descriptions of census methods for about 130 species of vertebrates are included, plus a few non-terrestrial species such as coastal whales and manatees where the census methods are useful for other species. There is a separate chapter on calculations and statistics.

Dunning, John B. *CRC Handbook of Avian Body Masses*. Boca Raton, FL: CRC Press, 1993. 371 p. $65.00. ISBN 0849342589.

Includes all known estimates of bird body mass found in the literature. The information provided for each entry includes Latin name, sex and number of individuals sampled, mean, standard deviation, and range of the estimate, collecting season, and citation to the original publication. A section lists body masses and composition for migrant birds in the eastern United States with their wet mass, dry mass, fat-free mass, and ash-free mass. The index is by genus.

Fowler, Jim and Louis Cohen. *Statistics for Ornithologists*. Tring England: British Trust for Ornithology, 1986. (BTO Guide no. 22). 176 p.

An introduction to statistics for amateur and beginning professional ornithologists. All the usual topics are covered, the main difference between this and any other statistical guide being the examples, which are all taken from the world of ornithology.

Fry, Frederic L. *Captive Invertebrates: A Guide to Their Biology and Husbandry*. Malabar, FL: Krieger, 1992. 135 p. $29.50. ISBN 0894645552.

Covers most invertebrate species used for scientific studies or kept as pets. Each group is described, then housing, water, nutrition, reproduction, and medical disorders are discussed. The author also includes an appendix listing commercial sources for invertebrates.

Goddard, Jerome. *Physician's Guide to Arthropods of Medical Importance*. Boca Raton, FL: CRC Press, 1993. 332 p. $110.00. ISBN 084935160X.

Handbook of Protoctista: The Structure, Cultivation, Habitats, and Life Histories of the Eukaryotic Microorganisms and Their Descendants . . . . Lynn Margulis, et al.

For complete annotation (p. 188), see Chapter 8, "Microbiology and Immunology."

Kaufman, M. H. *The Atlas of Mouse Development*. San Diego, CA: Academic Press, 1992. 512 p. $172.00. ISBN 0124020356.

Includes over 180 plates and numerous photographs and electron micrographs covering the development of the mouse from pre-implantation to term.

Measuring and Monitoring Biological Diversity: Standard Methods for Amphibians. Washington: Smithsonian Institution Press, 1994. (Biological diversity handbook series). 364 p. $49.00, $17.95 (paper). ISBN 1560982705, 1560982845 (paper).

As well as the standard methods for monitoring amphibian diversity, this handbook includes information on handling live amphibians, recording frog calls, preparing specimens, and vendors selling equipment for amphibian studies.

Nagorsen, D. W. and R. L. Peterson. *Mammal Collector's Manual: A Guide for Collecting, Documenting, and Preparing Mammal Specimens for Scientific Research*. Toronto: Royal Ontario Museum, 1980. 79 p. ISBN 0888542550.

Has methods of collecting mammals in the field, documentation, preservation, and shipping specimens.

Protocols in Protozoology. J. J. Lee and A. T. Soldo, eds. Lawrence, KS: Society of Protozoology, 1992. ISBN 0935868577 (v. 1).

This is the first volume of a projected series of protocols. The present volume provides protocols for isolation, culture, nutrition, and bioassay; ecological methods; fixation, staining, and microscopic techniques; molecular biological and genetic methods; and educational experiments. About 140 protocols are described.

Skalski, John R. and Douglas S. Robson. *Techniques for Wildlife Investigations: Design and Analysis of Capture Data*. San Diego: Academic Press, 1992. 237 p. $59.95. ISBN 0126476756.

Provides information on designing and carrying out mark-recapture studies, with special emphasis on proper statistical methods and randomization.

Survey Designs and Statistical Methods for the Estimation of Avian Population Trends. John R. Sauer and Sam Droege, eds. Washington, DC: U.S. Dept. of the Interior, Fish and Wildlife Service, 1990. (Biological report, 90(1)). 166 p.

The proceedings of a workshop on the analysis of avian population trends, held in 1988. The first part discusses various survey methods such as the Audubon Christmas bird counts, while the second discusses methods of statistical analysis of trends using these surveys. A final section offers examples using trends in scissor-tailed flycatcher populations.

Svendsen, Per and Jann Hau, eds. *Handbook of Laboratory Animal Science*. Boca Raton: CRC Press, 1994. 2 v. $95.00 (v. 1), $75.00 (v. 2). ISBN 084934378X (v. 1), 0849343909 (v. 2).

Volume 1 covers the selection and handling of animals in biomedical research (including experimental methods, alternatives to animal experiments, ethics, and legislation), while volume 2 covers animal models.

Histories

Birney, Elmer C. and Jerry R. Choate. *Seventy-five Years of Mammalogy (1919–1994)*. American Society of Mammalogists, 1994. (American Society of Mammalogists, Publication no. 11). 433 p. ISBN 0935868739.

Bridson, Gavin. *The History of Natural History: An Annotated Bibliography*. New York: Garland Pub., 1994. 740 p. $115.00. ISBN 0824023196.

Includes extensive bibliography on the history of zoology, in addition to material on the history of general biology and botany.

Farber, Paul Lawrence. *The Emergence of Ornithology as a Scientific Discipline: 1760–1850*. Boston: D. Reidel, 1982. (Studies in

the history of modern science, 12). 191 p. $82.00. ISBN
902771410X.

Concentrates on the development of ornithology as a science out of natural history.

International History of Mammalogy. Keir B. Sterling, general editor.
Bel Air, MD: One World Press, 1987. v. $25.00 (v. 1, paper).
ISBN 0910485003 (v. 1), 0910485011 v. 1, paper). v. 1: Eastern
Europe and Fennoscandia.

"The plan of this work is simple. It is to publish a series of chapters by authorities from every country of the world on the development of mammalogy in each of these nations during the modern scientific period."

Stresemann, Erwin. *Ornithology From Aristotle to the Present.*
Cambridge: Harvard University Press, 1975. 432 p. ISBN
0674644859.

Translation of *Entwicklung der Ornthologie* (1951). Covers the development of ornithology up to the 1950s, with emphasis on European activities. Ernst Mayr provided an appendix, "Materials for a History of American Ornithology," for this translation.

Nomenclature

Bulletin of Zoological Nomenclature. v. 1– , 1943– . London:
International Trust for Zoological Nomenclature. Quarterly.
$165.00. ISSN 0007-5167.

The official organ of the International Commission on Zoological Nomenclature. Contains notices prescribed by the International Congress of Zoology, announcements, opinions, new and revived cases, comments, and proposals.

Collins, Joseph T. *Standard Common and Current Scientific Names for
North American Amphibians and Reptiles*, 3rd ed. Society for the
Study of Amphibians and Reptiles, 1990. (Herpetological circular,
19). 41 p.

Includes publication date and describer, and approved scientific and common names as well as appendices covering Hawaiian and alien species.

*Common and Scientific Names of Aquatic Invertebrates from the United
States and Canada: Cnidaria and Ctenophora.* Stephen D. Cairns,
et al. Bethesda, MD: American Fisheries Society, 1991. (American
Fisheries Society special publication, 22). 75 p. $25.00. ISBN
0913235628, 0913235490 (paper).

Common and Scientific Names of Aquatic Invertebrates from the United States and Canada: Decapod Crustaceans. Austin B. Williams, et al. Bethesda, MD: American Fisheries Society, 1989. (American Fisheries Society special publication, 17). 77 p. $17.00. ISBN 0913235628, 0913235490 (paper).

Common and Scientific Names of Aquatic Invertebrates from the United States and Canada: Mollusks. Donna D. Turgeon, et al. Bethesda, MD: American Fisheries Society, 1988. (American Fisheries Society special publication, 16). 277 p. ISBN 0913235474, 0913235482 (paper).

Common and Scientific Names of Fishes from the United States and Canada, 5th ed. C. Richard Robins, et al. Bethesda, MD: American Fisheries Society, 1991. (American Fisheries Society, 20). 183 p. $32.00, $24.00 (paper). ISBN 0913235709, 0913235695 (paper).

Previous editions published as *A List of Common and Scientific Names of Fishes* . . . . Contains scientific and common names, occurrence, references for first description, and appendices on exotics and hybrid fishes.

Crocodilian, Tuatara, and Turtle Species of the World: A Taxonomic and Geographic Reference. F. Wayne King and Russell L. Burke, eds. Washington, DC: Association of Systematics Collections, 1989. 216 p. $29.00. ISBN 0942924150.

Contains species, taxonomic problems, type and location, distribution, comments, status, and common name(s).

Eschmeyer, William N. *Catalog of the Genera of Recent Fishes*. San Francisco: California Academy of Sciences, 1990. 697 p. $55.00. ISBN 09040228238.

Lists genera in alphabetical order with name, author, date, type specimen, remarks, and status. Separate sections list names by class and literature cited. Also available on the Internet at the Harvard University Biodiversity Gopher.

Frost, Darrel R. *Amphibian Species of the World: A Taxonomic and Geo-graphical Reference*. Lawrence, KS: Allen Press, 1985. 732 p. ISBN 0942924118.

In taxonomic order, with scientific name, authority, year of publication, type species, specimen, and location, distribution, and status.

International Commission on Zoological Nomenclature. *Code International de Nomenclature Zoologique. International Code of Zoological Nomenclature*, 3rd ed. London: International Trust for Zoological

Nomenclature, Berkeley: University of California Press, 1985. 338 p. ISBN 0520055462.

"Adopted by the XX General Assembly of the International Union of Biological Sciences."

Luca, Florenza de. *Taxonomic Authority List*. Rome: Food and Agriculture Organization of the United Nations, 1988. 465 p. ISBN 9251027722.

Created by the Aquatic Sciences and Fisheries Information System for use in preparing the *FAO Yearbook of Fishery Statistics* and the *Aquatic Sciences and Fisheries Abstracts*. Contains over 10,000 terms, in systematic and alphabetical lists, with the date and author of the term.

Nomenclator Zoologicus. v. 1–7, 1939–75. London: Zoological Society of London.

"Names of genera and subgenera in zoology from the 10th edition of Linneaus, 1758 to the end of 1965, with a bibliographical reference for the original description of each."

Official Lists and Indexes of Names and Works in Zoology. R. V. Melville and J. D. D. Smith, eds. London: International Trust for Zoological Nomenclature, 1987. 366 p. ISBN 0853010048.

Contains scientific names and titles of works which have been voted on by the International Commission on Zoological Nomenclature and published in the *Bulletin of Zoological Nomenclature* up to the end of 1985. Contains 4 main sections: Family-group names, generic names, specific names, and titles of works in addition to a systematic index and bibliographic references. Contains material from earlier editions of *Official Index of Rejected and Invalid Family Group Names in Zoology*, *Official Index of Rejected and Invalid Works in Zoological Nomenclature*, *Official List of Family Group Names in Zoology*, and *Official List of Works Approved as Available for Zoological Nomenclature*.

Periodicals

Acta Zoologica: An International Journal of Zoomorphology. v. 1– , 1920– . Pergamon. Quarterly. $290.00. ISSN 0001-7272.

Published by the Royal Swedish Academy of Sciences and the Royal Danish Academy of Sciences and Letters. Publishes "original research papers and reviews in the fields of animal organization, development, structure and function, from the cellular level to phylogenetic and ecological levels."

Alauda. v. 1– , 1929– . Brounoy, France: Museum National d'Histoire
    Naturelle. Quarterly. $53.00. ISSN 0002-4619.

The journal of the Scociété d'Etudes Ornithologiques de France. Text in French
with English summaries. Publishes original papers, most dealing with European
birds.

American Birds. v. 1– , 19– . New York: National Audubon Society.
    5/yr. $32.00. ISSN 0004-7686.

A birder's magazine, with features and columns discussing conservation, identification, and sightings of American birds.

American Malacological Bulletin. v. 1– , 1931– . American Malaco-
    logical Union. Biannual. $32.00. ISSN 0740-2783.

"The official journal publication of the American Malacological Union." Publishes "original, unpublished research, important short reports, and detailed
reviews dealing with molluscs."

American Zoologist. v. 1– , 1961– . Lawrence, KS: Allen Press.
    Bimonthly. $400.00. ISSN 0003-1569.

A publication of the American Society of Zoologists. "Uninvited manuscripts
are not considered. The *American Zoologist* publishes original, invited papers,
derived principally from Symposia sponsored by the American Society of Zoologists and its affiliates. The articles are mainly of a review or synthetic
nature."

Annales Zoologici Fennici. v. 1– , 1934– . Helsinki: Finnish Zoological
    Publishing Board. Quarterly. $30.00. ISSN 0003-455X.

In English, French, and German. Publishes original research on ecology, ecological physiology, faunistics, and systematics of animals in North European
countries, though research from other boreal regions is also considered.

Ardea. v. 1– , 1912– . Groningen, the Netherlands: Netherlands
    Ornithologists' Union. Semi-annual. $45.00. ISSN 0373-2266.

The journal of the Netherlands Ornithologists' Union. Publishes "manuscripts
reporting significant new findings in ornithology. Emphasis is laid on studies
covering ecology, ethology, taxonomy and zoogeography." In English and Dutch.

The Auk: A Quarterly Journal of Ornithology. v. 1– , 1884– .
    Washington, DC: American Ornithologists' Union. Quarterly.
    $60.00. ISSN 0004-8038.

The journal of the American Ornithologists' Union. *"The Auk* welcomes original reports on the biology of birds. Appropriate topics include the documentation, analysis, and interpretation of laboratory and field studies, theoretical or methodological developments, and reviews of information or ideas."

Bird Behaviour. v. 1– , 1977. Wallingford, CT: Bird Behaviour Press. Irregular. $30.00. ISSN 0156-1383.

"An international and interdisciplinary journal that publishes original research on descriptive and quantitative analyses of behaviour, behavioural ecology, experimental psychology and behavioural physiology of birds."

Bird Study. v. 1– , 1954– . Osney Mead, UK: Blackwell Scientific Publications. 3/yr. $99.00. ISSN 0006-3657.

The official journal of the British Trust for Ornithology. "Original papers on all aspects of field ornithology, especially distribution, status, censusing, migration, habitat and breeding ecology."

Bulletin of the Natural History Museum: Zoology Series. v. 1– , 1949– . London: British Museum of Natural History. Semiannual. ISSN 0007-1498.

Formerly the *Bulletin of the British Museum of Natural History: Zoological Series.* "Papers in the *Bulletin* are primarily the results of research carried out on the unique and ever-growing collections of the Museum, both by the scientific staff and by specialists from elsewhere who make use of the Museum's resources."

Bulletin of the Museum of Comparative Zoology. v. 1– , 1863– . Cambridge, MA: Museum of Comparative Zoology, Harvard University. Irregular. Price varies. ISSN 0027-4100.

Published 3–4 times per year. Each issue consists of a single lengthy article, usually on a taxonomic topic.

Bulletin of Zoological Nomenclature. v. 1– , 1943– . London: International Commission on Zoological Nomenclature. Quarterly. $135.00. ISSN 0007-5167.

"The Offical Periodical of the International Commission on Zoological Nomenclature. . . . At present the *Bulletin* comprises mainly applications concerning names of particular animals or groups of animals, resulting comments and the Commission's eventual rulings (Opinions). Proposed amendments to the Code are also published for discussion."

Canadian Journal of Zoology/Revue Canadienne de Zoologie. v. 1– ,
    1929– . Ottawa: National Research Council of Canada. Monthly.
    $401.00. ISSN 0008-4301.

Publishes "in English and French, articles, notes, reviews, and comments in the general fields of behaviour, biochemistry, physiology, developmental biology, ecology, genetics, morphology, ultrastructure, parasitology, pathology, systematics, and evolution."

Comparative Biochemistry and Physiology: Part A, Physiology. v. 1– ,
    1961– . Tarrytown, NY: Pergamon. Monthly. $2,020.00. ISSN
    0300-9629.

"This section deals specifically with cellular physiology, respiration, circulation, neurophysiology, sensory physiology, ecological physiology, organ-specifiic functions, etc., at all levels of organismal organization."

The Condor: A Journal of Avian Biology. v. 1– , 1899– . Lawrence,
    KS: Allen Press. Quarterly. $60.00. ISSN 0010-5422.

The journal of the Cooper Ornithological Society. "Devoted to the biology of wild species of birds."

Copeia. v. 1– , 1913– . Lawrence, KS: Allen Press. Quarterly.
    $90.00. ISSN 0045-8511.

Published by the American Society of Ichthyologists and Herpetologists. Publishes "results of original research performed by members in which fish, amphibians, or reptiles are utilized as study organisms."

Emu. v. 1– , 1901– . Moonee Ponds, Australia: Royal Australasian
    Ornithologists Union. Quarterly. ISSN 0158-4197.

The journal of the Royal Australasian Ornithologists Union. "The *Emu* prints original papers and short communications on the ornithology of the Australasian region."

Environmental Biology of Fishes. v. 1– , 19– . Dordrecht: Kluwer
    Academic. Monthly. $247.50. ISSN 0378-1909.

"An international journal which publishes original studies on ecology, life-history, epigenetics, behavior, physiology, morphology, systematics and evolution of marine and freshwater fishes and fishlike organisms."

Folia Primatologica: International Journal of Primatology. v. 1– ,
    1963– . Basel, Switzerland: S. Karger. 8/yr. $374.00. ISSN
    0015-5713.

In English, French, and German. Publishes in all areas of primatology, including full-length articles and short reports.

Herpetologica. v. 1– , 1936. Lafayette, LA: Herpetologists League.
    Quarterly. $70.00. ISSN 0018-0831.

Annual supplements published as *Herpetological Monographs*. Publishes "original papers dealing largely or exclusively with the biology of amphibians and reptiles; theoretical and primarily quantitative manuscripts are particularly encouraged. Contributors need not be members of the Herpetologists' Union."

Herpetological Review. v. 1– , 1967– . Hays, KS: Society for the
    Study of Amphibians and Reptiles. Quarterly. $60.00. ISSN 0018-
    084X.

Published by the Society for the Study of Amphibians and Reptiles. "A peer-reviewed quarterly that publishes, in English, articles and notes of a semi-technical or non-technical nature." An organ for news and opinion.

Ibis. v. 1– , 1859– . Osney Mead, UK: Blackwell Scientific. Quarterly.
    $180.00. ISSN 0019-0019.

"The international journal of the British Ornithologists' Union. . . . *Ibis* publishes original papers and comments in the English language, covering the field of ornithology."

International Journal of Primatology. v. 1– , 1980– . New York:
    Plenum. Quarterly. $275.00. ISSN 0164-0291.

The official journal of the International Primatological Society. "A multidisciplinary journal devoted to basic primatology, i.e., to studies in which the primates are featured as such," including laboratory and field work.

Invertebrate Reproduction and Development. v. 1– , 19– . Rehovot,
    Israel: Balaban. Bimonthly. $350.00. ISSN 0792-4259.

Continues *International Journal of Invertebrate Reproduction and Development*. "The journal publishes original papers and reviews with a wide approach to the sexual, reproductive and developmental (embryonic and postembryonic) biology of the Invertebrata."

Japanese Journal of Ornithology. v. 1– , 1915. Tokyo: Ornithological
    Society of Japan. Quarterly. ISSN 0913-400X.

Continues *Tori*. In Japanese and English. Publishes articles dealing with Asian birds.

Journal für Ornithologie. v. 1– , 1853– . Garmisch-Partenkirchen,

Germany: Deutsche Ornithologen-Gesellschaft. Quarterly. ISSN 0021-8375.

Articles published in German and English, with English summaries. Publishes articles on all aspects of ornithology, especially European birds.

Journal of Avian Biology. v. 25– , 1994– . Copenhagen: Munksgaard. Quarterly. $105.00. ISSN 0908-8857.

Published by the Scandinavian Ornithologists' Union, continues *Ornis Scandinavica*. "The journal presents original work on all aspects of ornithology."

Journal of Comparative Physiology A: Sensory, Neural, and Behavioral Physiology. v. 1– , 1924– . Berlin: Springer International. Monthly. $1,831.00. ISSN 0340-7594.

Topics covered include "physiological basis of behavior, sensory physiology, neural physiology, orientation, communication, locomotion, hormonal control of behavior."

Journal of Comparative Physiology B: Biochemical, Systemic, and Environmental Physiology. v. 1– , 1924– . Berlin: Springer International. 8/yr. $908.00. ISSN 0174-1578.

Covers "comparative aspects of metabolism and enzymology, metabolic regulation, respiration and gas transport, physiology of body fluids, circulation, temperature relations, endocrine regulation, muscular physiology."

Journal of Crustacean Biology. v. 1– , 1981. Lawrence, KS: Allen Press. Quarterly. $90.00. ISSN 0278-0372.

Published by the Crustacean Society. "Provides international exchange of information among persons interested in any aspect of crustacean studies."

Journal of Eukaryotic Microbiology. v. 1– , 1954– . Lawrence, KS: Society of Protozoologists. Bimonthly. $116.00. ISSN 1066-5234.

Formerly the *Journal of Protozoology*. Publishes "original research on protists, including lower algae and fungi, and covering all aspects of such organisms."

Journal of Experimental Zoology. v. 1– , 1904– . New York: Wiley-Liss. 18/yr. $2,050.00. ISSN 0022-104X.

"Published under the auspices of the American Society of Zoologists and the Division of Comparative Physiology and Biochemistry." Publishes "the results of original research of an experimental or analytical nature in zoology, including investigations of all levels of biological organization from the molecular to the organismal."

Journal of Field Ornithology. v. 51– , 1980– . Lawrence, KS: Association of Field Ornithologists. Quarterly. $45.00. ISSN 0748-4690.

Continues *Bird-Banding*. The journal "welcomes original articles that emphasize the description or experimental study of birds in their natural habitats." The abstracts are in English and Spanish.

Journal of Fish Biology. v. 1– , 1969– . London: Academic Press. Monthly. $623.00. ISSN 0022-1112.

"Covers all aspects of fish and fisheries research, both freshwater and marine."

Journal of Herpetology. v. 1– , 1968– . Athens, OH: Society for the Study of Amphibians and Reptiles. Quarterly. $60.00. ISSN 0022-1511.

"Publishes, in English, Articles and Notes on research relating to the study of amphibians and reptiles."

Journal of Ichthyology. v. 1– , 1968– . Silver Spring, MD: Scripta Technica. 9/yr. $996.00. ISSN 0032-9452.

Translation of the Russian *Voprosy Ikhtiologii* (ISSN 0042-8752). Covers all aspects of ichthyology and fisheries, both marine and freshwater.

Journal of Mammalogy. v. 1– , 1919– . Provo, UT: Department of Zoology, Brigham Young University. Quarterly. $35.00. ISSN 0022-2372.

The journal of the American Society of Mammalogists. Publishes technical papers on all aspects of mammalogy as well as news and commentary. Each issue of volume 75 (1994) includes short essays about the history of the society in commemoration of its 75th anniversary.

Journal of Molluscan Studies. v. 1– , 1893– . Oxford, UK: Oxford University Press. Quarterly. $165.00. ISSN 0260-1230.

"The Journal publishes papers dealing with all aspects of the study of molluscs. Shorter Research Notes are welcomed."

Journal of Nematology. v. 1– , 1969– . Hanover, PA: Sheridan Press. Quarterly. $70.00. ISSN 0022-300X.

The official journal of the Society of Nematologists. "Original papers on basic, applied, descriptive, or experimental nematology are considered for publication . . . . At least one author must be a member of the Society of Nematologists."

Journal of the Helminthological Society of Washington. v. 1– , 19– Lawrence, KS: Allen Press. Semiannual. $20.00. ISSN 1049-233X.

Published by the Helminthological Society of Washington. "A semiannual journal of research devoted to Helminthology and all branches of Parasitology."

Journal of Zoology. v. 1– , 1830– . Oxford, UK: Oxford University Press. Monthly. $795.00. ISSN 0952-8369.

Published by the Zoological Society of London. "The *Journal of Zoology* incorporates the Proceedings of the Zoological Society of London (founded in 1830) and the Transactions of the Zoological Society of London (founded in 1833). It contains original papers within the general field of experimental and descriptive zoology, and notices of the business transacted at the Scientific Meetings of the Society."

Malacological Review. v. 1– , 1968– . Ann Arbor: Museum of Zoology, University of Michigan. Annual. $33.00. ISSN 0076-3004.

The *Review* is affiliated with a number of malacological periodicals and societies and publishes "the results of original work, of either descriptive or experimental nature, devoted primarily ar exclusively to the study of mollusks, and will publish reviews of major research areas or malacological subjects."

Mammalia: Journal de Morphologie, Biologie, Systématique des Mammifères. v. 1– , 1936– . Paris: Museum National d'Histoire Naturelle. Quarterly. $143.00. ISSN 0025-1461.

"*Mammalia* publishes in French and English original notes and research papers dealing with all aspects of mammalian systematics, biology and ecology."

Netherlands Journal of Zoology. v. 1– , 1947– . Leiden: E. J. Brill. Quarterly. $87.00. ISSN 0028-2960.

"Publishes original research papers on topics from the fields of ecological and functional morphology, behavioural and physiological ecology, environmental physiology and biosystematics."

Physiological Zoology. v. 1– , 1928– . Chicago: University of Chicago Press. Bimonthly. $200.00. ISSN 0031-935X.

"Sponsored by the Division of Comparative Physiology and Biochemistry, American Society of Zoologists. . . . An outlet for research in environmental and adaptational physiology and biochemistry. Explores comparative physiology as wel as physiological ecology."

Primates. v. 1– , 1957– . Inuyama, Japan: Japan Monkey Centre. Quarterly. $239.00. ISSN 0032-8332.

"An international journal of primatology, whose general object is to provide facilities for the elucidation of the entire aspect of primates in common with man."

Systematic Biology. v. 1– , 1952– . Washington, DC: National Museum of Natural History. Quarterly. $40.00. ISSN 1063-5157.

Formerly *Systematic Zoology*. The journal of the Society of Systematic Biologists. Publishes "original contributions of theory, principles, and methods of systematics as well as evolution, morphology, zoogeography, paleontology, genetics, and classification."

The Wilson Bulletin: A Quarterly Magazine of Ornithology. v. 1– , 1889– . Lawrence, KS: Allen Press. Quarterly. $40.00. ISSN 0043-5643.

The journal of the Wilson Ornithological Society. "Publishes significant research and review articles in the field of ornithology."

Zeitschrift für Zoologische Systematik und Evolutionsforschung. v. 1– , 1963– . Hamburg: Paul Parey. Quarterly. ISSN 0044-3808.

"The journal publishes original articles on systematic zoology and evolutionary biology but not taxonomic descriptions of new species." Articles in German, English, or French.

Zoologica Scripta. v. 1– , 1971– . Tarrytown, NY: Pergamon Press. Quarterly. ISSN 0300-3256.

"An international journal published for the Norwegian Academy of Science and Letters and the Royal Swedish Academy of Sciences. . . . *Zoologica Scripta* publishes papers based on original research, and review articles, from the fields of taxonomy, systematics, phylogeny, and biogeography."

Zoological Journal of the Linnean Society. v. 1– , 18– . London: Academic Press. Monthly. $643.00. ISSN 0024-4082.

Published for the Linnean Society of London. "The *Zoological Journal* publishes original papers on systematic, comparative, functional and ecological zoology."

Zoological Science: An International Journal. v. 1– , 19– . Tokyo: Zoological Society of Japan. Bimonthly. $311.00. ISSN 0289-0003.

Formed by the merger of *Annotationes Zoologicae Japonenses* and *Zoological Magazine*. The official journal of the Zoological Society of Japan. "Devoted to the publication of original articles, reviews and communications in the broad field of Zoology."

Zoologischer Anzeiger. v. 1– , 1878– . Jena, Germany: Gustav Fischer Verlag. Irregular. $278.00. ISSN 0044-5231.

"Devoted to comparative zoology, with special emphasis on morphology, systematics, and biogeography. Thus it will act as a complement to *ZACS (Zoology: Analysis of Complex Systems)*" (see below).

Zoology: Analysis of Complex Systems. v. 98– , 1994– . Jena, Germany: Gustav Fischer Verlag. Irregular. $278.00. ISSN 0944-2006.

Formed by the union of three sections of *Zoologischer Jahrbücher: Abteilung für Allgemeine Zoologie und Physiologie der Tiere, Abteilung für Anatomie und Ontogenie der Tiere*, and *Abteilung für Systematik, Ökologie und Geographie der Tiere*. "*Zoology* provides a publishing forum for articles that emphasize integrative and comparative aspects of animal biology."

Zoomorphology: An International Journal of Comparative and Functional Morphology. v. 1– , 1924– . Heidelberg: Springer International. Quarterly. $573.00. ISSN 0720-213X.

"The journal will accept original papers based on morphological investigation of invertebrates and vertebrates at the macroscopic, microscopic and ultrastructural levels, including embryological studies."

Zoophysiology. v. 11– , 1979– . New York: Springer-Verlag. Irregular. Price varies. ISSN 0720-1842.

This monographic series "presents treatises on selected subjects of timely interest to animal physiologists." Recent volumes include *Vertebrate Flight, Diverse Divers*, and *Comparative Physiology of the Vertebrate Kidney*. Continues *Zoophysiology and Ecology*.

Reviews of the Literature

Advances in Parasitology. v. 1– , 19– . San Diego, CA: Academic Press. Annual. $95.00. ISSN 0065-308X.

Review articles covering all taxonomic groups of parasites.

Current Mammalogy. v. 1– , 1987– . New York: Plenum. Irregular. $110.00.

Includes review articles on a broad range of subjects dealing with mammalogy. Volume 2 was published in 1990.

Current Ornithology. v. 1– , 1983– . New York: Plenum. $85.00.
    ISSN 0742-390X.
Review articles covering all areas of ornithology.

Current Topics in Developmental Biology. v. 1– , 19– . San Diego,
    CA: Academic Press. Irregular. $79.00. ISSN 0070-2153.
"Provides the reader with a survey of major issues at the forefront of contemporary developmental biology."

Mammal Review. v. 1– , 19– . Oxford, UK: Blackwell Scientific
    Publications. Quarterly. $142.00. ISSN 0305-1838.
A publication of the Mammal Society. "It is not intended for the publication of the results of original research but rather as a secondary journal, carrying reviews of, and reports on, any aspects of mammalogy. It also serves to report the proceedings of the Society's Symposia, which are again of a review nature."

Reviews in Fish Biology and Fisheries. v. 1– , 19– . London: Chapman
    and Hall. Quarterly. $220.00. ISSN 0960-3166.
"Reviews are accepted in any field of fish biology where the emphasis is on the whole organism. Subjects covered include: physiology, evolutionary biology, taxonomy, zoogeography, behaviour, ecology, and exploitation."

Societies

American Association for Zoological Nomenclature
    c/o National Museum of Natural History, MRC 163, Smithsonian
    Institution, Washington, DC 20560.
Founded: 1983.   250 members.   For those interested in systematics. Publications: *AAZN Newsletter*.

American Cetacean Society
    P.O. Box 2639, San Pedro, CA 90731.
Founded: 1967. 2600 members. Laypeople and professionals interested in whales, dolphins, and porpoises. Publications: *WhaleNews* and *Whalewatcher*.

American Fisheries Society
    5410 Grosvenor Lane, Suite 110, Bethesda, MD 20814.

Founded: 1870. 8,400 members. Publications: *AFS Membership Directory and Handbook, Fisheries: Bulletin of the American Fisheries Society, Journal of Aquatic Animal Health, Transactions of the American Fisheries Society.*

American Malacological Union
    P.O. Box 30, North Myrtle Beach, SC 29582.
Founded: 1931. 750 members. For professionals and hobbyists interested in mollusks. Publications: *American Malacological Bulletin* and supplements and *Newsletter.*

American Ornithologists' Union
    National Museum of Natural History, Smithsonian Institution,
    Washington DC, 20560.
Founded: 1883. 5,000 members. Publications: *The Auk, Check-List of North American Birds, Membership List* (triennial), *Ornithological Monographs, Ornithological Newsletter.*

American Society of Ichthyologists and Herpetologists
    Department of Zoology, Business Office, Southern Illinois University,
    Carbondale, IL 62901-6501.
Founded: 1913. 3,600 members. Publications: *Copeia.*

American Society of Mammalogists
    Dept. of Zoology, 501 Widtsoe Bldg., Provo, UT 84602.
Founded: 1919. 3,600 members. Publications: *Journal of Mammalogy* and *Mammalian Species.*

American Society of Zoologists
    401 N. Michigan Ave., Chicago, IL 60651-4267.
Founded: 1890. 3,700 members. For professional zoologists. Publications: *American Zoologist.*

Association of Field Ornithologists
    c/o Elissa Landre, Broadmoor Wildlife Sanctuary, Massachusetts
    Audubon Society, 280 Eliot St., South Natick, MA 01760.
Founded: 1924. 1,400 members. Formerly Northeastern Bird-Banding Association. Publications: *Journal of Field Ornithology.*

Cooper Ornithological Society
    c/o Martin L. Morton, Occidental College, Biology Department, Los
    Angeles CA 90041.

Founded: 1893. 2200 members. Publications: *Condor*, *The Flock* (directory), *Studies in Avian Biology.*

The Crustacean Society
    c/o Denton Belk, 840 Mulberry Ave, San Antonio, TX 78212-3194.

Founded: 1980. 850 members. Publications: *The Ecdysiast* and *The Journal of Crustacean Biology.*

Herpetologists' League
    Texas Natural Heritage Program, Texas Park and Wildlife Department, 3000 Interstate Highway S., Suite 100, Austin, TX 78704.

Founded: 1936. 2,000 members. Publications: *Herpetologica* and *Herpetological Monographs.*

Society for the Study of Amphibians and Reptiles
    c/o Dr. Douglas Taylor, Miami University, Department of Zoology, Oxford, OH 45056.

Founded: 1958. 2,700 members. Formerly Ohio Herpetological Society. Publications: *Herpetological Circulars*, *Herpetological Review*, *Journal of Herpetology.*

Society of Nematologists
    c/o R. N. Huehel, Ph.D. USDA, NARS, Nematology Lab, Building 011A BARC-W, Beltsville, MD 20705.

Founded: 1961. 860 members. Publications: *Annals of Applied Nematology*, *Annual Meeting Presentations Abstracts*, *Directory*, *Journal of Nematology*, *Nematology Newsletter.*

Society of Protozoologists
    Box 199, Baruch College, 17 Lexington Ave., New York, NY 10010.

Founded: 1947. 1125 members. Publications: *The Abstracts*, *Directory of Members*, *Journal of Protozoology*, *Newsletter*, *Illustrated Guide to the Protozoa.*

Society of Systematic Biologists
    c/o National Museum of Natural History, NHB163, Washington, DC 20560.

Founded: 1948. 1.550 members. Formerly Society of Systematic Zoology. Publications: *Systematic Biology.*

Wilson Ornithological Society
    University of Michigan, Museum of Zoology, Ann Arbor, MI 48109-1079.

Founded: 1888. 1592 members. Publications: *Wilson Bulletin* and *Membership Directory* (triennial).

Texts/General Works

Alexander, R. McNeill. *Animals*. New York: Cambridge University Press, 1990. 509 p. $37.50 (paper). ISBN 0521343917, 052134865X (paper).

A survey of the animal kingdom; revised from Alexander's earlier works, *The Invertebrates* and *The Chordates*.

Barnes, Robert D. *Invertebrate Zoology*. 5th ed. Philadelphia: Saunders College Pub., 1987. 893 p. $58.75. ISBN 003008914X.

Undergraduates. Systematic treatment.

Barnes, R. S. K., P. Calow, and P. J. W. Olive. *The Invertebrates: A New Synthesis*, 2nd ed. Boston: Blackwell Scientific Publications, 1993. 488 p. $99.95, $42.95 (paper). ISBN 0632031255, 0632031271 (paper).

Undergraduates. Includes both systematic and functional treatment of invertebrates.

Bone, Q., N. B. Marshall, and J. H. S. Blaxter. *Biology of Fishes*, 2nd ed. Chapman and Hall, 1994. 288 p. $39.95. ISBN 075140022X.

Ecology and Classification of North American Freshwater Invertebrates. James H. Thorp and Alan P. Covich, eds. San Diego: Academic Press, 1991. 911 p. $59.95. ISBN 0126906459.

Provides taxonomic keys to the generic level; aquatic insects are largely excluded. Chapters on the remaining groups cover general biology, taxonomy, and ecology.

Gilbert, Scott F. *Developmental Biology*. 4th ed., rev. Oxford, UK: Sinauer, 1994. 900 p. $57.95. ISBN 0878932496.

Gill, Frank B. *Ornithology*, 2nd ed. New York: W. H. Freeman, 1994. 763 p. $55.95. ISBN 0716724154.

Hairston, Nelson G. *Vertebrate Zoology: An Experimental Field Approach*. New York: Cambridge University Press, 1994. 280 p. $34.95 (paper). ISBN 0521417031, 0521427126 (paper).

Hickman, Cleveland P., Larry S. Roberts, and Allan Larson. *Integrated*

Principles of Zoology, 9th ed. St. Louis: Mosby, 1993. $55.95. ISBN 080166375X.

For undergraduates.

Mayr, Ernst and Peter D. Ashlock. *Principles of Systematic Zoology*, 2nd ed. New York: McGraw-Hill, 1991. 475 p. ISBN 0070411441.

Meglitsch, Paul A. and Frederick R. Schram. *Invertebrate Zoology*, 3rd ed. New York: Oxford University Press, 1991. 623 p. $47.95. ISBN 0195049004.

Undergraduate text; taxonomic treatment of invertebrates.

Moyle, Peter B. and Joseph J. Cech. *Fishes: An Introduction to Ichthyology*, 2nd ed. Englewood Cliffs, NJ: Prentice Hall, 1988. 559 p. $58.00. ISBN 0133092113.

Physiology of Fishes. Edited by David H. Evans. Boca Raton: CRC Press, 1993. 592 p. $95.00. ISBN 0849380421.

For advanced students and practitioners.

Proctor, Noble S. and Patrick J. Lynch. *Manual of Ornithology: Avian Structure and Function*. New Haven: Yale University Press, 1993. 340 p. $40.00. ISBN 0300057466.

Designed as a lab manual for course in ornithology. In addition to discussion of avian anatomy, includes chapter on field techniques, including identifying, photographing, and banding birds, as well as how to prepare study skins.

Simpson, George Gaylord. *Principles of Animal Taxonomy*. New York: Columbia University Press, 1990. 247 p. $19.50. ISBN 023109650X (paper). Reprint of 1961 ed.

Vaughan, Terry A. *Mammalogy*, 3rd ed. Philadelphia: Saunders, 1986. 576 p. $53.25. ISBN 00350584744.

Welty, Joel Carl and Luis Baptista. *The Life of Birds*, 4th ed. New York: Saunders, 1988. 581 p. $53.25. ISBN 0030689236.

Wilson, E. O., D. Siegel-Causey, D. R. Brooks, and V. A. Funk. *The Compleat Cladist: A Primer of Phylogenetics Procedures*. Lawrence, KS: Museum of Natural History, 1991. (Special Publication, 19). 158 p. $14.95. ISBN 0893380350.

A workbook providing a guide to basic phylogenetic techniques.

Vermeij, Geerat J. *A Natural History of Shells*. Princeton: Princeton University Press, 1993. 207 p. $29.95. ISBN 069108596X.

A handy guide to the ecology, evolution, and natural history of the Mollusca, suitable for the general public and students alike.

Zug, George R. *Herpetology: An Introductory Biology of Amphibians and Reptiles*. New York: Academic Press, 1993. 527 p. $50.00. ISBN 0127826203.

For advanced undergraduates.

Treatises

Avian Biology. Donald S. Farner and James R. King, eds. New York: Academic Press, 1971– . v. $94.00 (v. 9). ISBN 0122494091 (v. 9).

This treatise is intended to update A. J. Marshall's *Biology and Comparative Physiology of Birds*. It will cover all aspects of bird biology. Currently up to volume 9, published in 1993.

Berthold, Peter. *Bird Migration: A General Survey*. New York: Oxford University Press, 1993. 239 p. ISBN 0198546920, 0198546912 (paper).

A general overview of current research on the topic of bird migration, intended for students and laypeople, as well as researchers.

Biochemistry and Physiology of Protozoa, 2nd ed. Michael Levandowsky and S. H. Hutner, eds. New York: Academic Press, 1979–1981. 4 v. ISBN 0124446043 (v. 4).

The first edition, published in 1951-64, was edited by A. Lwoff. This treatise attempts to cover the biochemistry and physiology of all the protozoa.

Biology of Crustacea. Dorothy E. Bliss, ed. New York: Academic Press, 1982-1985. 10 v. $960.00 (set). ISBN 0121064107 (v. 10).

Covers all areas of crustacean biology, including systematics, embryology, neurobiology, anatomy and physiology, ecology, and economic aspects.

Biology of the Reptilia. Carl Gans, ed. New York: Academic Press, 1969–1985. 15 v.

This treatise is "addressed to and designed for specialists who need a summary on the status of our knowledge in a particular system or process in the reptilia.

It is hence intended for people who have at least some minimal background in the areas covered."

Chemical Zoology. Marcel Florkin and Bradley T. Scheer, eds. New York: Academic Press, 1967–1979. 11 v. $1,460.00 (set). ISBN 0122610415 (v. 11).

Biochemical aspects of zoology. Each volume covers various topics in one group of animals, such as chemical aspects of hibernation or pigments of protozoa.

Fish Physiology. W. S. Hoar and D. J. Randall, eds. New York: Academic Press, 1969– . v. $99.00 (v. 13). ISBN 0123504376 (v. 13).

Each volume covers different aspects of fish physiology, such as the gills, locomotion, and reproduction. There are author, systematic, and subject indexes in each volume. Currently up to volume 13, *Molecular Endocrinology of Fish.*

Grasse, P. P. *Traite de Zoologie: Anatomie, Systématique, Biologie.* Paris: Masson, 1948– . 17 v. Price varies.

Encyclopedic treatise on zoology arranged systematically by taxonomic divisions. In French. Survey of the biology of animals; generous bibliographies.

Handbook of Marine Mammals. Sam H. Ridgway and Richard J. Harrison, eds. New York: Academic Press, 1981– . v. $89.00 (v. 5). ISBN 0125885059 (v. 5). v. 1: Walrus, sea lions, fur seals, and sea otter; v. 2: Seals; v. 3: Sirenians and baleen whales; v. 4: River dolphins and the toothed whales; v. 5: First book of dolphins.

Covers the biology and life history of marine mammals, as well as distribution and identification.

Handbuch der Zoologie: Eine Naturgeschichte der Stamme des Tierreiches, 2nd ed. Gegrundet von Willy Kukenthal. Berlin: W. de Gruyter, 1968– . Price varies.

Another multi-volume treatise covering the animal kingdom, similar to Grasse's *Traite de Zoologie*, above. In German. The first edition was completed in 1967; to date, the second contains volumes for the insects and mammals updating the first edition.

Malakhov. V. V. *Nematodes: Structure, Development, Classification, and Phylogeny.* Washington, DC: Smithsonian Institution Press, 1994. 286 p. ISBN 1560982551. Originally published in Russian in 1986.

Reviews morphology and development of nematodes and proposes a classification system for nematodes and related groups.

McLelland, John. *A Color Atlas of Avian Anatomy*. Philadelphia: Saunders, 1991. 127 p. $76.50. ISBN 0721635369.

Color photographs illustrating bird anatomy, chiefly chicken.

Microscopic Anatomy of Invertebrates. Frederick W. Harrison, ed. New York: Wiley-Liss. 15 v. 1991– . $185.00 (v. 10). ISBN 0471561177 (v. 10).

Presents microscopic anatomy of all invertebrate groups, from protozoa to the invertebrate members of the phyla Chordata. The emphasis is on functional morphology.

The Mollusca. Karl M. Wilbur, ed. New York: Academic Press, 1983–1988. 12 v. $150.00 (v. 12). ISBN 0127514120.

Comprehensive treatise discussing all major aspects of molluscan biology and paleontology.

Reproductive Biology of Invertebrates. K. G. Adiyodi and R. G. Adiyodi, eds. New York: Wiley-Liss, 1983– .

Covers all aspects of invertebrate reproduction, currently up to volume 6B, *Asexual Propagation and Reproductive Strategies*.

Videler, John J. *Fish Swimming*. New York: Chapman & Hall, 1993. (Fish and Fisheries Series, 10). 260 p. ISBN 0412409600.

Reviews the biophysics, kinematics, physiology and costs of fish swimming.